PHYSICS *of* NEUTRON STARS

Edited by
A. M. Kaminker

Nova Science Publishers, Inc.

PHYSICS *of* NEUTRON STARS

Art Director: Christopher Concannon
Graphics: Elenor Kallberg and Maria Ester Hawrys
Manuscript Coordinator: Roseann Pena
Book Production: Tammy Sauter, Benjamin Fung
and Michelle Lalo
Circulation: Irene Kwartiroff and Annette Hellinger

Library of Congress Cataloging–in–Publication Data

Physics of neutron stars / A. M. Kaminker (editor).
p. cm.
Includes bibliographical references and index.
ISBN 1-56072-164-2 : $97.00
1. Neutron stars. 2. Astrophysics. I. Kaminker, A. M.
QB843.N4P48 1994 94-1971
523.8'874--dc20 CIP

6080 Jericho Turnpike, Suite 207
Commack, New York 11725
Tele. 516-499-3103 Fax 516-499-3146
E Mail Novasci1@aol.com

Printed in the United States of America

CONTENTS

Physics of Neutron Stars
Edited by A. M. Kaminker

ISBN 1-56072-164-2

SUPERNOVA 1987A AND FORMATION OF ROTATING NEUTRON STARS

V.S. Imshennik and D.K. Nadyozhin
Institute of Theorectical and Experimental Physics

ABSTRACT

Explosion of Supernova (SN) 1987A gave much information to develop theoretical concepts concerning SN explosions. The work briefly presents this information. Of paramount importance now, two years after the explosion, is investigation of the relict star remnant in the site of SN 1987A explosion. Reasonably assuming that the star remnant is a neutron star, which is supported by neutrino signals registered, a two-stage collapse of the rotating core of the star-progenitor (PG) of SN 1987A is proposed.

Also presented and analyzed are numerical calculations of a collapse of an iron-oxygen core of the star of mass 2M$\odot$ at an initial solid state rotation with an angular velocity $\omega_o = 0.86 \cdot (GM)^{1/2}R_o^{-3/2}$. The numerical model is one-dimensional and is constructed by averaging the centrifugal force over a solid angle. A physical model for the calculations is in excellent agreement with a physical model used to calculate a collapse of a stationary star core, published earlier. This allowed a detailed comparison of these two calculations.

The second stage of the collapse is described with the help of verisimilar estimates, assuming a mechanism of rapid torque moment losses by a rotating neutron star due to, for example, magnetic field. This work allows a conclusion that the lower estimate of a rotational period of the final neutron star state for SN 1987A, i.e., of a pulsar period, is about a millisecond. For a millisecond pulsar to appear under such conditions, no other mechanisms of an accretion type is necessary in a tight binary star system.

Moscow 1989

1. Introduction

"This paper is based on two reports of the authors (with Dr. S.I. Linnikov contributing) made at the All-Union Seminar "Physics of Neutron Stars" in Leningrad, March 15 - 18, 1988. The information of these reports has been revised in this work from the viewpoint of an ever growing interest to the problem of a star remnant of SN 1987A explosion: a neutron star or even a black hole. Therefore, attention in the paper is focused on the collapse of the rotating core of a star, a blue supergiant Sunduleak-69°202 (Sk 1), a progenitor of SN 1987A (PGS 1987A).

To consider the problem we used both old (partially unpublished) results for our paper [1] and a number of qualitative estimates of a two-stage collapse. We believe that the explosion of a massive star Sk 1 as an explosion of a supernova (SN) is due to the so-called magneto-rotational mechanism [2]. At present this mechanism should be treated in a broader sense, emphasizing the effects of rotation itself on a collapsing star core. This idea was first suggested in an earlier review [3]. Here this idea is developed and qualified on the basis of the results of numerical simulation of the collapse process in rotating star cores and some qualitative estimates.

For observations, of primary importance is an expected rotational period of a neutron star, which can be derived in terms of a two-stage collapse of the rotating core of PGS 1987A. It should be mentioned that the two-stage collapse does not exclude formation of a black hole, but this paper will deal with a verisimilar outcome of the collapse: formation of a rotating neutron star. The outcome is supported by a coincidence of integrated characteristics of a neutrino pulse derived from the theory of a collapse resulting in a neutron star, with the data of two neutrino detectors KAMIOKANDE (KN) and IMB [4]. However, there are some problems connected with interpretation of the data obtained by the LSD instrument near Mont Blanc and angular anisotropy of events in the KN and IMB instruments, which do not exclude formation of a black hole and challenges the theory of elementary particles (see also [3] and references therein).

In work [3] a possibility of a millisecond pulsar formation was shown with the help of a simple "astronomical" estimate, the pulsar being, strictly speaking, a rotating neutron star with a rotational period about 2 ms in the site of an exploded blue supergiant in the explosion center of SN 1987A. Below, arguments are given, which follow from a two-stage collapse model of a rotating star core. These arguments also allow estimation of a rotational period of the relict neutron star as 1 ms.

Section 2 deals with major observational data of SN 1987S and their theoretical interpretation, which make this SN different from other explosions of SNs beyond the Galaxy, detected earlier, and therefore, attract great attention. Section 3 presents the results of numerical simulation for the collapse of the rotating star core and their discussion. In Section 4 the second stage of the star core collapse is described on the basis of qualitative estimates, assuming that the second stage realizes as a result of the rotating neutron star (formed in the first stage of the collapse) losing a major part of its torque. A hypotetic (probably, magnetic) mechanism of this fundamental process is discussed. It stands to reason that this mechanism is the underlying cause of the SN explosion, though explosion itself is still unidentified process, in which a small part of the energy released in collapse ($\leq 10^{54}$ erg) is transformed into an explosion energy $< 10^{52}$ erg) [3]. Investigations of SN 1987A can give an insight into this Nature mystery, and the present work is an effort in this direction.

2. Distinctive features of SN 1987A among other supernovas

That SN 1987A belongs to a biased galaxy, the Large Magellanic Cloud (LMC) is generally accepted. Its explosion was so bright that spectral measurements could be made with a high resolution while retaining a great value of the signal/noise ratio. In the SN 1987A spectra narrow absorption lines of interstellar environment between the Earth and The SN were identified and measured [5]. Their shifts with the receding velocities fall in an interval from several km/s to 240 - 290 km/s. Because the mean radiation velocity of the LMC receding is estimated to be 270 km/s, the implied velocities of SN 1987A receding strongly support its affiliation to the LMC. There are also some other facts confirming this [3]. The distance to the LMC is usually taken to be 50 kpc, to an accuracy of 10%. The explosion was detected in the star supercluster 30 Dor (NGC 2070), which is evidently close to the central region of the LMC. Hence, the distance to SN 1987A was taken to be $D = (50 \pm 5)$ kpc. For other supernovas the distance is not so well established.

A comprehensive investigation of SN 1987A was undertaken with ground-based radio-, IR, and optical telescopes; UV and X-ray telescopes aboard orbital stations; X-ray and gamma spectrometers in high altitude rockets and balloons. Spectra were recorded from the far ultraviolet (~ 1000 Å) to the deep IR regions (~ 50 000 Å), which were later supplemented with X-ray and gamma-spectra. These all-wave observations, whose importance for astrophysics was repeatedly emphasized by I.S. Shklovsky [6], provided a reliable determination of a bolometric brilliance curve for SN 1987A and the nature of the radioactive energy source at the star postmaximum stage of evolution. With the distance to SN 1987A known, electromagnetic radiation measurements in all spectrum regions were of absolute character. For the sake of precision, it should be noted that the bolometric brilliance curve was obtained by integrating the data of observations in the whole spectral range covered by ground-based telescopes from near UV to deep IR (from U to M in terms of photometric bands).

There are all grounds to believe that a well known star Sk1, a blue supergiant of B31a spectral class was a progenitor of SN 1987A. Its spectrum had no appreciable anomalies and, therefore, the Sk1 parameters can be safely given as $T_{eff} \approx 16\,000$ K, $L \approx 4.8 \cdot 10^{38}$ erg/s, $R \approx 47 R_{\odot}$, $M = (15 - 25) M_{\odot}$. In the closest vicinity of Sk1 there are two more stars, which most likely make up a triple star system in the LMC. The location of SN 1987A coincided with that of Sk1 to an accuracy of ~ $0''.1$, whereas the locations of the two other stars depart by $1''.5$ and $3''$ from the SN coordinates. However, a decisive argument to identify Sk1 as a progenitor of SN 1987A was its disappearance in the far ultraviolet ($\lambda \leq 2000$ Å), which was detected a week after the explosion [7]. 600 days later the luminosity of SN 1987A was almost similar to that of PGS 1987A [8], this must independently confirm the "death" of Sk1. It was tempting to assume that PGS 1987A was an invisible fourth star in the star system Sk. But analysis of the observations of even the first few weeks practically excluded this possibility [3]. Further observations confirmed the above conclusion.

Identification of PGS 1987A as a type II SN due to intensive hydrogen lines in its spectrum caused a crisis in generally accepted notions concerning type II SN. The hydrodynamics theory for the majority of type II sn firmly predicts the so-called extended structure of PGS with an initial radius $R \geq 10^3 R_{\odot}$ [9]. The radius of the first reliably observed PGS 1987A turned out to be much smaller, more than an order of magnitude. However, for PGS with a small initial radius (compact PGS) the hydrodynamics theory accurately predicts their behavior at the first stage of explosion. When applied to PGS 1987A the theory precisely described the bolometric brightness curve, the effective temperature, its shell speed at the photosphere level, and some photometric and spectral characteristics (see Ref. [3] and references therein). According to this theory, for the observations of the first stage ($t \leq 50$ days) of SN 1987A outburst to be adequately described, the main parameters of the explosion should be: $R = (30 - 60) R_{\odot}$, $M = (10 - 25) M_{\odot}$, $\varepsilon = (1 - 3) \cdot 10^{51}$ erg, where ε is the total explosion energy released instantaneously in the central area of the PGS. Here, M is not the whole mass of the PGS but rather a mass of the ejected SN shell, which is evidently equal to the PGS mass minus the mass of the collapsed star core ($\leq 2M_\odot$ for a neutron star formation and $\sim 10M_\odot$ for a black hole).

The main advantage of the hydrodynamics theory in interpreting the SN 1987A observations is that it provides an independent conclusion about a compact structure of the PGS, $R \leq 10^2 R_\odot$, which corresponds to the initial radius of the blue supergiant Sk1. We may safely state that all specific features of SN 1987A in comparison with typical type II SN can be explained if we assume its PGS initial structure to be compact. This assumption naturally requires "knowledge" of density and chemical composition distributions on the star radius. Figure 1 shows a bolometric brightness curve constructed by Utrobin [10] taking $R = 45R_\odot$, $M = 16\ M_\odot$, $\varepsilon = 2 \cdot 10^{51}$ erg for a polytrope = 3 at a hydrogen concentration $\chi = 0.7$ (curve 1). Curve 2 additionally includes the existence of a radioactive energy source of a mass ^{56}Ni $M_{Ni} = 0.08\ M_\odot$ and describes the outburst behavior at $t > 40$ days. These calculated curves closely coincide with the results of observations of the outburst first stage. Other parameters (such as the effective temperature T_{eff} and speed at the photosphere level etc.) are also in agreement with the observations. Comparison of the calculated and observational T_{eff} suggests a lower hydrogen concentration $0.1 < \chi < 0.7$ [10].

The hydrodynamics theory of SN explosions, for the sake of a PGS compact initial structure, predicts a momentary powerful pulse of UV- and X-ray radiation (UV pulse) with $\varepsilon_r \approx 2 \cdot 10^{47}$ erg, $\Delta t_r \approx 100$ s, $T_r \approx 2 \cdot 10^6$ K, which accompanies a shock wave reaching the PGS surface [11]. It is from this moment that the electromagnetic outburst of SN 1987A began. The spectrum of the UV pulse radically differs from the Planck's at the same temperature, because the Compton scattering of electrons dominates their absorption. As a result, radiation in some areas undergoes Compton scattering to give the Bose-Einstein spectrum. An accurate calculation of the photon spectra is still to be made, but the present hydrodynamics theory is sufficient to predict hydrodynamic and thermodynamic parameters of the substance and radiation. For example, a maximum speed for the substance was calculated to be about 60 000 km/s. The recorded maximum speed of the SN 1987A shell expansion is about 18 000 km/s. The discrepancy may be due to that the moment of the UV pulse was missed and the first SN 1987A spectra were recorded only 1.4 day later after the outburst. The UV pulse plays an important role in ionizing the star environment (see Ref. [3]) and its total energy constitutes about 2% of the SN 1987A brightness integral.

Many qualitative and even quantitative parameters of SN 1987A can be reasonably explained on the basis of the SN outburst hydrodynamics theory in terms of analytical solutions and asymptotic modes, starting from the theory of a shock wave outlet from the PGS inside (see [3] and references therein). To completely understand the nature of SN 1987A on asymptotic mode of free flying apart (at $t \geq 1$ day) and the stage of the cooling and recombination wave (at $t \geq 5$ days) should be involved. The asymptotic mode explains an extremely sharp decrease of the shell substance density on the radius $\rho \sim r^{-(9 \div 10)}$, which was confirmed by observations. The cooling and recombination wave allows a physical interpretation of the effective temperature $T_{eff} \approx 5000$ K stabilization from $t \geq 5$ days up to the SN shell optical transparence at $t \geq 200$ days. Free flying apart and cooling and recombination wave enable interpretation of an increase in luminosity L after passing the minimum ($t \approx 7$ days). The increase takes up to 40 days, when the radioactive energy source resulting from a two-stage decomposition $^{56}Ni \rightarrow {}^{56}Co \rightarrow {}^{56}Fe$ takes effect (see Fig. 1). A completely self-consistent explanation of SN 1987A can be given by numerical simulation of the SN shell ejection. This simulation should include all analytical and asymptotic solutions of the initial equations. The numerical simulation of the shell ejection goes on and promises new interesting conclusions.

From the astronomical viewpoint, a most important consequence of the PGS initial compact structure is a radical decrease of the luminosity in the maximum of the bolometric brightness curve (and photometric Brightness curves as well) as compared with typical SN II (to say nothing of type 1 SN) at the same total explosion energy. According to hydrodynamics theory this decrease results from a reduction of the electromagnetic radiation fraction: in extended models this fraction is about $3 \cdot 10^{-2}$, whereas in compact it is about 10^{-3}. Therefore, explosion in extended models is 2^m - 3^m brighter than in the compact model. Figure 2 demonstrates this decrease for the photometric band V, where SN 1987A is among typical type II SN. From a theoretical point of view, the above UV pulse may be regarded as compensating for the lowering of the brightness maximum. This pulse is either lacking or very weak in SN with an extended initial structure. Two of them are shown in Fig. 2, having plateau-like brightness curves. Later stages in the SN brightness curves are of different character because of the effect of an additional radioactive energy source. Noteworthy is that all the four type II SN in Fig. 2 have "tails" in the brightness curves, which strongly resemble the post-maximum stage of SN 1987A. This finding suggests that the assumption of a radioactive energy source in SN 1987A is also relevant for all type II SN.

SN 1987A reached its brightness maximum on 86 or 87th day (see Figs. 1 and 2), and a month later the brightness started fainting exponentially; the process lasted about 300 days. A characteristic time of fading turned out almost equal to the decay time of radioactive ^{56}Co-III, namely, 26 days. This fundamental fact was reported in Ref. [12], where the mass of initial radioactive ^{56}Ni was found to be $M_{Ni} = (0.078 \pm 0.015)M_\odot$ (the error is due to the uncertain distance to the LMC). The total decay energy in the series ^{56}Ni → ^{56}Co → ^{56}Fe of such a mass M_{Ni} is $1.46 \cdot 10^{49}$ erg. On the other hand, the total radiation energy, an integral of the bolometric brightness curve, was found to be $8 \cdot 10^{48}$ erg [12]. A substantial part of ^{56}Ni decay energy ($\sim 6 \cdot 10^{48}$ erg) was spent to support IR and optical SN 1987A radiation, the rest ($\sim 9 \cdot 10^{48}$ erg) served to increase the shell kinetic energy, insignificant as it was in comparison with the initial kinetic energy ($\sim 2 \cdot 10^{51}$ erg). And only a negligibly small fraction of this energy ($\leq 10^{46}$ erg) was emitted in X-ray and gamma ranges. This conclusion based on the observations was supported by theoretical models of SN 1987A shell glow.

A surprising coincidence of the SN 1987A bolometric brightness curve with the exponential law of ^{56}Co decay indicates that gamma-emission and gamma-induced hard X-ray radiation, which accompany radioactive decay, should not go beyond the SN shell in appreciable amounts. This theoretical prediction was confirmed by 20 - 300 keV X-ray radiation detected by KVANT in August 1987 [15] and by 4 - 30 keV radiation registered by "GINGA" satellite [16]. The hard X-ray spectrum ($\geq$ 20 keV) points to a dominating role of the Compton scattering of ^{56}Co gamma-radiation and allows determination of not only ^{56}Co distribution in the SN shell but also some of its hydrodynamic characteristics [17]. Deviations of the bolometric brightness curve from its exponential shape (by ~ 7%) were first registered in February 1988. In about 600 days almost 100% deviation was observed, i.e., the total energy flux for the bolometric magnitude of SN 1987A was about two times smaller than predicted for the exponential curve of radioactive decay [8]. The X-ray radiation, having its maximum in February 1988 (in the hard region $\sim 3 \cdot 10^{38}$ erg/s), fell off by the autumn of 1988, presumably because the effect of the flux growth due to the SN shell glow was giving way to the effect of smaller amounts of the ^{56}Co nuclide decay. Thus, X-ray astronomy has the first direct proof of the theoretically predicted possibility that appreciable amounts of radioactive ^{56}Ni can be synthesized in SN explosions.

^{56}Ni is the only nuclide, which can perceptibly influence a SN brightness curve, satisfying as it does all the criteria of a radioactive energy source, both by its synthesis and the abundance of the final product of its decay ^{56}Fe in space. Having the greatest binding energy among all alfa-particle nuclei, ^{56}Ni is in the center of thermonuclear synthesis and, therefore, it can be synthesized in great amounts. It is these amounts that help explain the abundance of ^{56}Fe in space. All previous candidates for the role (Ra, ^{254}Cf, ^{7}Be, ^{59}Fe - in the historical chronology) and had a suitable half-life, but were absolutely inadequate to account for the intensity of their synthesis and the abundance of their decomposition products. The main product of ^{56}Ni decay, nuclide ^{56}Co, was detected directly: its main gamma-emission lines were recorded at 847 and 1238 keV [18].

X-ray and gamma-observations clearly indicated that complicated processes of the material intermixing occur in the SN 1987A shell and its shape is biased from the spherical symmetry; the extent of this bias has found no quantitative explanation in the modern hydrodynamics theory of SN explosions as yet. It would be extremely beneficial if such observations could give information about the behavior of the innermost layers of the flying apart shell. These layers most likely belonged to the star core before the explosion, to its outer layers, to be precise. During the explosion they got loose from the collapsing part of the star core and scattered in the interstellar space. It was here that an unknown mechanism of the star explosion worked, causing a specific distribution of ^{56}Co in the SN shell and its deviations from the spherical symmetry. The modern hydrodynamics theory cannot explain a considerable delay in the expansion of the innermost layers of SN 1987A shell that is responsible for a rather long exponential mode of the bolometric brightness curve. As was mentioned above (see Fig. 1), the theory adequately describes the initial stage of SN 1987A explosion. With some improvements, the theory can be expected to describe the maximum explosion stage as well.

Though an exact explosion mechanism of such a massive PGS as the blue supergiant Sk1 remains undetermined yet, the theory rather unequivocally predicts formation of an iron core of a mass $(1.5 - 2)M_\odot$ in the PGS 1987A during its hydrostatically equilibrium evolution. This core was also predicted to collapse, in the long run, to form a neutron star or even a black hole (see Ref. [3] and references therein). The prediction of the core collapse, which evidently works as a trigger mechanism of the star shell ejection, was supported by neutrino signals recorded in underground neutrino observatories [13, 14, 19, 20]. The theory of the neutrino pulse from a collapsing iron star core developed in Ref. [21] adequately described two integrated parameters of a neutrino signal: the time dependence of the number of $\tilde{\nu}_e$-events and an averaged energy of these events for K !! and IBM instruments [4]. The work [21] deals with numerical calculations of a collapse of a rather massive iron-oxygen star core of a total mass $2M_\odot$ in terms of the neutrino heat conduction and a characteristic duration of about 10 s. This period also includes the beginning of the cooling of a hydrostatically equilibrium neutron star due to neutrino radiation. it was shown that no ejection of the star shell with the energy of a SN explosion occurred. The physical model proposed [21] was spherically symmetrical and involved a number of approximations in the state equations and transfer of a neutrino radiation. These approximations proved valid for a rather great entropy of a homologously collapsing central core of the star $S \geq 2k_B$/nucleon. For a massive core of $2M_\odot$ these approximations are well grounded. Their qualifications, when applying them to iron cores of smaller masses, $\leq 1.5M_\odot$, but of spherical symmetry, can hardly help establish an effective mechanism of SN explosion [22, 23]. More promising seem to be physical models, which include the effects of the collapsing core rotation, resulting in distortions of a strictly spherical symmetry. This approach led [2] to the hypothesis of a magneto-rotational mechanism of SN explosions. Some results of neutrino detectors (see Ref. [3]) can be interpreted in terms of the hypothesis of the rotating star core collapse, which may proceed in two stages [24].

Registration of neutrino signals from SN 1987A mean the beginning of an extragalaxy neutrino astronomy. This astronomy emphasized the role of underground neutrino observatories in the studies of the physics of star collapse and explosion. The explosion of SN 1987A can be regarded as a test of how we are ready to reliably record neutrino pulses from collapsing star cores in the Galaxy, whose probability is estimated to be high.

One more fact should be mentioned. The period from the moment, when neutrino signals were received by K II, IBM, and BST instruments on February 23, 1987 (7.36), to the first optical registration of SN 1987A outburst (10.40) is in agreement with a theoretically calculated time interval for a shock wave to travel from the center to the surface of Sk1, which is 0.5 - 2.5 h. This time, in its tern, depends on the structure parameters of PGS 1987A and the total energy of explosion $(1.5 - 3) \cdot 10^{51}$ erg. The finding, on the one hand, is an independent argument for that Sk1 can be identified as a PGS for SN 1987A. Moreover, it is a striking illustration of a trigger nature of a star core collapse for SN explosions. On the other hand, the LSD instrument could not fail to register the explosion if it had occurred at an earlier moment. Hydrodynamic calculations show that an explosive energy release at that earlier moment (say, 4.7 h before K II, IMB, and BST instruments started to operate) could not exceed a relatively small value $5 \cdot 10^{49}$ erg [25].

3. A numerical model for the collapse of a rotating iron-oxygen star core $2M_\odot$

A rather comprehensive physical model of a collapsing iron-oxygen star core was developed in Ref. [21], which however, did not take account of the core rotation. The model included the effects of neutrino transfer of energy and lepton charge in terms of neutrino heat conduction after the central area of the star became nontransparent for neutrinos. Before the neutrino nontransparence occurred, neutrino energy losses were taken into account as a three-dimensional radiation all over the collapsing star. The boundary between the neutrino nontransparent star core and the outer transparent accreting shell was a function of time and was determined during the problem solution. For the transparent shell, energy losses by the three-dimensional neutrino radiation and an additional warming up by diluted neutrino radiation from the surface of the nontransparent core (from the neutrinosphere level) were included. In the outer oxygen layer the model incorporated the kinetics of thermonuclear oxygen combustion. The kinetics of beta-processes in the transparent shell was described by the kinetic equilibrium approximation appropriate for a high entropy of massive star cores of $(1.5 - 2)M_\odot$ [26]. The collapse of a massive star core with $M_{Fe+O} = 2M_\odot$ ($M_{Fe} = 1.82M_\odot$, $M_O = 0.18M_\odot$) was calculated in detail. The calculations showed a practically complete lack of outer layer ejections with appreciable energy and mass, though all physical effects stimulating the explosion were taken into consideration in the model (see Ref. [21] for details).

In work [1] we calculated a star core collapse assuming a strong initial solid state rotation with an angular frequency $\omega_0 = 0.86 \cdot (GM / R_0^3)^{1/2}$ on the basis of the physical model developed in Ref. [21].*

*Footnote (Unfortunately, the initial angular velocity of a solid state rotation used in the calculations in work [1] was erroneous. In fact it is equal to 0.86 fraction of the angular frequency unit (rather than 0.80 used) $\Omega_0 = \sqrt{GM} / R_0^{3/2} = 1.78\ s^{-1}$. The angular frequency unit, in its turn, is a measure of the angular frequency in Fig. 1 in Ref. [1]. So in Fig. 1 ω_0 stands, in fact, for Ω_0. Table 3 in Ref. [1] also has an error: $\omega_c / \omega_0 = 5.8 \cdot 10^3$ instead of the given $\omega_c / \omega_0 = 1.69 \cdot 10^3$. The value $5.8 \cdot 10^3$ is in agreement with the plot in Fig. 1, where $\omega_c / \Omega_0 = 5 \cdot 10^3$.)

Application of such a complicated physical model to describe neutrino radiation became possible owing to retaining the spherically symmetrical approximation of the configuration. For this, the centrifugal force was averaged over a solid angle and all deviations from the spherical symmetry were neglected. The initial conditions were set as a polytrope with $n = 3$, i.e., $\rho_0 = \rho_3(m)$, $r_0 = r_3(m)$, except for the temperature, which satisfied the virial theory only after the rotation effects were taken into account. Thus, calculations started from a precise, hydrostatically equilibrium (the centrifugal force included), configuration at the boundary of stability. They resulted in a state of almost precise hydrostatic equilibrium in the form of a rapidly rotating hot neutron star, because no appreciable reset of the outer layers could be made, like in the case of a stationary core. The resulting initial configuration corresponded to the stability boundary and differed from that of a stationary star core: its radius was smaller, the density and temperature in the center were higher. Quantitatively these changes made about 10% (see Table 1 in Ref. [1]). Unlike the initial state, where the rotation contribution is small (the rotation energy was about 2% of the gravitational energy), their final state experiences a strong and dominating influence of the rotation (the ratio of these energies was about 40%).

When the Lagrange coordinate m is used, the hydrodynamics equations averaged over the solid angle have the form, before the stage of neutrino nontransparence:

$$\frac{\partial r}{\partial t} = v$$
$$\frac{\partial v}{\partial t} = -4\pi r^2 \frac{\partial P}{\partial m} - \frac{Gm}{r^2} + \frac{2}{3}\frac{\omega_0^2 r_0^4}{r^3} \qquad (1)$$
$$\frac{\partial r^3}{\partial m} = \frac{3}{4\pi\rho}$$

$$\frac{\partial E}{\partial t} + P\frac{\partial}{\partial t}\left(\frac{1}{\rho}\right) = \tilde{\varepsilon} \qquad (2)$$

where $\tilde{\varepsilon}$ is the total sum of all the sources and sinks for the energy due to thermonuclear energy release and three-dimensional neutrino radiation per unit mass. The internal specific energy E and pressure P are functions of T and ρ according to the state equation, which includes iron dissociation thermodynamics, relativistic electron degeneration, and the presence of positrons.

The star core nontransparent for neutrinos can be described by a more complicated system of equations. In the right side of (2) $\tilde{\varepsilon}$ is substituted by the divergency of an energy flux, which can be expressed with temperature gradients and neutrino chemical potential in terms of neutrino heat conduction approximation. In addition, one more equation should be added describing the diffusion of a lepton charge. And lastly, the pressure and specific energy of an equilibrium neutrino radiation is added to E and P.

An expression for the total energy conservation can be derived from equations (1) and (2):

$$\frac{\partial}{\partial t}\left(E + \frac{v^2}{2} - \frac{Gm}{r} + \frac{1}{3}\frac{\omega_0^2 r_0^4}{r^2}\right) = -4\pi\frac{\partial}{\partial m}(r^2 v P) + \tilde{\varepsilon} \qquad (3)$$

It can be seen that the energy of rotation is added to the left side of the expression. The initial form of equation (1) is due to the use of the conservation law of the torque at a given Lagrange coordinate m:

$$j = v_\varphi r \sin\Theta = v_{\varphi 0} r_0 \sin\Theta = \omega_0 r_0^2 \sin^2\Theta$$

with an arbitrary polar angle Θ before averaging and including a material particle travelling along a spherical radius r. The radial projection of the centrifugal force can than be expressed with the torque j:

$$F_r = v_\varphi^2/r = j^2/r^3 \sin^2\Theta = \left(\omega_0^2 r_0^4/r^3\right)\sin^2\Theta$$

The centrifugal force is averaged over a solid angle, resulting in the term

$$\bar{F}_r = \frac{2}{3}\cdot\frac{\omega_0^2 r_0^4}{r^3}$$

in the right side of equation (1).

It follows that the angular velocity ω at any moment and for any Lagrange layer is connected with ω_0 and r_0 by a simple relation

$$\omega = \omega_0 \left(\frac{r_0}{r}\right)^2 \qquad (4)$$

Of interest are some changes in the parameters of the neutrino nontransparence onset in the rotating star core (see Table 2 in Ref. [1]). The density in the center ρ_c exceeds that in a stationary core by 2.5 times, whereas the temperature T_c decreases by 20%. Thus, the difference between these two models still remains small though a bit more appreciable than in the early moment.

Now we will proceed to a more detailed discussion, than in Ref. [1], of the basic results of calculations for the given model of a rotating iron-oxygen star core. The numerical model comprised 150 Lagrange zones, whose number was reduced to 74 after the shock wave front reached the star surface. With this division, the cooling of the neutron star by neutrino radiation was calculated for some time (about 2 s) until the virial theorem to be valid with an accuracy to five meaningful figures (~ $2 \cdot 10^{-5}$). Figure 3 shows a (r - t) dependence for some typical Lagrange zones of the star. The dependence includes the most interesting, dynamic part of the collapse in the time interval from t = 0.4 s to 0.8 s (the reference point was chosen so that t_0 = 17.2465 s should be added to t to obtain the total time from the true initial moment of the model calculation). In this time interval, which is about 0.05 s (between t = 0.5 and 0.55 s), the major part of the star core collapses and the density increases by about 10^4 times. This time interval, that can be called a characteristic collapse time, is similar to that in the stationary core collapse [26]. The similarity is quite understandable, because the parameters of the collapsing cores do not yet differ substantially. Moreover, the parameters are compared for the moment, when the neutrino nontransparence sets in at the end of the dynamical stage of the collapse and the material densities become as high as 10^{13} g/cm^3. After the dynamic stage (at t > 1 s) the star core continues to contract but at a lower rate. This contraction proceeds when the accreting shock wave (rectangular crosses in Fig. 3) and neutrinosphere (asterisks in Fig. 3) are already present.

The shock wave (SW) arises at t ≈ 0.53 s, m ≈ 0.6M$_\odot$ and r ≈ $5 \cdot 10^6$ cm (the leftmost asterisk near the Lagrange trajectory m = 0.553M$_\odot$). At the dynamic stage of the collapse, a homologous central core appears with a mass 0.6M$_\odot$. The SW front first moves rapidly outwards: at t ≈ 0.56 s it crosses the trajectory m = 1.56M$_\odot$ at r = $1.8 \cdot 10^7$ cm. Taking its coordinate origin at r = $5 \cdot 10^6$ cm, the mean velocity of its propagation at the dynamic stage would be D ≈ $4 \cdot 10^8$ cm/s, whereas the rate of the material collapse is $v_{ac} \approx -3 \cdot 10^9$ cm/s (Fig. 4). This behavior is in contrast with that in a stationary star core, where the SW front practically does not propagate along the radius [26]. At t > 0.56 s the SW velocity is twice as small and only at t ≈ 1.1 s it reaches the star surface at r ≈ $2.5 \cdot 10^8$ cm and causes a weak star shell ejection. The ejection involves only two outer layers with a total mass and a kinetic energy $E_{kin} = 3 \cdot 10^{46}$ erg, which is negligibly small as compared to the SN outburst energy.

It may be concluded that the SW was still of the some accreting nature as in the stationary case [26]. This conclusion is supported by the velocity curves v(m) shown in Fig. 4 for different moments at the dynamic collapse stage (at t ≤ 1 s). At instant 6 (t = 0.53 s) the material behind the SW front does not move at any appreciable velocity, $|v| \le 2 \cdot 10^8$ cm/s at the accretion rate before the SW front of $v_{ac} \approx -2.4 \cdot 10^9$ cm/s. According to the continuity equation, the SW front propagation velocity is

$$D = -\frac{\rho_0}{\rho - \rho_0} v_{ac} + \frac{\rho}{\rho - \rho_0} v \qquad (5)$$

where v is velocity behind the SW front. At v = 0, $D = -\frac{\rho_0}{\rho - \rho_0} v_{ac} > 0$, i.e., the front velocity is several time $\left(\frac{\rho}{\rho_0} - 1 = \frac{2}{\gamma - 1} \approx 6\right)$ smaller than the accretion rate. If

material compression continues behind the SW front (see Fig. 9 and 10 in Ref. [26]), that is, v < 0, then the second term in (5) can make the D value zero. On the contrary, high positive velocities v in (5) would increase the value of D. This would give rise to an expanding SW capable of initiating a considerable ejection of outer layers and trigger the mechanism of hydrodynamic bounce for the SN explosion [22, 26]. The case v ≈ 0, considered here, can be regarded as an intermediate between an expanding and accretion SW. At D < 0 the accretion SW, according to (5), would obey the condition

$$v \le \frac{\rho_0}{\rho} v_{ac} \qquad (6)$$

that is a rather high (in an absolute magnitude) velocity of final compression v. Figure 4 demonstrates small positive velocities of the outer zones for all times t > 0.425 s, which correlate with their above mentioned ejection.

In Fig. 3 the position of the neutrinosphere is worthwile attention. It appears at t ≈ 0.52 s when m = 0.3 M$_\odot$, then it almost coincides with the position of the Lagrange zone m = 0.965 M$_\odot$, though its radius remains constant r = (4 - 3) ·

10^6 cm. In a rotating star core the neutrinosphere soon begins lagging behind the SW front, whereas in a stationary case it closely follows the front up to the star core surface [26].

Two directions can be distinguished in the development of the theory of the hydrodynamic bounce. Originally, a simple, so-called prompt, mechanism of the shell ejection was developed. Recently a new, so-called delayed, mechanism was proposed in Ref. [28]. But this mechanism has also been modified since new results were obtained in Ref. [22, 23]. These works suggest that a convection-instable area with a negative entropy gradient could arise behind the accretion SW. A large scale convection then develops within a characteristic time of ~ 0.1 s accompanied by energy transfer to the SW front. If this energy comes from the neutrino nontransparent layers, i.e., from under the neutrinosphere, then convection is a much more effective way of energy transfer than neutrino heat conduction. The multi-dimensional nature of the delayed mechanism of a collapsing star shell ejection may quite well be due to the convective energy transfer (because convection is a three-dimensional motion). This idea, however, requires extremely complicated numerical calculations to become well grounded [22]. Figure 3 shows that in the case considered here the neutrinosphere rapidly starts to fall behind the SW front (in ~ 0.1 s) and neutrino radiation would play the role attributed to it above. Weak (diluted) neutrino radiation would pass through the SW front without affecting its dynamics. If convection arises it would occur in a wide gap between the SW and neutrinosphere, where no appreciable energy density of neutrino radiation could exist.

Two variables can be regarded as major characteristics for the rotation of a collapsing star core: an angular velocity ω/Θ_o and the ratio of an averaged centrifugal force to the gravity force α:

$$\alpha = \frac{2}{3G} \cdot \frac{\omega_0^2 r_0^4}{rm} \qquad (7)$$

Figures 5a, b, c and 6a, b, c present the profiles of these variables for 17 instants. The first 8 profiles correlate with those in Fig. 4, the rest represent the cooling stage of a neutron star due to neutrino radiation. In the case considered here ω_o = 1.53 s^{-1} at M = 2M$_\Theta$ and R_o = 4.38 · 10^8 cm. The curves in Fig. 5a show how much the rotation differs from that predicted by the low of solid state rotation: the star surface rotates with its initial velocity, whereas in the center the velocity increases by 6 · 10^3 times. The boundary between the fast and the slow rotation is close to the SW front as long as the SW is strong enough (up to ~ 1.8M$_\Theta$). With time, ω increases in the center and decreases near the surface (Fig. 5a). The rotation period in the center remains minimal all the time and is equal to $\tau_c = 2\pi / \omega_c \approx 0.7$ ms.

The behavior of α representing the dynamic rotation effects in (7) during the collapse is far from commonplace. As can be seen from Fig. 6a the value of α behind the SW front, at the dynamic collapse stage (moments 5 and 6), increases to its initial maximum value at the surface $\alpha_s \approx 0.43$. Then, right behind the front of the SW reaching the surface (moments 7 and 8) an absolute maximum is formed with $\alpha \approx 0.6$. At the cooling stage of a neutron star (Fig. 6c) α values change considerably: they increase in the region of fast rotation (up to J = 60, m = 1.85M$_\Theta$) and decrease in the slow rotation region. In the former region it may be as high as ~ 0.8, in the latter it drops to become smaller than the initial α_s. These plots show that the hydrostatic equilibrium of a neutral star in its final state is due to centrifugal forces rather than to a pressure gradient.

Shown in Fig. 7 are the profiles of some physical variables ω, α, ρ, T, and r for a final stage of a neutron star, after a hydrostatic equilibrium is established with a relatively high accuracy (~ 2 · 10-5) ensured by the virial theorem. The calculations were discontinued at an arbitrary instant t ≈ 2.9 s but after some cooling of the neutron star occurred, so that T_c became two times smaller than its maximum value ~ 10^{11}, during cooling. The neutrinosphere receded by that time to m ≈ 0.3M$_\Theta$ (compare with Fig. 3). Figure 7 represents the structure of a hydrostatically equilibrium neutron star with a strong rotation. The rotation is of differential nature due to the preceding collapse (see Figs. 5 and 6). Further cooling of the star by neutrino radiation can slightly change the relation between the contributions from the centrifugal forces and those due to a pressure gradient. However, the contribution of the pressure-induced forces to the hydrostatic equilibrium should remain appreciable enough because of nonideality effects and nuclear matter degeneration.

The approximation with averaging the rotation effects over a polar angle Θ can give a relation, which would provide useful estimates and help interpret and check the results of numerical calculations, remaining valid at any values of m and t. According to (7) radius r at a given α in an arbitrary moment t can be expressed with its initial value r_o, initial angular frequency ω_o, and mass m. Under hydrostatic equilibrium conditions with $\partial v/\partial t = 0$ in (1), the value α unequivocally determines the force due to a pressure gradient

$$4\pi r^2 \frac{\partial P}{\partial m} = -(1-\alpha)\frac{Gm}{r^2} \quad (8)$$

Let us differentiate expression (7) with respect to m (at ω_0 independent of m). Using the relationship of the material density ρ and $\partial r/\partial m$ from (1) and making some simple transformations, we obtain an expression for ρ through α and its logarithmic derivative ($\partial \log \alpha / \partial \log m$) at arbitrary m and t:

$$\frac{\rho}{\rho_0} = \frac{\frac{27}{32}\left(\frac{Gm}{\omega_0^2 r_0^3}\alpha\right)^3}{1 - \frac{\pi \rho_0 r_0^3}{m}\left(1 + \frac{\partial \log\alpha}{\partial \log m}\right)} \quad (9)$$

Now let us use (9) to describe the star center, neglecting the derivative ($\partial \log \alpha / \partial \log m$) in it. With $m = \frac{4}{3}\pi r_0^3 \rho_{0c}$ (where "c" is the index for all values in the center) near the center, we obtain a simple relation:

$$\frac{\rho_c}{\rho_{0c}} = \frac{27}{8}\beta_0^3 \alpha_c^3 \left(\frac{\Omega_0}{\omega_0}\right)^6 = 5.37 \cdot 10^5 \alpha_c^3 \left(\frac{\Omega_0}{\omega_0}\right)^6 = 1.33 \cdot 10^6 \alpha_c^3 \quad (10)$$

The value β_0 in the expression is equal to the ratio of the density in the center ρ_{0c} to a mean density $\bar{\rho}_0 \equiv 3\,M / 4\pi\, R_0^3$ of the star core before collapse. In the case of a polytrope gas ball with n = 3, β_0 = 54.18. The second part of (10) was calculated using the value, and the final result was obtained taking $\omega_0 = 0.86\ \Omega_0$. Substitution of $\alpha_c = 0.683$ for the final state of a rotating neutron star (Fig. 7) gives $\rho_c/\rho_{0c} = 4.23 \cdot 10^5$, which then gives $\rho_c = 2.58 \cdot 10^{14}$ g/cm^3 at $\rho_{0c} = 6.11 \cdot 10^8$ g/cm^3 determined by the initial conditions. This ρ_c value is in complete agreement with a calculated ρ_c in Fig. 7. The derivative $(\partial \log \alpha / \partial \log m)_c$ is really small ($\sim 10^{-2}$) so its contribution can be neglected in the estimation.

The expression (9) can be used for similar estimates at any m, both in the final (Fig. 7) and all intermediate configurations, if the value of ($\partial \log \alpha / \partial \log m$) is estimated from the data in Fig. 6a, c and Fig. 7. Simple relation (10) also shows that the star core rotation, though obeying the law of solid state rotation, cannot prevent the central density ρ_c from increasing up to the density of nuclear matter ($\sim 3 \cdot 10^{14}$ g/cm^3) in spite of the fact that α_c did not become equal to unity owing to the neutrino-induced cooling. It means that there remains a considerable contribution of the forces due to a pressure gradient. In formula (10) an ultimately possible initial rotation, i.e., $\omega_0 = \Omega_0$, would give a coefficient of $\sim 0.4 \approx (0.86)^6$ for the final density ρ_c. Consideration of the case $\alpha_c = 1$ resulting from further neutrino cooling would hardly be justifiable because of the effects of nonideality and nuclear matter degeneration. Nevertheless, the consideration provides support for that the density in the center of a rotating neutron star formed during the collapse of the star core turns out to be close to the density of nuclear matter, the initial rotation of the star core obeying the solid state laws.

The Table below, partly based on the data from Table 3 in Ref. [1], presents the major characteristic values for the final state of a rotating neutron star (column 1). Column 2 of the Table gives integrated values for different kinds of energy, including neutrino energy losses. Column 3 presents corresponding values for a stationary star, which were calculated in [21] at t ≈ 2.4 s after the magnetosphere became neutrino nontransparent. In the time scale of this work t ≈ 2.4 s corresponds to t ≈ 2.9 s. This discrepancy is due to that in [21] the moment when the nontransparence sets in was taken as t = 0, while in this work it corresponds to t ≈ 0.52 s. The values in Column 3 slightly differ from the characteristics of the star final state reported in [21]. An explanation may be that they belong to an earlier time moment when accretion was still go on and the accretion SW was only approaching the neutron star surface. How small this difference is can be seen from Column 4 that presents the characteristics of the final state calculated in [21] at t ≈ 5.1 s. Comparison of the data from Columns 3 and 4 leads to the conclusion that after the accretion stage is over, the choice of a time moment to characterize the final stage is of no importance at all, because the star parameters change with relatively long characteristic times of the star cooling by neutrinos.

The Table shows that the configuration of a rotating neutron star is about 10 times less dense, its radius (linear dimensions) and gravitation energy are 3 times greater, whereas its temperature and internal energy are 10 times smaller. Its energy of rotation, however, increases due to collapse by $7.8 \cdot 10^2$ times as compared to the initial value and exceeds its internal energy. The rotation, which acquire a differentiated character with a maximum angular rate in the center, greatly contributes to the star hydrostatic equilibrium, though the configuration of a rotating neutron star may considerably deviate from the spherical symmetry.

The configuration of a hydrostatically equilibrium rotating neutron star, resulting from a gravitational collapse, may be unstable toward transformation into a three-axial figure. This instability is generally possible at a rather great ration $\Delta = \varepsilon_{rot} / |\varepsilon_g|$. Speaking about stability, we consider only axially symmetric configurations, because such configurations are the basis of approximate model, in which centrifugal forces are averaged over the polar angle with respect to a chosen axis of rotation. Tassoul in his monograph [29] stated that classical results for uniform McLoren spheroids can be applied to differentially rotating polytropes that depend, but slightly, on the torque distribution law and the polytrope index. In McLoren spheroids a centenary instability with respect to the transformation into a three-axial figure appears at $\Delta > 0.1375$, whereas dynamic instability (leading to rod-shaped configurations) at $\Delta > 0.2738$. Our calculations give a configuration with a high value of $\Delta \approx 0.42$. It greatly exceeds the stability range for axially symmetric configurations. Given the above instability does occur it will cause changes in our calculations. Evidently, a more complicated configuration should result with a three-axial figure. The configuration, nevertheless, will be hydrostatically equilibrium provided the torque is retained: the centrifugal forces will equilibrate gravitational. The gravitational collapse of such a three-axial configuration will recommence only after a real loss of local torques. A conclusion may be that the two stage collapse model would still be valid even if the instability is responsible for the transformation into a three-axial figure. The process of local torque losses, however, may be of a different nature.

Work [27] reports one of the latest two-dimensional calculations of the gravitational collapse with rotation. The initial conditions for a star central core of mass $1.6M_\odot$ were taken from the evolution calculations, then a rather strong rotation with $\omega_{Co} \approx 4\ s^{-1}$ was superimposed on them, which is in discord with the hydrostatic equilibrium of the initial state. This is not of importance because the energy of rotation remains small (see Table 2 in Ref. [27], where the value of ε_{rot} is evidently misprinted: it is an order of magnitude greater). Using a simplified physical model, the authors of work [27] made calculations, with a detailed difference scale, of the gravitational collapse almost to the final stage, when the accretion SW provided $m \approx 1.3\ M_\odot$. Table 3 in Ref. [27] presents some characteristics of a rotating neutron star at the moment when the calculations were discontinued (the ε_{rot} values in the Table are also an order of magnitude greater): from the moment of bounce, which is in our case $t \approx 0.53$ s, and for a period of 18 ms before the neutrino cooling effect became appreciable. Qualitatively, our calculations correlate with the model A from Ref. [27] with the above-mentioned initial angular speed. The density in the center $\rho_c \approx 2.1 \cdot 10^{14}$ g/cm^3 is in a fair agreement with ρ_c from the Table; the values of the angular speed $\omega_c \approx 1.2 \cdot 10^4\ s^{-1}$ and the ratio $\omega_c / \omega_0 \approx 3 \cdot 10^3$ are close, though not in excellent agreement. The energy of rotation $9 \cdot 10^{51}$ erg is 6 times smaller than the value in our calculations. These discrepancies is evidently due to a more intensive initial rotation, which has a differentiated character from the very beginning of the calculations. The results on two-dimensional effects reported in Ref. [27] are very interesting and deserve separate discussion. In spite of a strong deviation from the spherical symmetry with $r^{pole}/r^{eq} \sim 0.5$, they are not in disagreement with our calculations in which the centrifugal forces were averaged over the polar angle.

Let us now consider neutrino radiation. The Table shows that the total energy of neutrino losses and the neutrino luminosity are about 10 times smaller than at $\omega_o = 0$. By the time when the calculations were over, the neutrinosphere practically did not exist (see also Figs. 3 and 7 and the discussion above). The neutrinosphere parameters $\rho_{\nu ph}$ and $T_{\nu ph}$ were close to ρ_c and T_c, its radius $R_{\nu ph}$ very small. Figure 8, borrowed from Ref. [1] (Fig. 2 therein), shows the time dependence of neutrino luminosity in comparison with a stationary star (the data from Ref. [21]). The characteristic energies of emitted neutrinos were emphasized in Ref. [1] to be two times smaller in a rotating star core. It should be noted that mean neutrino "optical" thicknesses $\bar{\tau}_{\nu\nu}$ beneath the neutrinosphere are rather small. Under such conditions the neutrino distribution in energies should be approximated by the equilibrium Fermi-Dirack distribution with caution.

4. Millisecond pulsar

The numerical model of the collapse of a rotating star core based on equations (1) and (2) includes the local law of the specific torque conservation $j = \omega r^2 \sin^2\Theta = \omega_0 r_0^2 \sin^2\Theta$ or, after averaging over the polar angle Θ,

$$\bar{j} = \frac{2}{3}\omega r^2 = \frac{2}{3}\omega_0 r_0^2.$$

It follows that the angular rate of rotation ω of any Lagrange spherical layer at any instant is connected with its initial angular speed ω_o by relation (4).

The law of the torque conservation can be violated owing to different causes. Any shear viscosity (that of neutrino including) can cause the specific torque redistribution inside a differentially rotating neutron star. Neutrinos are only capable of influencing the torque distribution inside the central nontransparent core, where they can actively interact with the material. The torque of the material beneath the neutrinosphere is equal to

$$j_{\nu ph} = \frac{2}{3}\int_0^{M_{\nu ph}} \omega r^2 dm = \frac{2}{3}\omega_0 \int_0^{M_{\nu ph}} r_0^2 dm \qquad (11)$$

where $M_{\nu ph}$ is the mass covered by the neutrinosphere. For further estimations let $M_{\nu ph} \approx 1\,M_\Theta = 0.5M$ (Fig. 4). In the second right side of (11) the ratio from (4) is used and the integral is taken over the initial solid state rotating configuration. the total torque of the initial configuration (a polytrope gas ball with the index n = 3) at $M = 2M_\Theta$, $R_o = 4.38 \cdot 10^8$ cm, and $\omega_o = 1.53\ s^{-1}$ can be written as

$$J_0 = \frac{2}{3}\omega_0 \int_0^M r_0^2\, dm = \frac{2}{3} 0.113\, M R_0^2 \omega_0 = 8.8 \cdot 10^{49} erg\, s \qquad (12)$$

The torque of the central half of the mass ($M_{\nu ph} = 0.5$ M) is then equal to 0.17 J_o, therefore,

$$J_{\nu ph} = 1.3 \cdot 10^{-2} M R_0^2 \omega_0 = 1.5 \cdot 10^{49}\ erg\, s \qquad (13)$$

The estimate of the characteristic time t_J of the torque redistribution due to the neutrino viscosity, which can be presented by a rough formula, of dimension considerations only, is

$$t_J \sim \frac{c^2 J_{\nu ph}}{L_{\nu\tilde{\nu}} R_{\nu ph}^2\, \omega_{\nu ph}} = \frac{c^2 J_{\nu ph}}{L_{\nu\tilde{\nu}} R_{\nu pho}^2 \omega_0} = 1.3 \cdot 10^{-2}\, \frac{c^2 M}{L_{\nu\tilde{\nu}}} \left(\frac{R_0}{R_{\nu pho}}\right)^2 \qquad (14)$$

where $L_{\nu\tilde{\nu}}$ is the total neutrino luminosity and $R_{\nu ph}$ is the radius corresponding to the mass $M_{\nu ph}$ in the initial configuration. Substituting $L_{\nu\tilde{\nu}} \sim 10^{52}$ erg/s (Fig. 8) and $R_{\nu ph} \approx 1.24 \cdot 10^8$ cm (from the original data) into (14) we obtain $t_J \sim 100$ s. This time exceeds by more than an order of magnitude the time interval, after neutrino nontransparence sets in, covered by our calculations ($\Delta t \approx 2.4$ s). Hence, neutrino radiation cannot substantially violate condition (4) of local torque conservation used in our calculations.

The neutrino radiation, leaving the neutrinosphere, carries the torque away. If we assume that in the coordinate system related to the neutron star surface, neutrino radiation of any surface element is isotropic *,

*Footnote(In fact only $I_\nu(\Theta, \varphi) = I_\nu(\Theta, \varphi + \pi)$ is sufficient, where $I\nu$ is the neutrino intensity radiation, Θ is the angle between the normal to the neutrinosphere surface and the direction of neutrino motion, and φ is the azimuthal angle in the tangential plane to the neutrinosphere in the spot of this element location. Expression (15) remains valid when the symmetry is not spherical if a value of $< \omega_{\nu ph} R_{\nu ph}^2 >$ properly averaged over the neutrinosphere is substituted into it. In this case the flux of neutrino energy is different for different sections of the neutrinosphere, and, in spite of its local isotropy character, the star neutrino radiation is anisotropic.),

then in the spherically symmetric approximation:

$$\dot{J}_{\nu ph} = -\frac{2}{3} \frac{L_{\nu\tilde{\nu}}}{c^2} \omega_{\nu ph} R_{\nu ph}^2 \qquad (15)$$

that is, the characteristic time $J_{\nu ph} / \dot{J}_{\nu ph}$ of the torque loss is of the same order of magnitude as t_J from (14). Thus, the torque losses from the neutrinosphere are of minor importance for our calculations.

It should be noted that the torque losses from the neutrinosphere in the case of local isotropy of neutrino radiation do not directly result in a decrease of the angular speed of the neutrinosphere $\omega_{\nu ph}$. These losses are due to that neutrons carry away mass and thus decrease the inertia momentum only. Changes in $\omega_{\nu ph}$ may only be caused by the torque redistribution in the layers beneath the photosphere. When neutrino radiation lacks a local isotropy, $\omega_{\nu ph}$ can change too. For example, the predominant neutrino radiation in the direction (or against the direction) of the rotation would decrease (or increase) the rate of the torque rotation $\sim \left(\omega_{\nu ph} R_{\nu ph} / C\right)^{-1}$ times as compared to the estimate (15). In our case it would be ≈ 10 times.

Thus, neutrino radiation cannot substantially change the torque derived in our calculations for the rotating star central core. For the gravitational collapse to renew, other mechanisms of the torque loss should work. The magneto-rotational mechanism first advanced by Bysnovaty-Kogan [2] looks promising.*

*Footnote(The role of magnetic field in rapid, of typically hydrodynamic time, torque loss and explosion of outer layers of a star has been considered in numerical models [31]. However, doubts remain whether the initial magnetic field would prevent the star core from a rather rapid solid state rotation before it losses its stability with respect to collapse (see the initial conditions in Section 3). Such a field, probably of hollow-shaped structure, would cause the torque redistribution between the star core and its outer layers at later stages of the star evolution. We cannot unambiguously dispel the doubts, but would like to point to the following facts. The initial magnetic field can be generated by a well-known mechanism due to the Hall effect in the stellar plasma $\sim [\nabla \rho \cdot \nabla p]$. The effect is certainly substantial in the baroclinic configuration during the collapse of a rotating star core. In this case magnetic field arises in the process of the collapse itself and does not prevent the initial rotation from starting.)

Differentiated rotation rapidly makes the frozen-in magnetic field stronger by, say, transforming the initial hollow-shaped field into the toroidal, doubling its stress in one revolution. The resultant magnetic ponderomotive force forms an expanded SW accompanied by the ejection of the star core outer layers and the surrounding outer shell. The force would decelerate the rotation and relieve the torque in the central part of the star core, thus disturbing the hydrostatic equilibrium. In its turn the disturbance of the hydrostatic equilibrium should bring about the renewal of the star core collapse. We already discussed this scenario as applied to SN 1987A in Ref. [24].

In this work we shall give additional estimates of expected parameters of a neutron star rotation in its final state after a second collapse. The second collapse became possible because of the torque loss by a rapidly rotating neutron star (Fig. 7 and the Table).

We shall, first, assume that the residual rotation (after the torque loss) does not prevent formation of a neutron star, which is not different from that calculated in Ref. [21] for a stationary case. The assumption is appropriate because major characteristics of neutrino signals from SN 1987A are in satisfactory agreement with theoretical calculations, Ref. [21], and experimental observations [19, 20], see [4]. So we take $\alpha_{fin} \approx 0.1$ for the final state according to (7). The subscript "fin" will denote parameters for the final neutron star close to the parameters of the model in Ref. [21] (Column 4 in the Table). The subscript "in" will denote the parameters of our initial rotating neutron star after a rapid loss of the torque. According to (7), during the second collapse $\alpha \sim r^{-1} \sim \rho^{1/3}$, i.e., we use the principle of the local torque conservation for a rapid hydrodynamic collapse. From this we can obtain the required initial value α_{in}, taking from the Table the density ratios $\rho_{fin} / \rho_{in} \sim \rho_{cfin} / \rho_{cin} \approx 34/2.6 \approx 13$ $= \sim$ 34/2.6 $= \sim$ 13. Hence,

$$\alpha_{in} \approx \alpha_{fin} \, (\rho_{cin} / \rho_{cfin})^{1/3} \approx 0.04$$

The estimate shows that the tabular value of α (Column 2) should be decreased about 20 times, i.e., $\alpha_{in} \approx 0.04$ instead of $\alpha \approx 0.8$. This decrease can be attributed to some physical mechanism of the torque loss by a rotating neutron star. This mechanism works in characteristic time periods that exceed both the period of the star rotation and the characteristic time of the primary collapse, considered in Section 3 and Ref. [1].

According to (7) α is proportional to the square of the angular speed, $\alpha \sim (j^2/r) \sim j^2 \sim \omega^2$, so ω_{in} should be decreased 5 times as compared to ω_c: $\omega_{cin} \approx \omega_c \, (\alpha_{in}/\alpha_c)^{1/2} \approx 2 \cdot 10^3 \ s^{-1}$. The rotation slowed down by the deceleration mechanism would certainly lead to the secondary collapse to form a slowly rotating neutron star similar to the model proposed in Ref. [21]. Now the angular rotation rate in the final state remains to be estimated. The low of the torque conservation predicts $\omega \sim r^{-2} \sim \rho^{2/3}$, i.e., ω_{cfin} will be approximately equal to $\omega_{cfin} \approx \omega_{cin} \, (\rho_{cfin} / \rho_{cin})^{2/3} \approx 2 \cdot 10^3 \cdot 5.5 \approx 10^4 \ s^{-1}$. Thus, the final central period of rotation turns out to be $\sim$ 0.6 ms. From the differentiated nature of rotation, a rotation period of the neutron star surface - an expected pulsar period - can be several times longer ($\sim 1 \div 3$ ms).

All the foregoing estimates can be presented as a single dependence on the characteristic values:

$$\omega_{cfin} = \omega_c \left[(\alpha_{fin} / \alpha_c) \, (\rho_{cfin} / \rho_{cin}) \right]^{1/2} \qquad (16)$$

In (16) the factor in the square brackets is about unity if α_{fin} is taken equal 0.1, which small enough to consider ω_{cfin} as an upper limit for the angular speed of the final state of a neutron star. Our estimates show that approximately $\omega_{cfin} \approx \omega_c$, in other words, the value of ω_c obtained from the numerical model is almost equal to the estimate of the upper limit of the angular speed in the final state. The finding is quite obvious: the hydrostatic equilibrium of a rotating star, i.e., an intermediate equilibrium state, differs but only slightly from the final state of the hydrostatic equilibrium of a stationary star, in the central density, for example, only 13 times.

Any change in α causes changes in the rotation energy: if α decreases by 20 times, the value of ε_{rot} will decrease 20 times as well. Thus, the required torque decrease correlates with the energy released by the explosion $\varepsilon_{ex} \sim \varepsilon_{rot} \sim 5 \cdot 10^{52}$ erg, according to the data in the Table. This value of energy release is about an order of magnitude greater than the explosion energy of a typical SN. It does not follow, however, that the ejection of the shell occurs, because the gravitational energy of all the star is great ($- 1.2 \cdot 10^{53}$ erg). Even the gravitational energy of the outer, over the neutrinosphere, shell 1 $M_\odot$ turns out to be ~ ε_{ex} and equals -4.4 · 10^{52} erg. However, the energy released is sufficient to form a powerful SW followed by its spreading into the outer layers of smaller density. Under these conditions the shell could be ejected with an energy accounting for the SN outburst, that is, of about 10^{51} erg. These qualitative agreements should be verified by using, say, a two-dimensional numerical model of a secondary collapse of a neutron star.

The period of a neutron star rotation of about 1 ms could have been derived if the initial angular speed ω_o were decreased 5 times, that is, at $\omega_0 / \Omega_0 \approx 0.17$. This would correspond to the rotation period before the first collapse, about 21 s. The result (the calculations for the collapse were, in fact, made at $\omega_0 / \Omega_0 = 0.1$) would practically be in agreement with the calculations in Ref. [21] for $\omega_o = 0$. In the final state, however, the period of rotation will be again ~ 0.6 ms (in the calculations with $\omega_0 / \Omega_0 = 0.1$ it was found equal ~ 1 ms). Of course, the rotation energy ε_{rot} will be 25 times smaller, about 2 · 10^{51} erg, and equal to ε_{ex} of a SN, i.e., to a minimum energy value for the shell ejection. In the whole range of the initial angular speed variations $0.86 \geq \omega_0 / \Omega_0 \geq 0.17$ the result should be the same in our two-stage scenario of the collapse, namely, formation of a neutron star rotating with a period of about 1 ms.

5. Conclusion

In February 1989 the discovery of a pulsar with a period of ~ 0.5 ms was reported [30] to be made on January 20, 1989, in the optical region on the site of SN 1987A outburst. Unfortunately, this extremely important discovery was not confirmed by further observations either by authors themselves (on January 31) or by other researchers. Perhaps, the pulsar only "peeped out" for a short period of time through optically thick clouds, which consist of the ejected shell material of SN 1987A. Or it may faded, for example, because some circumstellar material had deposited onto it. In the latter case we can expect to see the pulsar again but with a somewhat different rotation period. The authors of Ref. [30] observed the pulsar for 7 hours and its period was measured with an amazing precision, to the 7th meaningful figure. However, any decisive theoretical conclusions would seem now premature before the existence of a pulsar on the site of SN 1987A outburst is unequivocally confirmed.

The above calculations of the collapse of a rotating star core with = $2M_\odot$ in terms of a numerical model and qualitative estimates of a secondary collapse of the same star core together with a satisfactory theoretical interpretation of the registered neutrino signals from SN 1987A seem to support our conclusion made in work [3]: a millisecond pulsar could appear on the site of SN 1987A outburst in the Large Magellanic Cloud, being a direct result of the collapse of the rotating star core. At least this work shows that the low-limiting estimate of the pulsar period, which does not yet prevent a collapse with a registered powerful neutrino pulse, is about a millisecond.

The authors would like to express their gratitude to Drs. D.A. Varshalovich, A.D. Kaminker, and D.G. Yakovlev for fruitful discussions and their proposal to write this work. Our thanks are due to Dr. S.I. Blinnikov for interesting and useful discussions, to Dr. V.P. Utrobin for presenting Fig. 1 and Ms. N.A. Vulikh for the assistance in the preparation of this work.

Parameters	$\frac{\omega_0 R_0^{3/2}}{\sqrt{GM}} = 0.86$ (t = 2.9 s)	$\omega_0 = 0$ (t = 2.4 s)	$\omega_0 = 0$ (t = 5.1 s)
ρ_c (g/cm^3)	$2.6 \cdot 1014$	$2.2 \cdot 10^{15}$	$3.4 \cdot 10^{15}$
T_c (K)	$6.2 \cdot 10^{10}$	$9.6 \cdot 10^{11}$	$1.1 \cdot 10^{12}$
$\rho_{\nu ph}$ (g/cm^3)	$4.5 \cdot 10^{13}$	$2.1 \cdot 10^{13}$	$7.6 \cdot 10^{13}$
$T_{\nu ph}$ (K)	$2.2 \cdot 10^{10}$	$6.3 \cdot 10^{10}$	$6.5 \cdot 10^{10}$
$R_{\nu ph}$(cm)	$1.4 \cdot 10^{6}$	$1.5 \cdot 10^{6}$	$1.1 \cdot 10^{6}$
α_c	0.68	0	0
α_s	0.09	0	0
α_{max}	0.89	0	0
ω_c / ω_0	$5.8 \cdot 10^{3}$	0	0
ω_c (s^{-1})	$8.8 \cdot 10^{3}$	0	0
ε_g (erg)	$-1.2 \cdot 10^{53}$	$-5.3 \cdot 10^{53}$	$-6.6 \cdot 10^{53}$
ε_{in} (erg)	$3.9 \cdot 10^{52}$	$3.3 \cdot 10^{53}$	$4.1 \cdot 10^{53}$
ε_{rot} (erg)	$5.2 \cdot 10^{52}$	0	0
ε_{kin} (erg)	$9.2 \cdot 10^{46}$	$8.4 \cdot 10^{45}$	$7 \cdot 10^{42}$
$\varepsilon_{\nu\tilde{\nu}}$(erg)	$3.3 \cdot 10^{52}$	$1.9 \cdot 10^{53}$	$2.1 \cdot 10^{53}$
$L_{\nu\tilde{\nu}}$(erg s^{-1})	$2.8 \cdot 10^{51}$	$2.5 \cdot 10^{52}$	$1.7 \cdot 10^{52}$

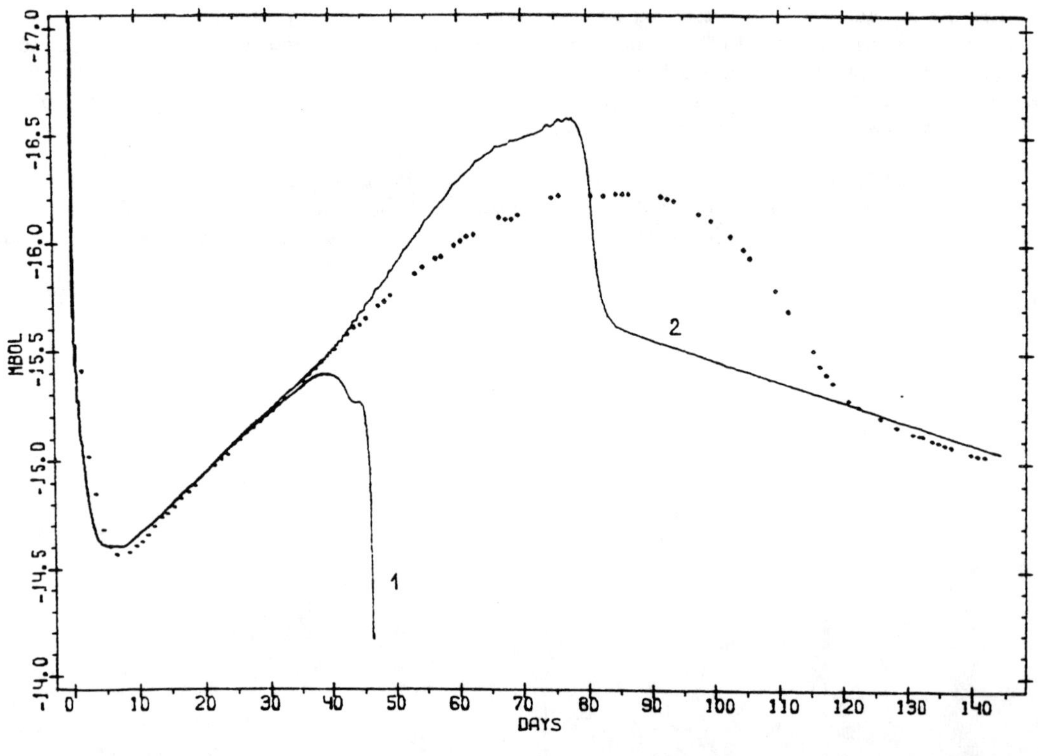

Fig. 1. Theoretical (solid lines) [10] and observed (dots) [32] bolometric curves of SN 1987A brightness.

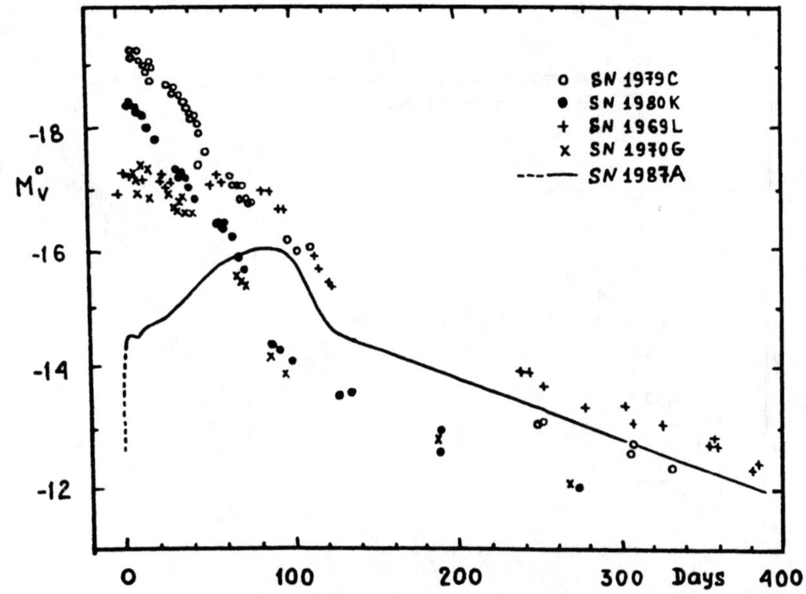

Fig. 2. SN 1987A brightness curve in the photometric region as compared to brightness curves of other SN of type II (from Ref. [33]). Supernovas SN 1979C and SN 1980K belong to the "linear SN", whereas SN 1969L and SN 1970 G to the "plateau-shaped SN."

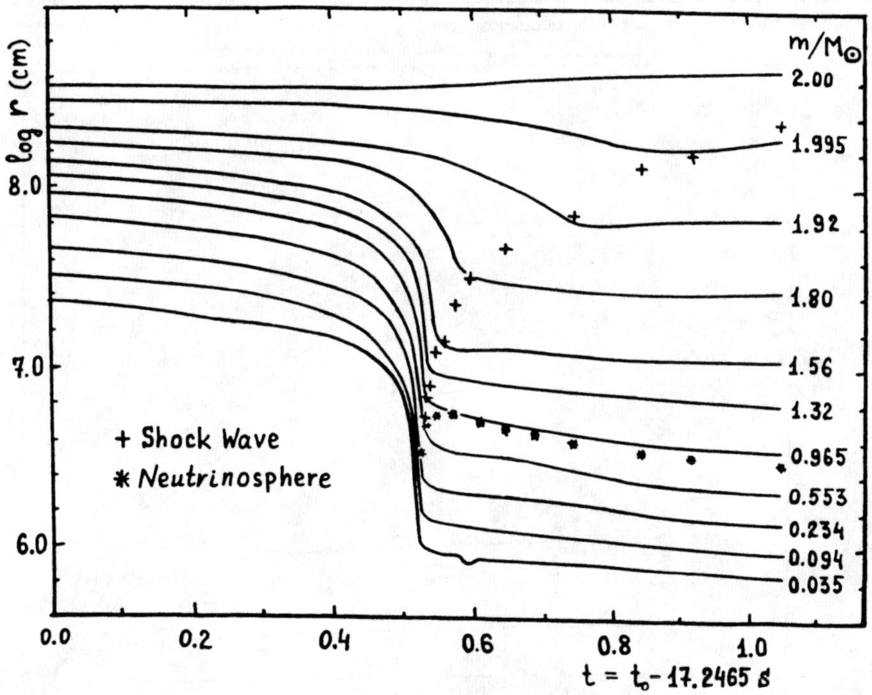

Fig. 3. Time dependence of the radii of different Lagrange layers for the collapse of a rotating iron-oxygen star core of a mass 2M⊙; t_0 is the time of the hydrodynamic calculation start.

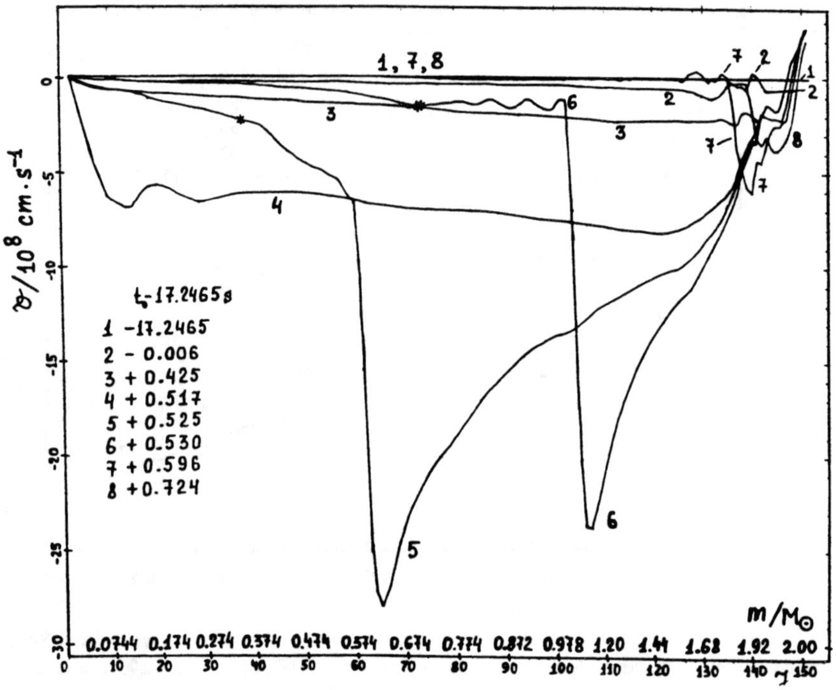

Fig. 4. Distribution of speed in the Lagrange coordinates: in mass M and The number of calculation zone J, in different instants. The asterisk shows the position of the neutrinosphere.

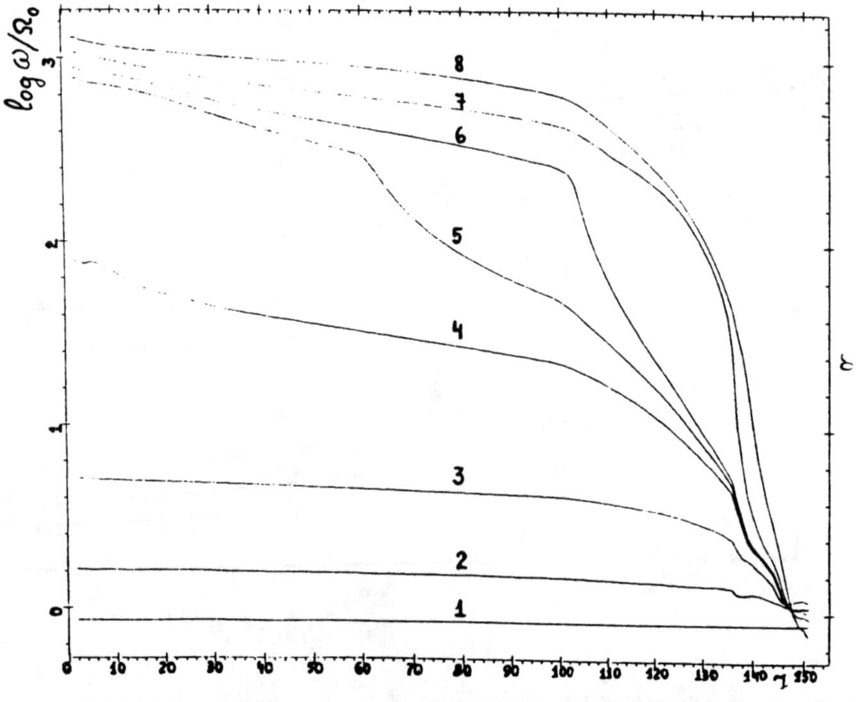

Fig. 5a. The angular speed distribution at the initial stage of the collapse in the Lagrange coordinate J (the corresponding scale M is in Fig. 4) for the instants from Fig. 4 (Ω_o = 1.78 s^{-1}).

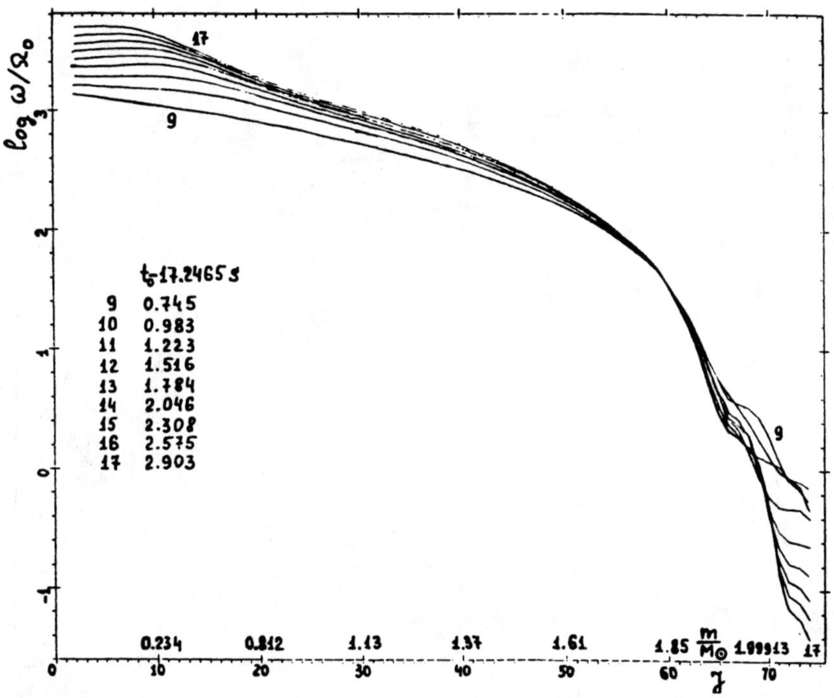

Fig. 5b. The angular speed distribution in the later instants, which correspond to the cooling of the central rotating neutron star (note the scale change in J and M0.

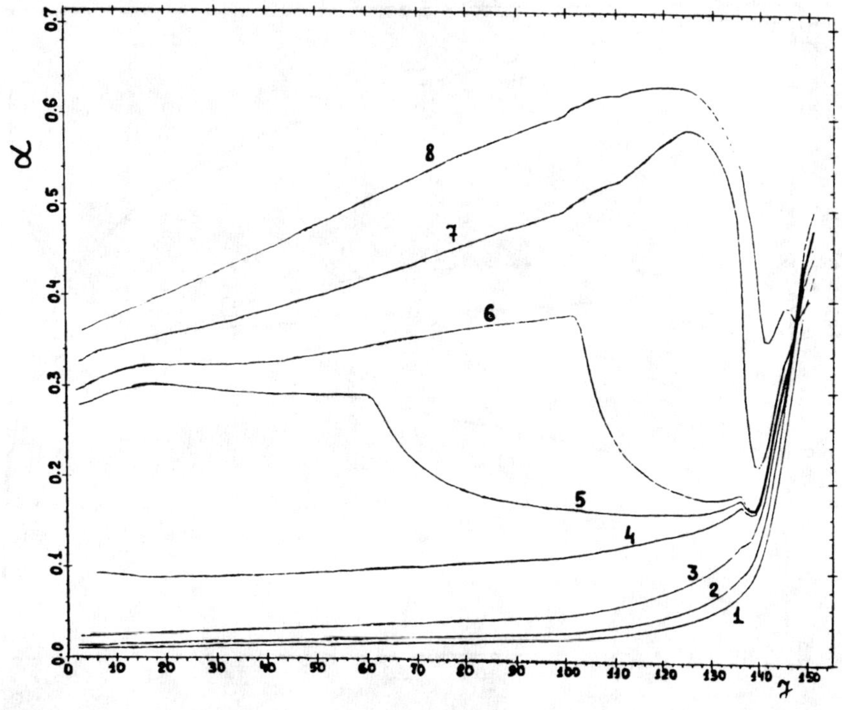

Fig. 6a. The ratio of the α centrifugal force to the force of gravitational attraction as a function of the Lagrange coordinate for the instants and the scale M as in Fig. 4.

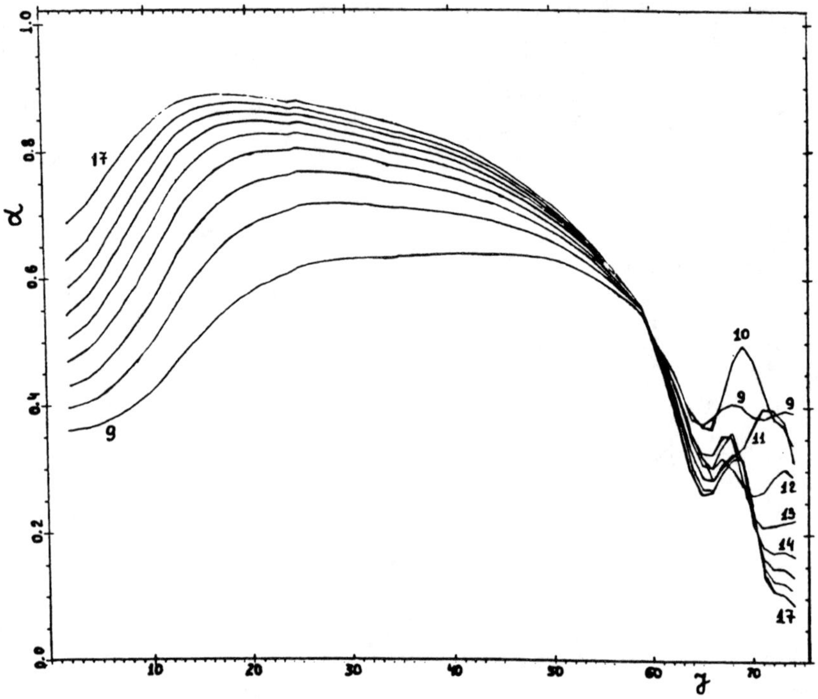

Fig. 6b. The same as in Fig. 6a but for the instants shown in Fig. 5b.

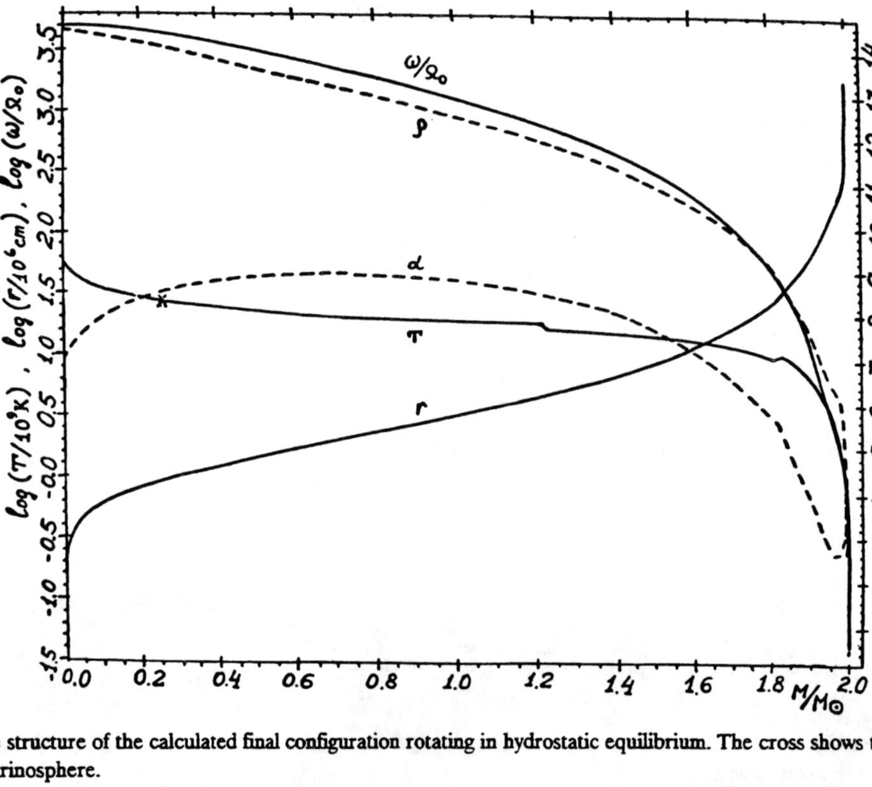

Fig. 7. The structure of the calculated final configuration rotating in hydrostatic equilibrium. The cross shows the location of the neutrinosphere.

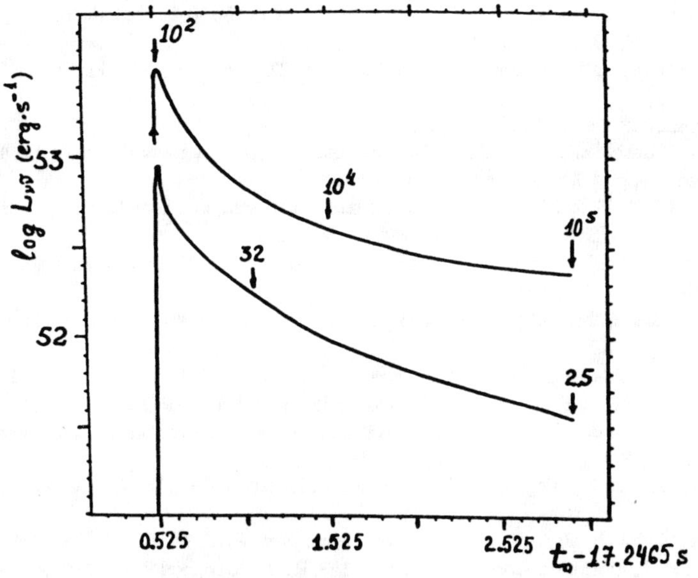

Fig. 8. Neutrino brightness curves for the collapse of a stationary (the upper curve) and rotating (the lower curve) star core of a mass 2M⊙. Over the curves are the values for neutrino "optical" thickness of the whole star core.

References

1. V.S. Imshennik and D.K. Nadyozhin, PAZh, 1977, v. 3, no. 8, pp. 353 - 358.
2. G.S. Bisnovaty-Kogan, AZh, 1970, v. 47, p. 813.
3. V.S. Imshennik and D.K. Nadyozhin, UFN, 1988, v. 156, no. 4, pp. 561 - 651. Sov. Sci. Rev. Sec. E. Astrophys. Space Phys. Rev., ed. R.A. Sunyaev, 1989, v. 7.
4. S.I. Blinnikov, V.S. Imshennik, and D.K. Nadyozhin, Astrophys. and Space Sci., 1988, v. 150, pp. 273 - 290.
5. A. Vidal-Madjar, P. Andreani, S. Cristiani, R. Ferlet, T. Lanz, and G. Vladilo, Astron. and Astrophys., 1987, v. 177. P. Andreani and A. Vidal-Majar, Nature, 1988, v. 333, p. 432. P. Andreani, R. Ferlet, and A. Vidal-Madjar, Nature, 1987, v. 326, p. 770.
6. I.S. Shklovskii, Stars: their Birth, Life and Death, 3rd ed., Nauka, Moscow, 1985.
7. G. Sonneborn, B. Altner, and R.P. Kirshner, Astrophys. J. Lett., 1987, v. 323, p. L35.
8. R.M. Catchpole et al., Spectroscopic and Photometric Observations of SN 1987A. Paper : Days 386 to 616, 10 December 1988.
9. E.K. Grassberg, V.S. Imshennik, and D.K. Nadyozhin, Astrophys. and Space Sci., 1971, v. 10, p. 28.
10. V.P. Utrobin, PAZh, 1989, v. 15, no. 2, pp. 99 - 108.
11. V.S. Imshennik and D.K. Nadyozhin, PAZh, 1988, v. 14, no. 12, p 1059.
12. R.M. Catchpole, P.A. Whitelock, M.W. Feast, J.W. Menzies, et al., Mon. Not RAS, 1988, v. 231, p. 75P.
13. K. Hirata, T. Kajita, M. Koshiba, et al., Phys. Rev. Lett., 1987, v. 58, p. 1490.
14. R.M. Bionta, G. Blewitt, C.B. Bratton, et al., Phys. Rev. Lett., 1987, v. 58, p. 1494.
15. R. Sunyaev, A. Kaniovskii, V. Efremov, et al., PAZh, 1987, v. 13, p. 1027; Nature, 1987, v. 330, p. 227.
16. T. Dotani, K. Hayashida, H. Inoue, et al., Nature, 1987, v. 330, p. 230.
17. S.A. Grebenev and R.A. Sunyaev, PAZh, 1988, v. 14, p. 675.
18. S.M. Matz et al., Nature, 1988, v. 331, p. 416.
19. V.L. Dadykin et al., Pis'ma ZhETF, 1987, v. 45, p. 464.
20. E.N. Alekseev et al., Pis'ma ZhETF, 1987, v. 45, p. 461.

21. V.S. Imshennik and D.K. Nadyozhin, ZhETF, 1972, v. 68, p. 1548. D.K. Nadyozhin, Astrophys. and Space Sci., 1977, v. 49, p. 399; 1977, v. 51, p. 283; 1978, v. 53, p. 131.
22. W.D. Arnett, The Origin and Evolution of Neutron Stars, eds. D.J. Helfand and J.-H. Huang: IAU Symposium N 125, 1987, pp. 278 - 280.
23. A. Burrows, Astrophys. J. Lett., 1987, v. 318, p. L63.
24. S.I. Blinnikov, V.S. Imshennik, and D.K. Nadyozhin, Soviet Scientific Reviews/ Section E, Astrophysics and Space Physics Reviews: 1988, v. 6, part 3, Supernovae, pp. 266 - 278.
25. E.K. Grasberg and D.K. Nadyozhin, Physics of Neutron Stars. Formation, Structure and Evolution, Ioffe Institute of Physics and Technology, the USSR AS, Leningrad, 1988, pp. 84 - 88.
26. V.S. Imshennik and D.K. Nadyozhin, Soviet Scientific Reviews/ Section E, Astrophysics and Space Physics Reviews, 1983, v. 2, pp. 76 - 161.
27. R. Mo@"nchmeyer and E. Mu@"ller, Preprint MPA 374, Core Collapse with Rotation and Neutron Star Formation, 1988, June.
28. J.R. Wilson, in Numerical Astrophysics, eds. J. Centrella, J. eblanc, and R. Bowers, Jones and Barlett, Boston, 1985.
29. J. -L. Tassoul, Theory of Rotating Stars, Princeton University Press, Princeton, New Jersey, 1978.
30. J. Middelditch, C. Pennypacker, D. Morris, et al., IAU Circular No. 4735, February 8, 1989. For discussion see Lindley D., Nature, 1989, v. 337, p. 595.
31. N.V. Ardelyan, G.S. Bisnovaty-Kogan, and Yu.P. Popov, AZh, 1979, v. 56, p. 1244. N.V. Ardelyan et al., AZh, 1987, v. 64, pp. 495, 761.
32. J.W. Menzies, R.M. Catchpole, and van G. Vuuren, Mon. Not. RAS, 1987, v. 227, p. 39P. R.M. Catchpole, J.W. Menzies, A.S. Monk, et al., Mon. Not. RAS, 1987, v. 229, p. 15P. R.M. Catchpole, P.A. Whitelock, W.W. Feast, et al., Mon. Not. RAS, 1988, v. 231, p. 75P.
33. M. Hamuy, N.B. Suntzeff, R. Gonzalez, G. Martin, Astron.
J., 1988, v. 95, p. 63.

Physics of Neutron Stars
Edited by A. M. Kaminker

ISBN 1-56072-164-2

SUPERNOVA 1987A: FORMATION OF HARD X-RAY AND GAMMA EMISSION SPECTRA

S.A. Grebenev and R.A. Sunyaev
Institute of Space Research, USSR Academy of Sciences, Moscow

Spectra of hard X-ray and γ-emission from Supernova 1987A, resulting from the radioactive decay of ^{56}Co and ^{57}Co in the shell, have been calculated together with the emission spectra from the star remnant: a neutron star or an accretting black hole in its center.

The exponential decrease of the bolometric brightness of Supernova 1987A (SN 1987A) in the optical and infrared regions, $L_B = L_{56} \cdot \exp(-t/t_{56})$, where $L_{56} \approx 1 \cdot 10^{42}$ erg/s and $t_{56} \approx 112$ days [1, 2] is a major evidence that about 0.075 $M_\odot$ of radioactive ^{56}Ni- ^{56}Co was synthesized in the supernova explosion. Beginning with the 120th day after the explosion, thermal radiation from the SM became completely determined by the energy released in ^{56}Co decay with a characteristic time of 111.3 days. Detection of hard X-ray emission from SM 1987A by the international orbital observatory ROENTGEN on the module KVANT [3, 4] and by the Japanese satellite GINGA [5] supports this conclusion. X-ray emission is a result of repeated Compton scatterings of gamma-photons, emitted in cobalt decay, by the shell electrons [6 - 8]. Photons, that leave the shell without being scattered, are responsible for an intensive line gamma emission from SN 1987A recorded by several observers [9 - 14].

Calculations [6 - 8, 15 - 21] made in terms of a radioactive model of X-ray and gamma emission from SN 1987A were found to be in a fair agreement with the observations and enabled some inference about the structure of the SN shell. The early X-ray emission detected from the SN may be to the following three causes: (1) the shell symmetry is far from being spherical [7], (2) the SN shell is extremely patchy [7], and (3) the material of the outer and inner layers undergoes constant intermixing to cause redistribution of radioactive elements all over the shell [18 - 21].

The results of the 18-month observations of a hard X-ray region of 15 - 200 keV of SN 1987A made by the observatory ROENTGEN on the module KVANT [3, 4, 22, 23] strongly restrict the parameters of possible models. In particular, they allow estimation of the character and extent of possible intermixing of radioactive cobalt in the shell [23]. In this report mechanisms of hard X-ray and gamma emission formation in SN 1987A are considered in detail in terms of a model with an optimal ^{56}Co distribution.

1. Model of the shell.

Density distribution $\rho(r)$ and the rate of the flying apart of the shell material v(r) are taken from the work by Arnett [24]. The shell consists of an oxygen-silicon (90% O and 10% Si) core of a mass $1M_\odot$, a helium layer of a mass $2.5M_\odot$ and an outer hydrogen-helium shell containing $10M_\odot$ of matter. The abundance of the elements of the iron group was supposed to be typical of the LMC and equal to 0.364 of that in the Sun. Figure 1 shows the distribution of $0.075M_\odot$ ^{56}Ni - ^{56}Co - ^{56}Fe, synthesized in the explosion, in the SN shell. At early stages of expansion, about 20% radioactive cobalt was assumed to be redistributed almost uniformly all over the shell. The ^{56}Co distribution used here is in the best agreement (in terms of the chosen model for the distribution of the shell material velocities and density) with the evolution of hard X-ray emission from the SN according to the results of the module KVANT (see Fig.1 in Ref. [23]). The data available are, however, insufficient to unequivocally determine ^{56}Co abundance in the shell outer layers, whose emission was important at early stages of expansion before systematic observations of SN 1987A by ROENTGEN observatory. Neither can it be determined exactly in the innermost layers, where ^{56}Co redistribution affects but slightly the observed X-ray emission. This ambiguity results in great errors in Fig. 1. The calculations involved the synthesis ~ $0.01M_\odot$ of non-radioactive ^{58}Ni. Knowledge of the abundance of this isotope is important when the intensity of the shell fluorescence is calculated in K-@a lines of Ni 7.5 keV. The distance to the LMC is taken equal to 55 kpc.

2. ^{56}Co X-ray and gamma emission. More than 96% of the energy released in ^{56}Co decay is carried away by gamma-quanta with energies 3.26 MeV (12%), 2.60 MeV (18%), 2.02 MeV (11%), 1.77 MeV (16%), 1.36 MeV (4%), 1.24 MeV (67%), 1.04MeV (14%), 847 keV (100%), and 511 keV (40%). Shown in the brackets is the probability of a photon birth with a given energy per one decay act [25]. The fate of all gamma-photons is more or less similar. The main elementary process of photon interactions with the shell material is Compton scattering by electrons (both free and bound in atoms). Photons very rapidly, within several scatters, decrease their energy down to about 100 keV. The spectrum observed is a result of diffusion: repeated Compton scattering of photons in the shell. The photon energy decreases at every scattering act by an electron at rest due to the recoil effect. Thus, a flux of photons forms directed downwards along the frequency axis. Photons experiencing to more than 20 scatterings ~ $m_e c^2/h\nu$ acquire X-ray energies $h\nu$ ~ 30 keV. The process is characterized by a wide distribution of photons according to the number of scattering they experience, therefore the spectrum of the outing emission should be countinuous and broad [26]. At energies smaller than ~ 20 keV the SN

emission spectrum is determined by photoabsorption on the K-shells of heavy elements, those of the iron group first of all, including radioactive ^{56}Co and its decay product ^{56}Fe. About one third of the hard photons absorbed appear in the fluorescence lines of Fe 6.4 keV and Ni 7.5 keV.

Monte-Carlo [27] spectra of X-ray emission outgoing from the SN shell 180, 270, 340 and 580 days after the explosion are shown in Fig.2 together with the experimental data of the HEXE and Pulsar X-1 instruments of the ROENTGEN observatory aboard the Mir station [3, 4, 22, 23] and the data recorded in a balloon experiment by the group from Caltech [10]. The shell expansion is accompanied by a decrease of its optical thickness through Thomson scattering $\tau_T \sim t^{-2}$. Accordingly, photons in the shell experience a fewer number of scatterings, $u \sim \tau_T^2 \sim t^{-4}$. We believe that the evolution of the spectra in Fig. 2 with time can be due to this process together with the exponential (the characteristic time is 111.3 days) decrease in the radioactive cobalt amount in the shell (by the time this paper is being written, more than 99% of the cobalt formed in the explosion has decomposed). At a later stage of expansion the number of scatterings experienced by the photons in the shell is not sufficient for them to reach X- ray energies. In particular the sharp drop at $h\nu \leq$ 40 keV in the SN 1987A spectrum (Fig.2) on the 580th day may be accounted for by an exponential break in the photon distribution by the number of scatterings taking place. In the emission spectra of earlier stages such a drop results from the photon death during photoionization of K-electrons of Fe group elements.

Photons that did not undergo scattering in the shell, i.e. photons of direct escape form a powerful line gamma emission of SN 1987A. Emission in the region of two most prominent lines of ^{56}Co, at 847 and 1238 keV, was recorded by the orbital station SMM [9] and during some balloon experiments [10 - 14]. Fig.3presents experimental data available at present and the results of theoretical calculations in terms of the model involving ^{56}Co intermixing (solid lines) and without it (dotted lines). It can be seen that the latter model is in rough contradiction with the observations. Early detection of gamma emission in the model involving ^{56}CO intermixing can be accounted for by low abundance of ^{56}Co in the shell outer layers, whose optical depth with respect to Thompson scattering is small. The velocities of expansion of these shell layers may be as high as 2000 - 5000 km/s, therefore the gamma lines of the shell have a strong Dopler broadening and are shifted towards higher energies (Fig. 4).

3. Optical and infrared luminosity of SN 1987A. The energy, that X-ray and gamma photons lose in Compton scattering because of the electron recoil effect, transforms, in the long run, into thermal emission infrared and optical shell emission. An evidence of this is an exponential decrease of the bolometric luminosity of SN 1987A with a characteristic time of about 112 days, beginning with the 120th day after the explosion [1, 2, 28 - 29]. In the first months after X-ray emission begins, hard X-ray and gamma photons leaving the shell carry away less than 1% of the ^{56}Co decay energy. As the shell becomes more and more transparent, hard X-rays take away with them a greater fraction of brightness, so that the bolometric luminosity curve starts to deviate from the exponential law [7, 21]. According to calculations (Fig. 5) this deviation is as great as 50% on the 640th day after the explosion.

The major fraction of the bolometric luminosity is carried away by gamma quanta with energies from 0.5 to 3.2 MeV. It was found that 30% of the luminosity at the moment of the hard X-ray and gamma emission detection and 70% of it 2 or 3 years later are carried away in the gamma lines of direct escape [21]. It is easy to explain because a hard ($h\nu \sim$ 1 MeV) photon loses about a half of its initial energy right after the first scattering because of the recoil. Thus, optical and infrared observations of SN 1987A allow an independent estimate of the photon flux of direct escape in cobalt gamma lines.

4.Emission of radioactive ^{57}Co. Some other radioactive isotopes are synthesized in a supernova explosion in addition to ^{56}Ni - ^{56}Co. Among them are ^{57}Ni - ^{57}Co, ^{22}Na, ^{44}Ti - ^{44}Sc, and their emission may greatly contribute to the shell emission [30, 31]. In the case a Supernova 1987A, radioactive ^{57}Co (a ^{57}Ni nuclide) with a lifetime of 391 days is most likely to be detected in the X-ray region [7, 17, 32 - 34]. Figure 6 shows the results of calculations for the X-ray radiation yield formed in the shell during ^{57}Co decay (in addition to ^{56}Co emission). The calculations included ^{57}Co lines with energies 136 keV (10.6%), 122.1 keV (85.9%) and 14.4 keV (9.2%). Shown in brackets is the probability of a photon birth of a given energy per one scatter [25]. The distribution of ^{57}Co in the shell was assumed coincident with that of ^{56}Co. The magnitude of an X-ray flux formed in the shell is proportional to the total amount of ^{57}Co from the explosion. The ratio of ^{57}Fe and ^{56}Fe abundance (the products of ^{57}Co and ^{56}Co decay) in the Solar System is η_{57} = 0.024 [34]. However, this value may not remain unchangeable in all supernova explosions. Calculations of an explosion neucleosynthesis for SN 1987A [36] predict the ratio of ^{57}Co to ^{56}Co yields twice as large as that in the Solar System. Its value η_{57} = 0.05 was used in the calculations presented in Fig.6. The detection of X-ray and gamma emission due to ^{57}Co decay, and the line of 122 keV from SN 1987A in the first place, would help to determine the abundance of ^{57}Co in the shell more precisely.

its presence beneath the SN shell even still earlier as a deviation in the optical curve of its brightness. Indeed, each act of ^{57}Co decay liberates only 144 keV of energy, i.e. 26 times less than in the case of ^{56}Co. However, the contributions of these isotopes into the shell warming up become equal very soon because ^{57}Co decay period is t_{57} = 391 days and that of ^{56}Co is much shorter t_{56} = 111.3 days.

$$t = \frac{t_{56} t_{57}}{t_{57} - t_{56}} \ln \left(\frac{26}{\eta_{57}} t_{57} / t_{56} \right) = 1296 - 156 \cdot \ln \left(\frac{\eta_{57}}{0.024} \right) \text{ days} \quad (1)$$

It is in the moment that the deviation in the optical curve of the SN brightness from the exponential law with a characteristic time of about 111 days may be expected. Since then the curve starts to obey the exponential law with a characteristic time of about 391 day. In fact it occurs even earlier (see Fig. 7). The optical luminosity of the shell "consumes" only a part of the neuclide decay energy, the other part is carried away by X-ray and gamma photons (see paragraph 3). ^{57}Co isotope is known to mainly emit in the soft gamma region for which the shell remains nontransparent much longer than for gamma lines of ^{56}Co, therefore the energy of ^{57}Co decay transforms into heat to a far greater extent than that of ^{57}Co.

It is worth while remarking that however insignificantly small are the amounts of radioactive ^{22}Na and ^{44}Ti in the shell as compared to ^{56}Co concentration it is the energy from their decay that will determine the optical and infrared luminosity of the SN in 3 or 4 years after the explosion (Fig. 7). The decay period of the radioactive ^{22}Na is 1370 days: 92% ($\approx$ 2.19 MeV) of the energy in every decay act is released as photons with energy 511 keV (the birth probability is 180%) and 1275 keV (99.9%) [25]; the abundance of ^{22}Na in comparison with that of ^{56}Co is $\eta_{22} \approx 0.001$ or about $3 \cdot 10^{-5}$ M$_\odot$ [31]. The decay period of ^{44}Ti is 68 years and 79% ($\approx$ 2.28 MeV) of the energy in each decay act is released as photons with energies 68 keV (88%), 78 keV (95%), 511 keV (189%), 1157 keV (99.9%) and 1499 keV (0.9%). The last three lines belong to ^{44}Sc decay, a nuclide of ^{44}Ti, whose decay period is 5.7 h [25]. The abundance of ^{44}Ti compared to ^{56}Co is $\eta_{44} \approx$ 0.0017 or about $1 \cdot 10^{-4}$ M$_\odot$ [31, 36]. The contribution of these isotopes into the shell thermal radiation is comparable with that of ^{56}Co in 1200 days after the explosion, in the case of ^{22}Na, and in 1345 days, in the case of ^{44}Ti, according to formulas similar to (1).

5. Emission of the star remnant.

The explosion of SN 1987A most likely occurred to form a star remnant: a neutron star or a black hole as is evidenced by the neutrino outburst detected in the moment of the explosion [37, 38]. Unfortunately, first expectations [39] to detect a pulsar with a luminosity of 10^{41} - 10^{42} erg/s beneath the shell failed to realize. The fact that no deviations of the optical luminosity curve for SN 1987A from the exponential law with a characteristic time ~ 111 days [40] occurred as long as 600 days after the explosion, excludes the possibility of an additional to ^{56}Co source of energy beneath the shell. This would restrict a possible luminosity of the star remnant to about $3 \cdot 10^{38}$ erg/s (Fig. 8).

Indeed, in the first years after the explosion practically all the remnant emission should transform into the thermal optical and infrared radiation of the shell, irrespective of the emission nature and the energy range of the luminosity [21]. In X-ray and gamma ranges the shell is nontransparent because of the photoabsorption during the ionization of K-electrons of heavy elements and Compton scattering. If the remnant intensively emits in the region of superhard gamma rays (for example, if a young pulsar of the type of the pulsar in the Crab is in the center of the SN shell), the absorption of gamma radiation may be due to two processes: (1) photobirth of e^+ - e^- pairs on the electrons and core of the shell and (2) $\gamma-\gamma$ interaction with optical and infrared photons in the SN shell, which is responsible for the absorption of gamma-quanta with energies 10^{11} - 10^{15} eV [21, 41]. The luminosity of the remnant in the radio and infrared region should also transform into thermal radiation; the brightness emission temperature here exceeds $m_e c^2$ as a result of Compton induced warming up of the plasma [42].

An additional to ^{56}Co source of energy beneath the shell, if present, and the emission from the star remnant in the first place would be detected by observing the optical and infrared curves of the SN brightness. However, the distortions in them caused by the shell becoming more and more transparent for hard rays would hardly allow an unequivocal determination of the source nature: optical data do not give information about the real rate of the bolometric luminosity decrease of the SN. Moreover, optical and infrared observations of SN 1987A at later stages of the shell expansion become complicated because of the presence of two bright stars with a total luminosity ~ 10^{38} erg/s [43] at an angular distance of about 1" - 3" from the SN. Therefore, observations of SN 1987A in other spectral regions become of paramount importance.

The observatory ROENTGEN on the module KVANT and the observatory GRANAT, planned for launching in 1989, are to seek for the hard X-ray emission from the star remnant. The shell drastically changes the emission spectrum of the remnant. The dominating role in the spectrum formation 2 or 3 years after the explosion belongs to the photoabsorption on the K-shells of the Fe group elements. Its effect will continue long after SN shell becomes optically thin with respect to Thomson scattering. The process prevents the shell from becoming transparent in the standard X-ray region for many years [33]. Figure 9 shows the results of calculations of the spectra of hard X-ray emission formed in the shell. These calculations were made assuming the remnant spectrum in the X-ray energy range to be similar to the emission spectrum of the pulsar in the Crab nebula, $I_\nu(\nu) \sim \nu^{-2.1}$. The source spectrum was normalized so that its luminosity in the range of 1 - 1000 keV at a distance of 55 kps were equal to $3 \cdot 10^{38}$ erg/s. The real X-ray luminosity of this source ($\sim 3 \cdot 10^{37}$ erg/s, a total luminosity of the plerion in the Crab) was mach lower. The calculations show that an accreting neutron star(or a black hole) or a young pulsar in the LMC could be detected in the nearest years only if their luminosity would be close to or exceed 10^{38}. It should be noted that almost critical Eddington or even supercritical accretion can be expected under the conditions predicted for the center of the SN shell near an accreting object [33]. An example of such an object in which supercritical accretion occurs may be a well known X-ray pulsar SMC X-1 emitting in a hard region [44]. On the other hand, object SS433 shows that under supercritical accretion conditions the major fraction of luminosity may not be in the hard X-ray region. Only complex observations in the X- ray region and bolometric measurements in optical, infrared and ultraviolet regions will enable identification the nature of the central object. If a young pulsar is in the shell center, its total optical and infrared luminosity may be much greater than that in the X-ray range.

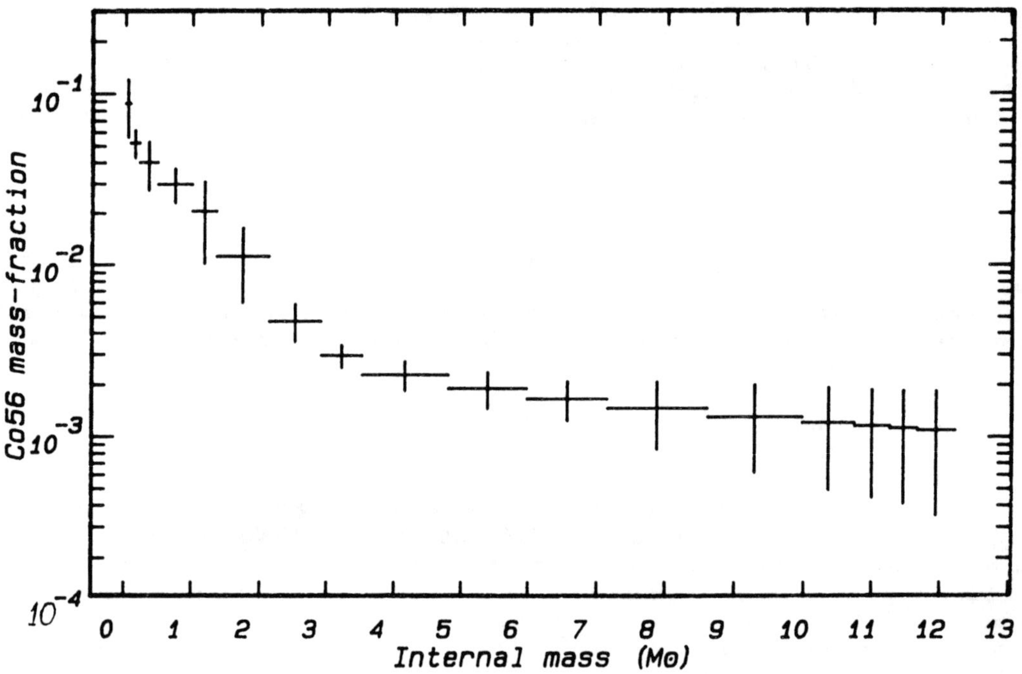

Fig.1. Distribution (in mass fraction) of radioactive ^{56}Co in the supernova shell in the moment of explosion derived from the observational data on the evolution of hard X-ray emission of 1987A, obtained by the observatory ROENTGEN on the module KVANT [23].

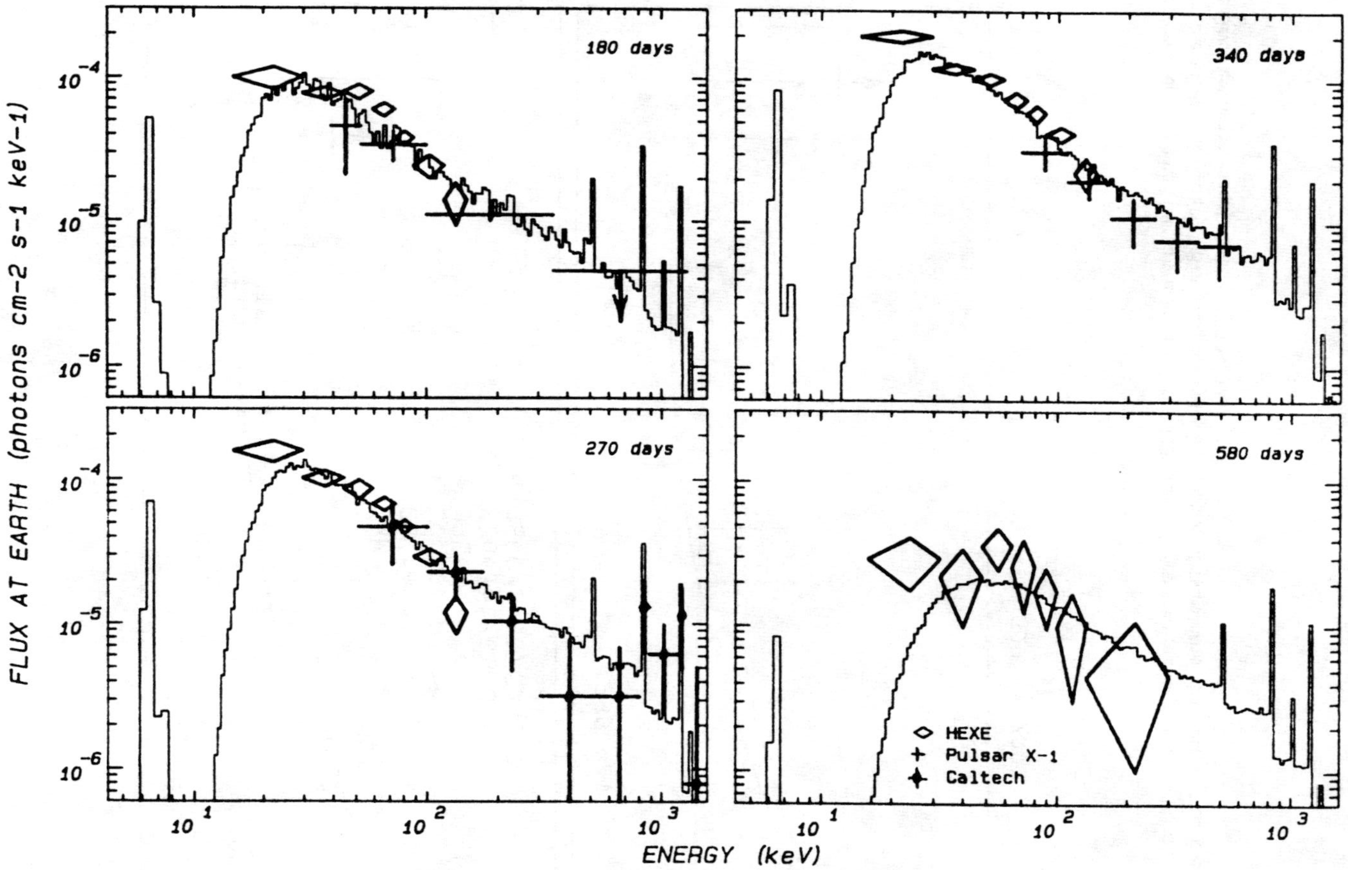

Fig.2. Spectra of the emission formed in the SN 1987 shell during radioactive ^{56}Co decay 180, 270, 340 and 580 days after the explosion. Monte Carlo calculations (histograms) and experimental data [3, 4, 10, 22, 23]. Energy resolution of the calculations in γ-lines of ^{56}CO is (in keV): 0.4 (blend of K-α lines of iron 6.4 keV, cobalt 6.9 keV and nickel 7.5 keV), 18 (511 keV), 29 (847 keV), 35 (1.04 MeV), 42 (1.24 MeV), 47 (1.36 MeV).

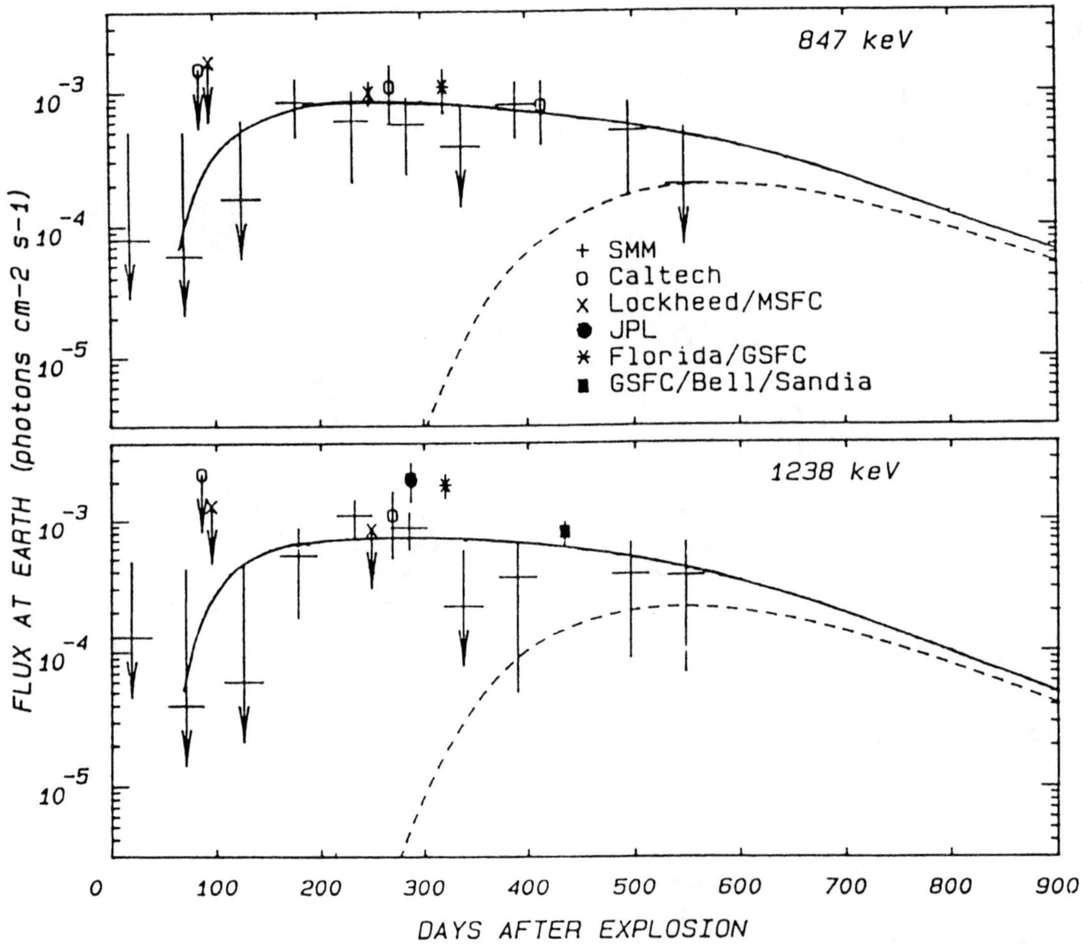

Fig.3. Dependence of the photon flux in γ-lines of 847 and 1238 keV ^{56}Co on the time elapsed since the explosion: solid lines are for the models of the SN 1987A shell involving intermixing of radioactive elements; dotted lines - without intermixing. Given are observational results for SN 1987A.

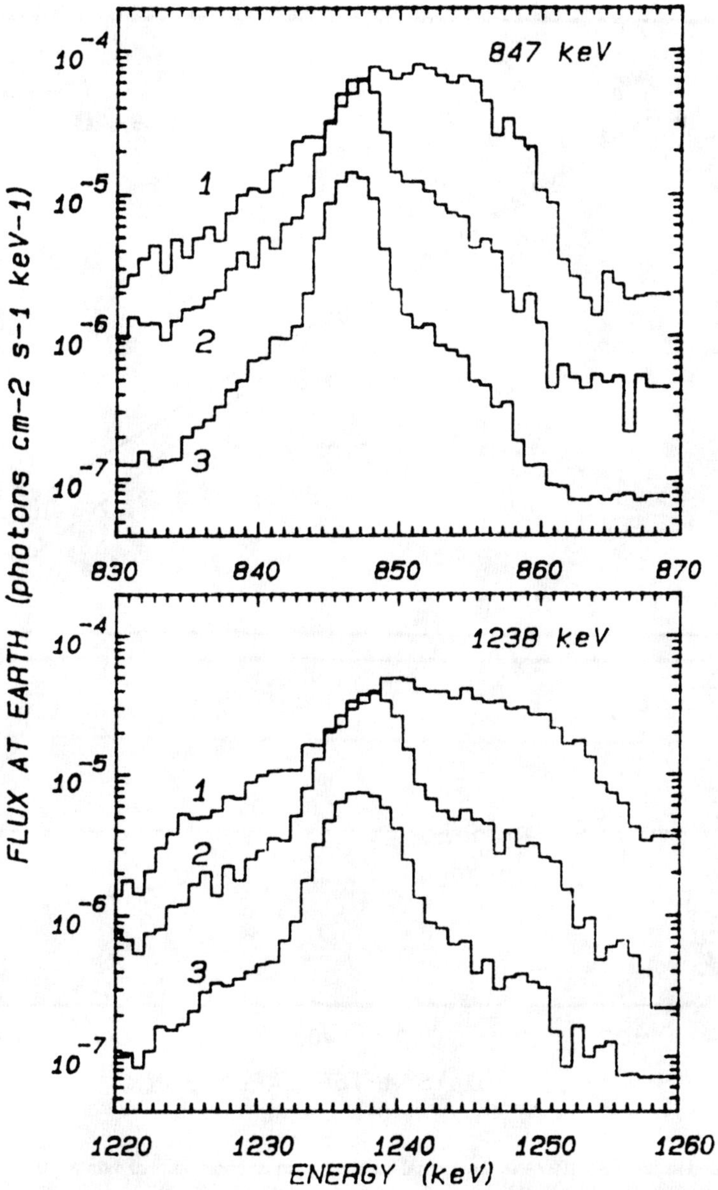

Fig.4. Profiles of γ-lines of 847 and 1238 keV ^{56}Co from SN 1987A for the model with radioactive elements intermixing on the 340th (1), 645th (2) and 900th (3)day after the explosion (Fig.1.). The background is determined by the emission in harder cobalt lines, subjected to Compton scattering in the shell. The lines have a strong Dopler broadening and are shifted towards greater energies because part of cobalt is in the outer, rapidly expanding zones of the shell. Infrared shift of the lines due to the LMC flying away at a speed of 270 km/s has been taken into account.

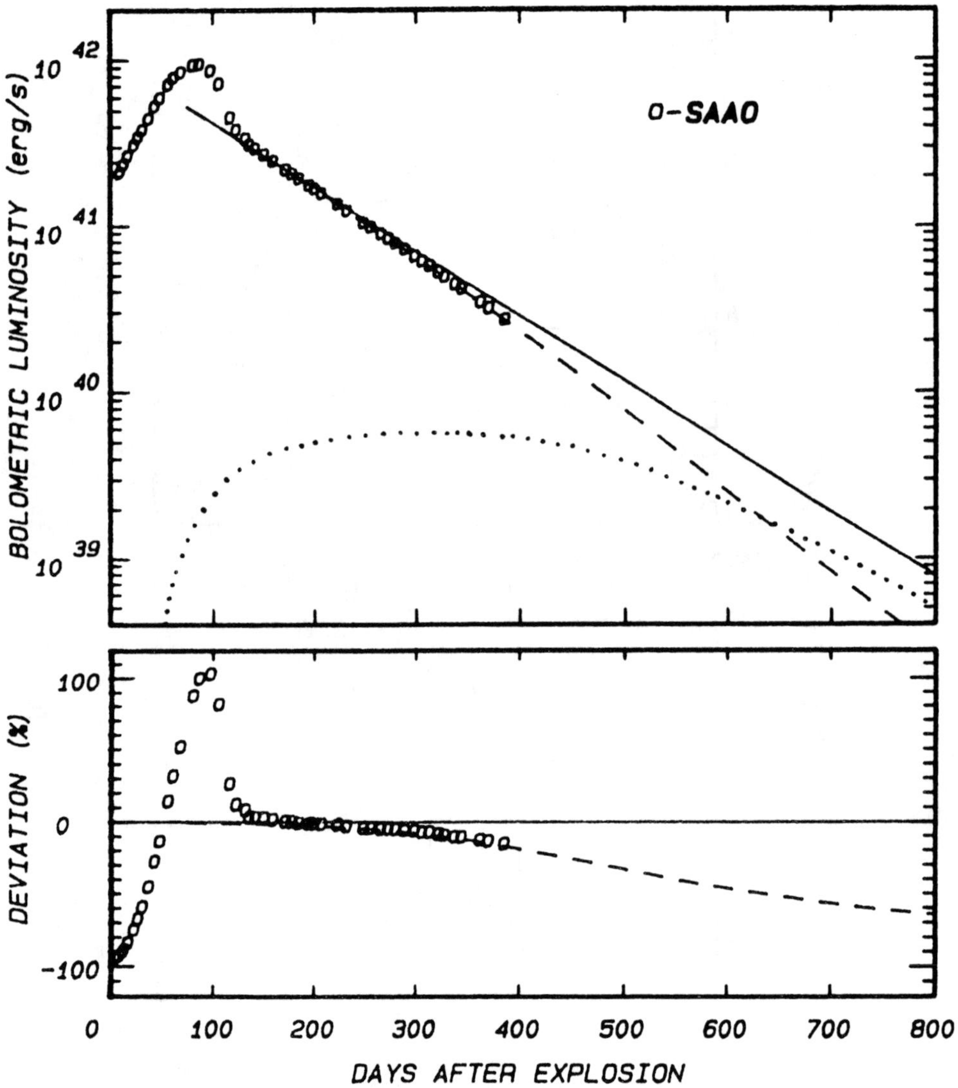

Fig.5. Bolometric luminosity of SN 1987A in the optical infrared regions: observational data obtained in SAAO [1, 28 - 29] and theoretical predictions. In the interval between the 120th and 240th days after the explosion the optical curve of the SN brightness coincided with the curve of energy release during 0.075M⊙ ^{56}Co decay with a characteristic time of 111.3 days (solid lines). Beginning with the 240th day the optical luminosity decreased rapidly (dashed lines). An explanation may be that a considerable fraction of the energy from Co decay is carried away by X- ray and γ-lines (dotted line) so that not all this energy transformed into thermal radiation by that time. From the Figure at the bottom it follows that even before the 120th day after the explosion the integrated (with respect to time) luminosity of the SN was primarily determined by the energy released in the decay of radioactive ^{56}Ni and ^{56}Co (straight line).

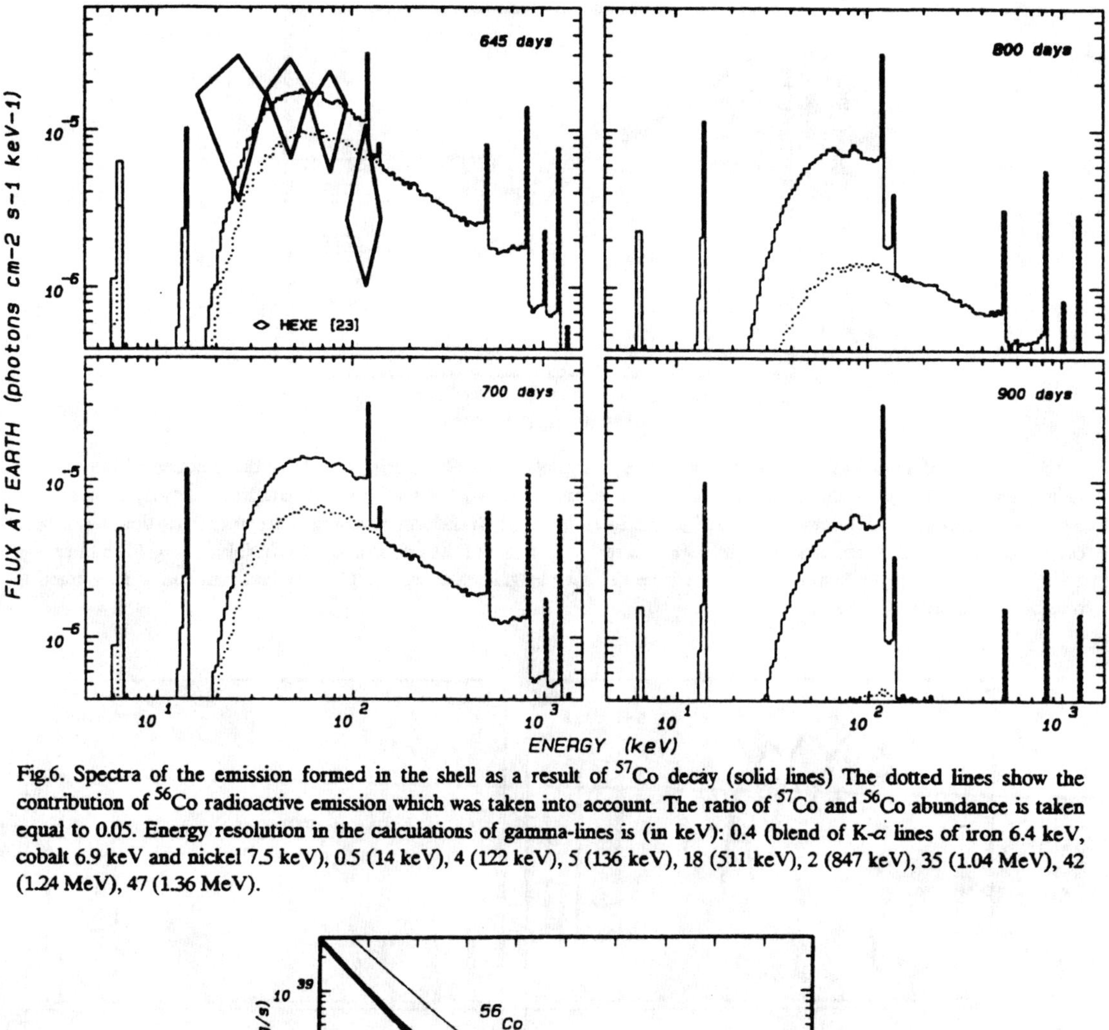

Fig.6. Spectra of the emission formed in the shell as a result of ^{57}Co decay (solid lines) The dotted lines show the contribution of ^{56}Co radioactive emission which was taken into account. The ratio of ^{57}Co and ^{56}Co abundance is taken equal to 0.05. Energy resolution in the calculations of gamma-lines is (in keV): 0.4 (blend of K-α lines of iron 6.4 keV, cobalt 6.9 keV and nickel 7.5 keV), 0.5 (14 keV), 4 (122 keV), 5 (136 keV), 18 (511 keV), 2 (847 keV), 35 (1.04 MeV), 42 (1.24 MeV), 47 (1.36 MeV).

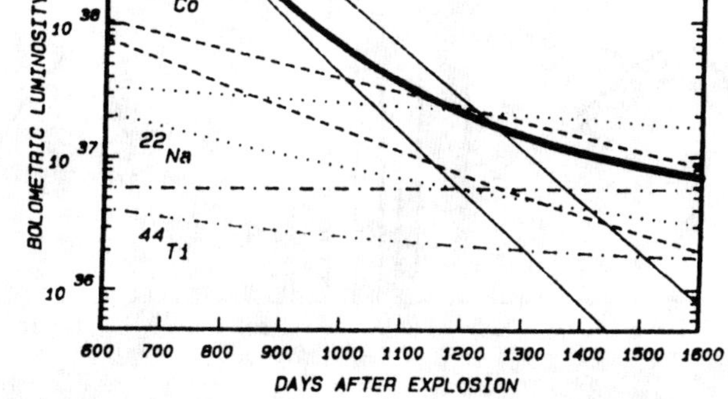

Fig.7. Deviation of the bolometric optical and infrared curve of the SN 1987A brightness from the exponential law with the fading time ~ 11days (a thick solid line). The thin solid lines are the shell luminosity due to ^{56}Co decay, dashed lines - for ^{57}CO, dotted lines for ^{22}Ni, dashed-dotted lines for ^{44}Ti (the upper lines are for bolometric luminosity, the lower lines for the luminosity transformed into the shell thermal emission).

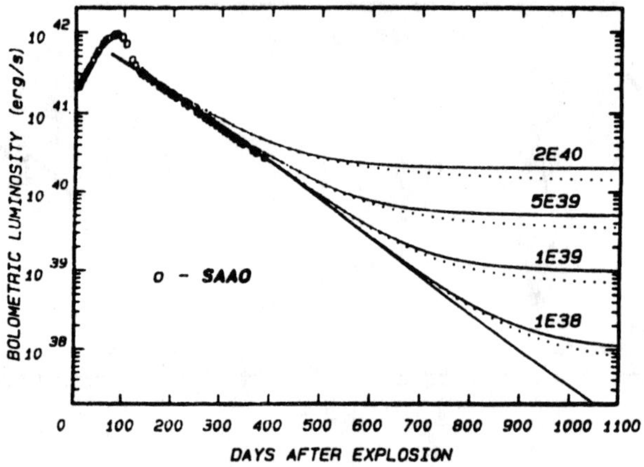

Fig.8. Deviation of the bolometric optical and infrared curve of the SN 1987A brightness from the exponential law with the fading time ~ 111 days due to the emission of the star remnant beneath the shell (the remnant luminosity is shown in erg/s over the curves) The solid lines correspond to the total luminosity transformation into thermal energy; the dotted lines correspond to the pulsar emission, with its spectrum resembling that of the pulsar in the Crab Nebula (Fig. 9) in the range 1 - 5000 keV; part of the luminosity escapes from the shell as gamma-emission. The remnant luminosity is assumed to remain unchanged for 2 - 3 years.

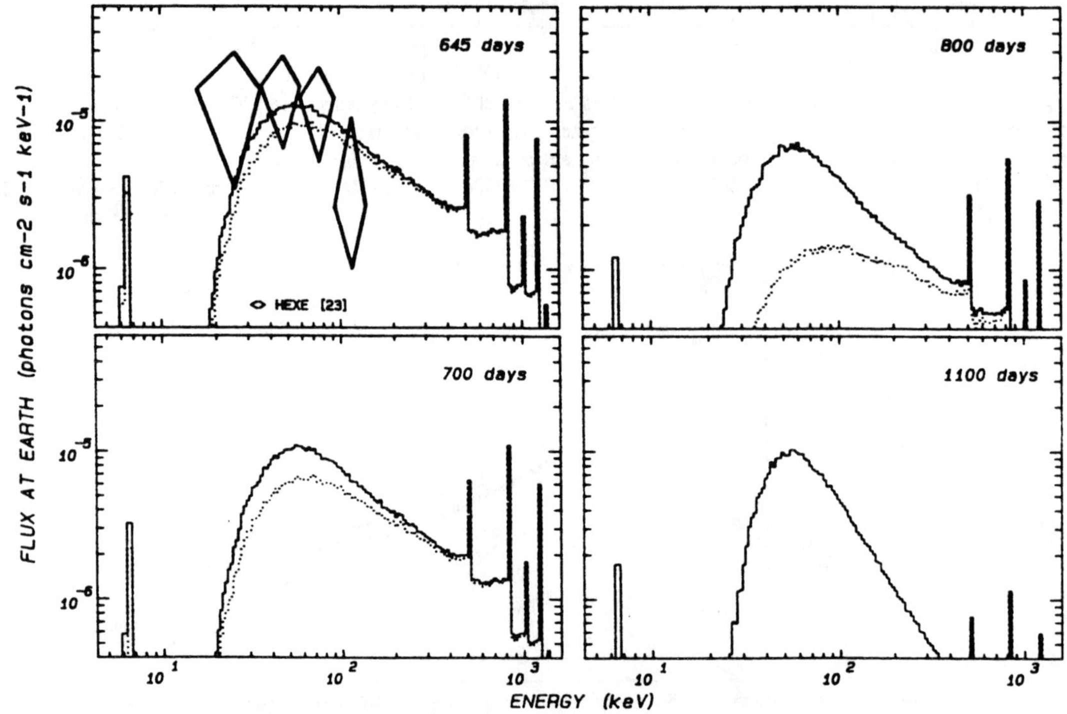

Fig .9. Spectra of emission formed in the shell due to the emission of an X-ray source in its center with the spectrum resembling that of the plerion in the Crab (solid lines). The contribution of radioactive ^{56}Co emission is shown by dotted lines. The luminosity of the central source in the range 1 - 1000 keV was taken equal to $3 \cdot 10^{36}$ erg/s.

References

1. R.M. Catchpole, J.W. Menzies, A.S. Monk et al., Monthly Not. Roy. Astron. Soc, 1987, v. 229, p. 15P.
2. M. Hamuy, N.B. Suntzeff, R. Gonzales, G. Martin., Astron. J., 1988, v. 95, p. 63.
3. R. Sunyaev, A. Kaniovskiy, V. Efremov et al., Nature, 1987, v. 330, p. 227.
4. R.A. Sunyaev, A.S. Kaniovskiy, V.V. Efremov et al., Pis'ma AZh, 1987, v. 13, p. 1027.
5. T. Dotani, K. Hayashida, H. Inoue et al., Nature, 1987, v. 330, p. 230.
6. R. McCray, J.M. Shull, P. Sutherland, Astrophys. J. Letters., 1987, v. 317, p. L73.
7. S.A. Grebenev and R.A Sunyaev, Astrophys. J. Letters., 1987, v. 13, p. 945.
8. Y. Xu, P. Sutherland, R. McCray, R.P. Ross, Astrophys. J., 1988, v. 327, p. 197.
9. S.M. Matz, G.H. Share, M.D. Leising et al., Nature, 1988, v. 331, p. 416 (SMM).
10. W.R. Cooke, D.M. Palmer, T.A. Prince et al., Astrophys. J. Letters, 1988, v. 334, p. L87 (Caltech).
11. W.A. Mahoney, L.S. Varnell, A.S. Jacobson et al., Astrophys. J. Letters, 1988, v. 334, p. L81 (JPL).
12. W.G. Sandie, G.H. Nakano, L.F. Chase et al., Astrophys. J. Letters, 1988, v. 334, p. L91 (Locheed/MSFC).
13. A.S. Rester, G. Eichorn, R.R. Coldwell et al., IAU Circ., N 4535, 1988 (Florida/GSFC).
14. S. Barthelmy et al., IAU Circ., N@4593, 1988 (GSFC/Bell/Sandia).
15. N. Gehrels, C.J. MacCallum, M. Leventhal, Astrophys. J. Letters, 1987, v. 320, p. L19.
16. O.S. Bartunov, S.I. Blinnikov, .V. Levakhina, and .K. Nadyozhin, Pis'ma AZh, 1987, v. 13, p. 744.
17. K.W. Chan, R.E. Lihgenfelter, Astrophys. J. Letters, 1987, v. 318, p. L51.
18. M. Itoh, S. Kumagai, T. Shigeyama et al., Nature, 1987, v. 330, p. 233.
19. P.A.Pinto, S.E. Woosley, Astrophys. J., 1988, v. 329, pP.820.
20. T. Ebisuzaki, N. Shibazaki, Astrophys. J. Letters, 1987, v. 327, p. L5.
21 S.A. Grebenev and R.A. Sunyaev, Pis'ma AZh, 1988, v. 14, p. 675.
22. R.A. Sunyaev, V.V. Efremov, A.S. Kaniovskiy et al., Pis'ma AZh, 1988, v. 14, p. 579.
23. R.A. Sunyaev, V.V. Efremov, A.S Kaniovskiy et al., Pis'ma AZh, 1989, v. 15, in press.
24. W.D. Arnett, Astrophys. J., 1988, v. 331, p. 377.
25. Skhemy Raspadov Radionuklidov. Energiya i Intensivnost' Izlucheniya, Energoatomizdat, Moscow, 1987, Book 1. Part 1.
26. R.A. Sunyaev and L.G. Titarchuk, Astron. Astrophys., 1980, v. 86, p.121.
27. L.A. Pozdnyakov, I.M. Sobol', R.A. Sunyaev, Itogi Nauki i Tekhniki (Ser. Astronomiya), VINITI, Moscow, 1982, v. 21, p. 238.
28. R.M. Catchpole, P.A. Whitelock, M.W. Feast et al., Monthly Not. Roy. Astron. Soc, 1988, v. 231, p. 75P.
29. P.A Whitelock, R.M. Catchpole, J.W. Menzies et al., Monthly Not. Roy. Astron. Soc, 1988, v. 234 ,p. 5P.
30. D. Kleiton, Nuclear Astrophysics (eds. Ch. Barnes, D. Kleiton and D. Shramm), Russian translation, Mir, Moscow, 1986, p. 359.
31. S.E. Woosley and T.A. Weaver. Nuclear Astrophysics (eds Ch.Barnes, . Kleiton and D. Shramm), Mir, Moscow, 1986, p. 359.
32. S.E. Woosley, P.A. Pinto, P. Martin, T.A. Weaver, Astrophys. J., 1987, v. 318, p. 664.
33 S.A. Grebenev and P.A. Sunyaev, Pis'ma AZh, 1988, v. 14, p. 1065.
34. P.A. Pinto, S.E. Woosley, Nature, 1988, v. 333, p. 534.
35. A. J. U. Chameron, Nuclear Astrophysics (ed. Ch. Barnes, D. Kleiton and D. Shramm). Russian translation, Mir, Moscow, 1986, p. 33.
36. M. Hashimoto, K. Nomoto, T. Shigeyama, Astron. Astrophys, in press.
37. R.M. Bionata et al., Phys. Rev. Lett., 1987, v. 58, p. 1494.
38. K. Hirata, T. Kajita, M. Koshiba et al., Phys. Rev. Lett., 1987, v. 58, p. 1490.
39. J.P. Ostriker, Nature, 1987, v. 327, p. 287.
40. G. Sonneborn, 1988, private communication.
41. R.J. Protheroe, Nature, 1987, v. 329, p. 135.
42. E.B. Levich and P.A. Sunyaev, Astron. Zh., 1971, v. 48, p. 461.
43. V.S. Imshennik and D.K. Nadyozhin, Uspekhi Fiz. Nauk, 1988, v. 156, p. 561.
44. M.M. Basko and R.A Sunyaev, Monthly Not. Roy. Astron. Soc, 1976, v. 175, p. 395.

Physics of Neutron Stars
Edited by A. M. Kaminker

ISBN 1-56072-164-2

HARD X-RAY RADIATION FROM SUPERNOVA 1987A

R.A. Sunyaev, A.S. Kaniovskiy, V. V. Efremov, S.M. Yumin, A.V. Kuznetzov, V.M. Loznikov, A.S. Melioranskiy, A.V. Prudkoglad[1], K . Peppin, V. Pitch, J. Anglehouser, I . Trumper, V. Voges[2], , A. Kendzorra, M. Bezler, R. Stauberg[3]
[1]Institute of Space Research, USSR AS
[2]Max Planc Institute , Institute of Space Physics, 8046 Garshing, FGR
[3]Astronomic Institute of Tubingen University, Tubingen, 7400, FGR

The results of observations of Supernova 1987A in a hard X- ray region by the observatory "Roentgen" aboard the module "Kvant" are presented.

1. Detection of x-ray radiation from Supernova 1987A. (SN 1987A).

On August 10, 1987 the X-ray telescope HEXE of the Roentgen observatory aboard the module KVANT detected hard X-ray radiation in the region of 20 - 100 keV at about 10 standard deviations from the location of Supernova 1987A (SN 1987A) in the Large Magellanic Cloud (LMC). Since then SN 1987A has become a major object of observations for the observatory. Later in August 1987 another observatory instrument, Pulsar X - 1, recorded X-ray emission from SN 1987A in the region of 50 - 300 keV. During observations of SN 1987A the instrument detected another already known X-ray source LMC X - 1. In order to eliminate a possible influence of this source, a series of off-set sessions was carried out in August with the observatory instruments 1 degree away from SN 1987A. At this guidance the source LMC X - 1 was beyond the viewing area of HEXE. The results obtained led to the conclusion that X-ray radiation detected belongs to SN 1987A. The observations with the HEXE instrument allowed the position of the X-ray source to be localized (Fig. 1). The position coincided with that of SN 1987A within one standard deviation to an accuracy of 6 angular minutes. Other known source in the LMC did not fall within 3 standard deviations from the localized source [1, 2].

2. Evolution of X-ray luminosity.

Observations of the Roentgen observatory in 1987 - 1988 help to follow the X-ray luminosity evolution of SN 1987A. Fig. 2 shows the evolution of the X-ray radiation from the SN with time in three "colours": 15 - 45, 45 - 105, and 105 - 200 keV.

In the period from August 1987 to February 1988 the observatory recorded a gradual growth of the X-ray flux from SN 1987A by about 40% in the region 45 - 105 keV. This finding is in a good agreement with theoretical models which assume that the main energy source in the SN shell is ^{56}Co decay. These models are also supported by an exponential decrease of bolometric optical and IR luminosity of 1987A with a characteristic time of 114 days, in agreement with the period of ^{56}Co decay [3, 4].

Unfortunately, we do not have accurate data concerning the observatory guidance from June to October, so the results of this period may be 10 - 30% lower. It is, however, evident that the X-ray luminosity of SN 1987A began gradually decrease since May 1988. This is confirmed by the measurements made in November and December 1988 when the location of the space station was precisely known. By December 1988 the emission from SN 1987A dropped 6 times as compared to the maximum value of January 1988. This decrease is also in accord with the theory of radioactive decay of ^{56}Co. However, in addition to the energy release by ^{56}Co, the effect of ^{57}Co isotope with the decay period 391 days becomes appreciable.

3. The source spectrum.

After X-ray radiation from SN 1987A was detected, the source was under constant observations by the Roentgen observatory. The total time of observation is given in Table 1.

In January 1988 the most intensive spectrum of SN 1987A was recorded. Fig. 3 shows the photon spectra recorded by the HEXE instrument in August and January and by Pulsar X - 1 in August and then in December and January. The spectral shape is in a good agreement with the theoretical predictions of the model including Compton scattering of γ-lines from ^{56}Co.

Fig.4 shows the SN spectra, in instrument readings, recorded by HEXE in January to December 1988. A gradual evolution of the intensity and shape of the spectra with time is evident.

4. Localization.

The complete set of data obtained from August 1987 to February 1988 confirmed the localization of the source in the region 45 - 105 keV so that a possible contribution of the LMC X-1 source could be estimated. The localization was determined assuming a linear increase of the flux from SN 1987A [4]. The source location coincided with that of SN 1987A to an accuracy of 1 standard devotion. At the same time, the 3σ limit on the flux from the LMC X - 1 was found to be $1.6 \cdot 10^{60}$ phot x $cm^2 s^{-1}$ keV^{-1}, which is 28% of the flux from the major X-ray radiation source.

5. Discussion.

The results obtained are in contradiction with simple models of the SN shell, where ^{56}Co is concentrated on the inner boundary of the shell. In these models, which use the velocity distributions in the SN shell [5], emission appears much later in comparison with real experimental data [6, 7].

The early detection of the SN 1987A emission may be due to some factors: a non-spectrical character of the SN 1987A shell [8], the availability of strong inhomogeneities in the shell [8], or strong mixing of cobalt in the shell [9, 10, 7]. Figure 2 shows the evolution of a radiation flux from the SN with time in terms of the model according to which about 80% of Co is concentrated in the central area while the rest is almost uniformly distributed in the outer layers of the SN shell (see also Fig. 5). This model distribution of cobalt was obtained on the basis of the X-ray luminosity evolution recorded by the HEXE instrument of the orbital observatory Roentgen (Fig. 2) [11]. The model allows explanation of a 180-day decay in the appearance of the X- ray flux from the SN after the explosion: it is the time when the observatory Roentgen first detected X-ray radiation from SN 1987A.

At later stages of the evolution the contribution of ^{57}Co into the energy balance of the shell becomes substantial. The X-ray emission spectrum from SN 1987A recorded in September and October requires other sources of energy in addition to ^{56}Co to be properly interpreted. If we assume that no other sources besides ^{56}Co and ^{57}Co exist, then the observed X-ray luminosity should require 1.8 ^{56}Co amount as compared to its abundance on the Earth. It should also be noted that calculations of neucleo-synthesis in the stars with a mass of about 25M⊖M@o [12] and similar calculations for an expected progenitor star of SN 1987 [13] predict an abnormally high abundance of ^{57}Co and ^{56}Co. At present the observatory Roentgen continues observations of SN 1987A to obtain more information about this interesting object.

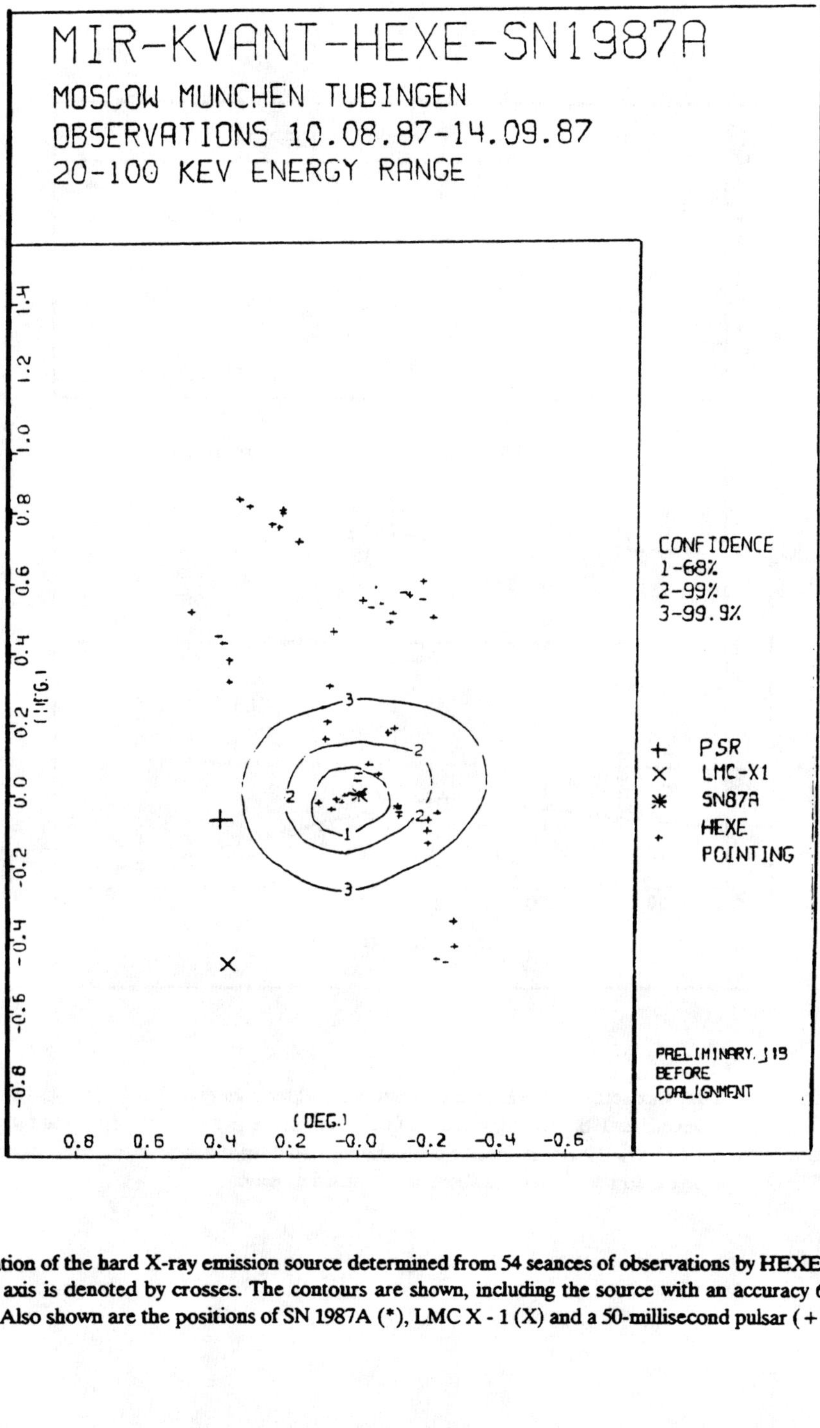

Fig. 1. Localization of the hard X-ray emission source determined from 54 seances of observations by HEXE. Guidance of the instrument axis is denoted by crosses. The contours are shown, including the source with an accuracy 68% (1), 99% (2), 99.9% (3). Also shown are the positions of SN 1987A (*), LMC X - 1 (X) and a 50-millisecond pulsar (+).

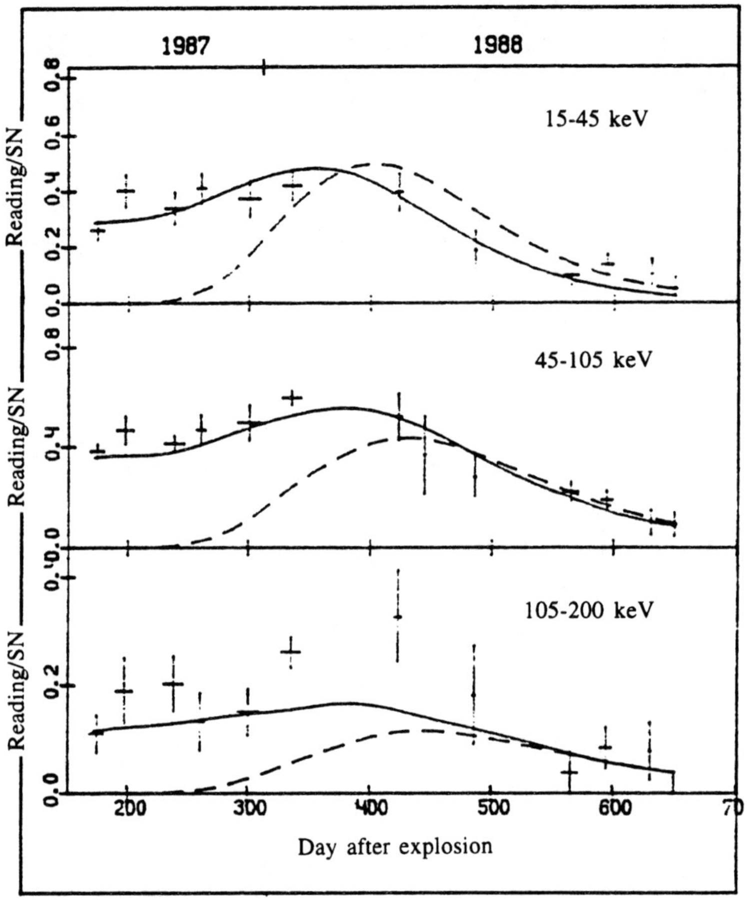

Fig. 2. Evolution of the X-ray flux according to HEXE observations in three energy region 15 - 45, 45 - 105, and 105 - 200 keV from August 1987 to December 1988. The curves of the flux evolution are given, which correspond to the models with ^{56}Co distribution in the center (dotted lines) and with a complicated ^{56}Co distribution determined from experimental data (solid lines). The error in the experimental points is about one standard deviation.

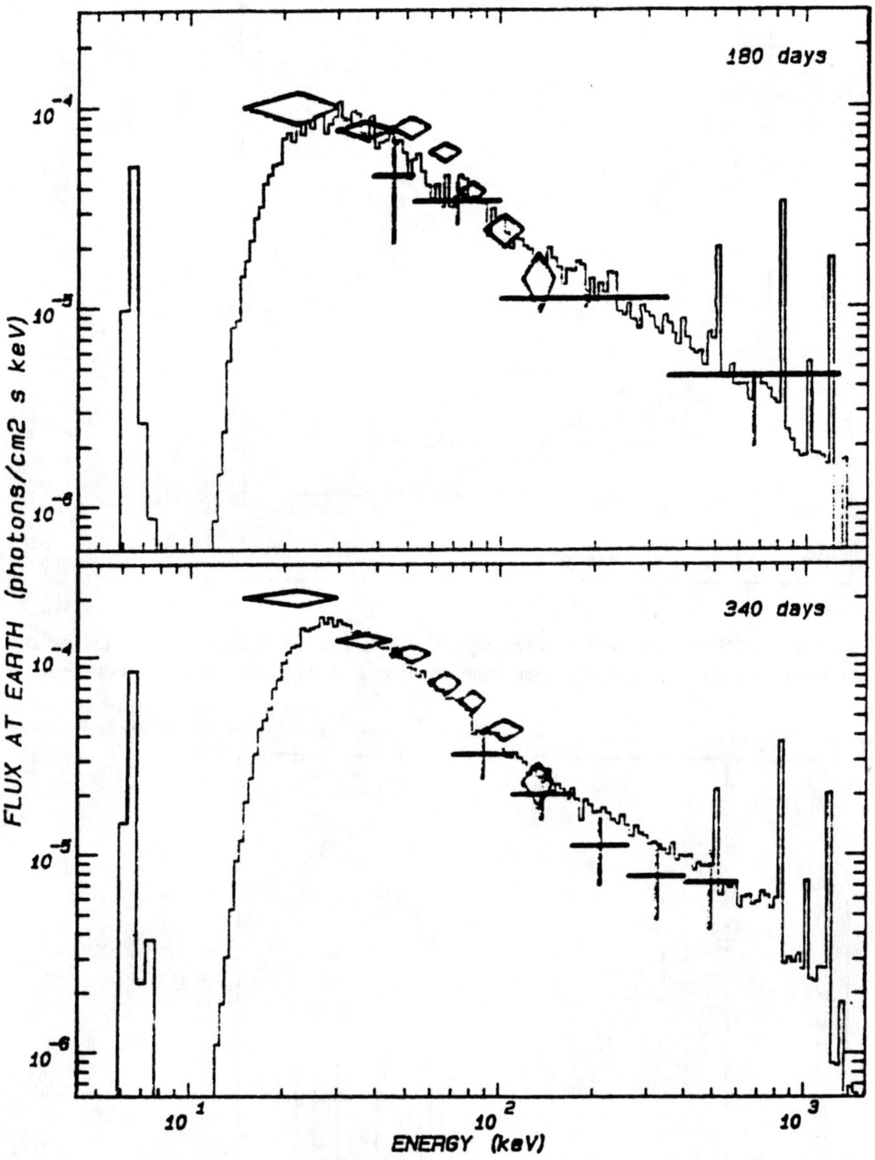

Fig. 3. The spectra of SN 1987A according to observations by the observatory Roentgen in August 1987 (the upper plot) and in January 1988 (the lower plot). Rhombs represent the results of HEXE, crosses - the data of Pulsur X - 1 instruments (the errors correspond to 1σ, the upper limits 3σ). The histograms are in agreement with the results of Monte-Carlo calculations in terms of the model with complicated ^{56}Co distribution in the SN shell (see Fig. 5) 180 days and 340 days after the explosion.

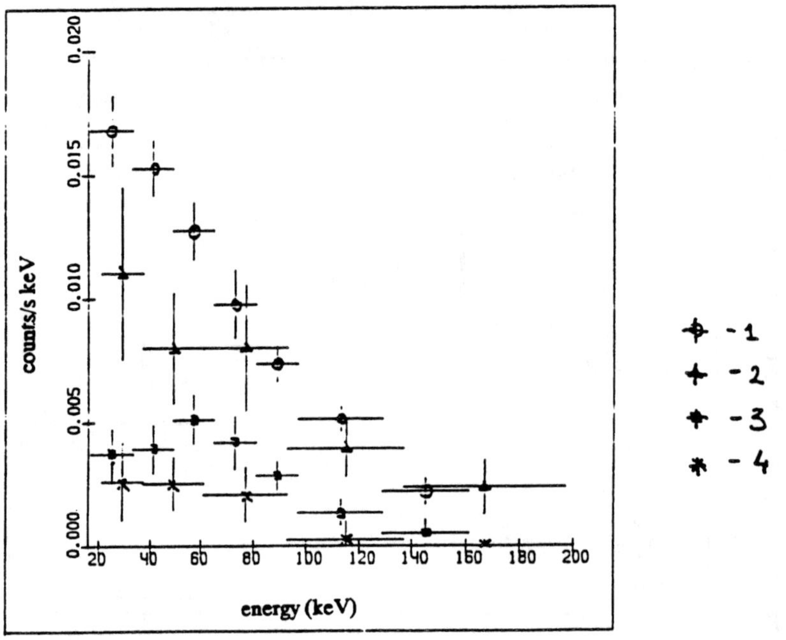

Fig. 4. Observations of SN 1987A (count spectra) in January (1), in April (2), in September-October (3) and December (4) 1988. Gradual decrease of the X-ray emission from the supernova is seen, shift to a harder X-ray spectrum is noticeable.

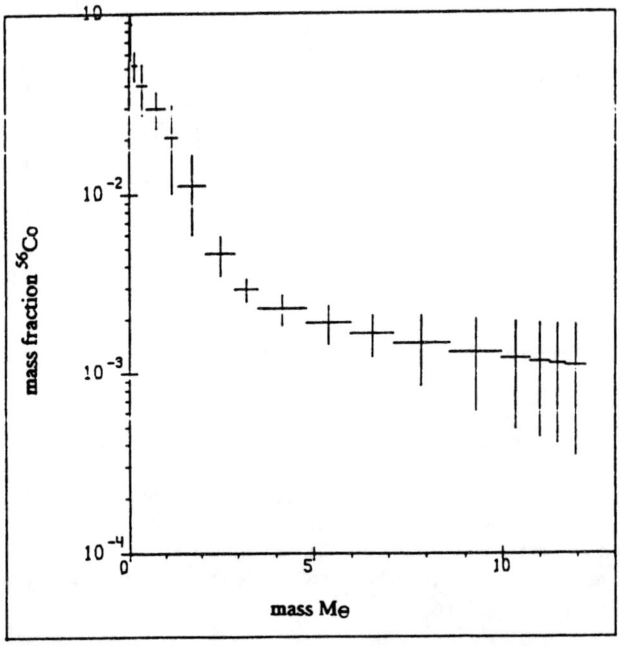

Fig. 5. Distribution (mass fraction) of ^{56}Co in the SN shell according to the Arnett shell model [5] determined from the results of the observations by HEXE.

Supplement: mass fraction ^{56}Co. mass Mϴ.

TABLE 1

Time of observation		SN 1987A s
1987		
	August	25000
	September	20000
	October	26000
	November	13000
	December	22000
1988		
	January	18000
	April	7900
	May	3100
	June	6400
	September	6200
	October	10000
	November	12000
	December	11300

REFERENCES

1. R. Sunyaev, A. Kaniovskiy, V. Efremov et al., Nature, 1987, v. 330, p. 227.
2. R.A. Sunyaev, A.S. Kaniovskiy, V.V. Efremov et al., Pis'ma AZh, 1987, v. 13, p. 1027.
3. R.M. Catchpole, J.W. Menzies, A.S. Monk et al., Monthly Not. Roy. Astron. Soc. 1987, v. 229, p. 15P.
4. M. Hamuy, N.B. Suntzeff, R. Conzales, and G. Martin, Astron. J., 1988, v. 327, p. 197.
5. W.D. Arnett, Astrophys. J., 1988, v. 331, p. 377.
6. Y. Xu, P. Sutherland, R. McCray, and R.P. Ross, Astrophys. J., 1988, v. 327, p. 197.
7. S.A. Grebnevand and R.A. Sunyaev, Pis'ma AZh, 1988, v. 14, p. 675.
8. S.A. Grebnev and R.A. Sunyaev, Pis'ma AZh, 1987, v. 13, p. 945.
9. P.A. Pinto and S.E. Woosley, Astrophys.J., 1988, v. 329, p. 820.
10. T. Ebisuzaki and N. Shibazaki, Astrophys. J. Letters, 1987, v. 327, p. L5.
11. R.A. Sunyaev, V.V. Efremov, A.S. Kaniovskiy et al., Pis'ma Zh, 1989, v. 15, in press.
12. S.E. Woosley and T.A. Weaver, Nuclear Astrophysics (edited by Ch. Barnse, D. Clayton, and D. Shramm), Mir, Moscow, Russian translation, 1986, p. 359.
13. M. Hashimoto, K. Nomoto, and T. Shigeyama,Astron. & Astrophys., in press.

Physics of Neutron Stars
Edited by A. M. Kaminker

ISBN 1-56072-164-2

X-RAY PULSARS

V.G. Kurt and E.Yu. Shafer
Institute of Space Research, USSR Academy of Sciences, 117810, Moscow

Presented are the data on the pulse period, orbital period, mass and radii for some of the 26 X-ray pulsars studied. Three of these pulsars are single, the rest 23 are binaries. The "Astron" satellite, five years in orbit, has been observing.14 pulsars, for which pulse periods were determined and brightness curves were constructed in the energy range from 2 to 25 keV. Also given are data on period changes per century and spectra of some pulsars. All known parameters for the 26 X-ray pulsars are summarized in Table 1.

Table 1. Parameters of X-ray pulsars in binaries

Pulsar	P,s	P_{orb}, day	R_{opt}, R⊖	M_{opt}, M⊖	M_x, M⊖	Spectral class	$V_{visible}$	Eclipse	D, kpc
1	2	3	4	5	6	7	8	9	10
A0538-66	0.069	16.66	–	–	–	B2IVe	13^m-15^m	?	50
SMC X-1	0.714	3.892	16.5±4	18	1±0.35	B0I	13.3	+	65
Her X-1	1.24	1.70	4.05±3	2	1.45±0.4	A9V-F0V	13.2	+	~5
4U0115+63	3.61	24.31	–	>5	–	Be	14-16	–	3.6
V0332+53	4.37	34.25	–	–	–	B	–	?	–
Cen X-3	4.84	2.087	12.2±2.5	19	1.7±0.6	06III-08III	13.4	+	10
1E2259+586	6.98	<0.08	–	–	–	–	–	?	3.6
4U1626-67	7.68	0.029	–	0.1	0.5±4.7	?, B-V=0m,12 U-B=-1m,2	18.5	–	–
2S1553-54	9.26	30.7	–	–	–	–	–	?	–
LMC X-4	13.5	1.408	9±4	17	1.7±1.5	07III-V	14.0	+–	50
2S1417-62	17.6	>15	–	–	–	F	16.2	?	–
OA01653-40	38.2	–	–	–	–	–	–	?	–
A0535+26	104	110.5	–	25	–	09.7IIIe	9.1	–	1.8
4U1728-24 (GX1+4)	122	>15	–	–	–	–	18.7	?	10?
4U1258-61 (GX304-1)	272	135?	–	6÷10	–	B2V	13.9	?	2.4
4U0900-40 (Vel X-1)	283	8.965	31±3	24	1.85±0.3	B0.51	6.9	+	1.4
4U1145-61	292	187?	–	15	–	PI,e	9.0	?	1.5
IE1145.1-6141	297	>12	–	–	–	B11	12	?	5
A1118-61	405	–	–	12-15	–	C9.5IV	12.1	?	5
4U1907+09	438	8.4	–	–	–	0.B	16.4	–	0.7
4U1538-52	529	3.73	16±4.5	19	$1.9^{+1.3}_{-0.9}$	B0I	14.5	+	7
2S1223-624 (GX301-2)	696	41.4	40±4	>30	–	B1.51a	10.8	–	2
4U0352+30	835	580?	–	20	–	09.5III-IV,e	6.0	?	0.35

X-ray pulsars are fairly well studied objects in X-ray astronomy. This, however, does not mean that we understand and can explain all of the observed events. But our understanding of how X-ray pulsars are built, and what are the mechanisms of their X-ray radiation generation is more advanced than, say, our understanding of those for g- flashes. This is mainly due to the availability of optical identification and, as a result of it, to a greater amount of data obtained with the help of ground-based optical telescopes and infrared instruments as well as satellite- based ultraviolet instruments.

X-ray pulsars were first detected in 1971 by an UHURU satellite and have been object of investigations of all X-ray-equipped satellites. Of special importance are the results of their localization in the sky with the help of SAS-3, HEAO-A satellites and Einstein Observatory. Determination of their exact coordinates allowed identification of most of them with visible stars, enabled a detailed photometry of the stars, recording their spectra, etc.

More than 25 X-ray pulsars have been detected by now, and about 20 of them were identified with optical objects [1]. Astronomers of this country have greatly contributed to the understanding of the pulsars nature. Worth mentioning are the pioneer works by Ya.B. Zeldovich and O. Guseinov, I.S. Shklovskii, N.I. Shakura, R.A. Sunyaev. A better feeling of that Ya.B. Zeldovich and I.S. Shklovskii are no longer with us in this world of ours cannot but emphasize the importance of this Conference. These distinguished scientists paid special attention to X-ray pulsars and compact objects, in general, the more so that X-ray pulsars have been observed in this country from the only satellite. It is "Astron", put in a high-apogee orbit on March 23, 1983, and still working there. All (or almost all) illustrative data concerning pulsars were obtained by observations from the "Astron." The characteristics of the instruments aboard the satellite are the following. The detectors area is 1760 cm^2, spectrum region 2 ÷ 25 keV, the number of spectral channels 11, time resolution 0.305 s and 2.27 ms, angular resolution 3° (by the diagram halfwidth), guidance accuracy and stabilization $\pm 1' + 2'$, duration of a session 3 - 4 h and the maximum sensitivity per 1-hour exposure 1 ÷ 0.1 m Crab [2 - 4].

So, what do we know about pulsars? First of all that they emit X-ray pulses with an almost constant period (Fig. 1). The shortest pulse period was found in a remnant of the Supernova 1054 in the center of the Crab Nebula. It was equal to 0.033282608 by the year 1982 [5]. This period, however, not typical of the objects which are discussed in this report. Therefore, a truly short-period object is the pulsar AO 538-66 in the Large Magellanic Cloud (LMC) with P = 0.069 s. The pulsar with the longest period is 4U 0352 + 30 (x Per); its period is 834 s. The difference in the periods is really enormous, about 4 orders of magnitude. Radiopulsars have a different range of pulse periods: from 1.557806449014 ms (PSR 1937-16) [6] to 4.30877928 s (PSR / 1845-15) [7, 8].

Usually measured are a period of X-ray pulsars P, its derivative $\dot{P}$, an orbital period (if a pulsar belongs to binaries) P_{orb}, a derivative of this parameter $\dot{P}_{orb}$, a visible stellar magnitude and color spectrum of its optical partner, as well as brightness and eclipse curves, if the latter occur. It should be noted that a high precision of P and $\dot{P}$ determination helps to isolate the harmonic component, due to the Doppler effect, of the pulsar motion in orbit with respect to the barycenter of a binary system. The rates of motion here are great and may be as high as 400 km/s. A mathematical apparatus developed for the analysis of brightness curves allowed determination of many parameters of the binary systems, mass components M_x and M_o being the most important among them, or at least their mass functions $f(M) = M_0^3 \sin^3 i / (M_x + M_0)^2$.

The problem of distance estimations to the pulsars is very complicated. The distance can be estimated from the visible and absolute values of the optical component, a "drop" in the E < 3 - 4 keV region of the spectrum, and optical spectrum of the visible component. Nevertheless, the distances have not properly been estimated, except for some very simple cases such as the pulsar in the Crab Nebula and the sources in the SMC (Small Magellanic Cloud) and the LMC. These distances must naturally be very large, i.e., more than 2 kpc (except for xPer, Vel X-1 and the Crab Nebula). The orbital periods of the pulsars are in the range from 0.691 h (41 min) to 580^d.

Only three pulsars out of almost thirty known are single. They are Tau X-1, LMC 0540-69, and MSH 15-52.

Table 2
Single X-ray Pulsars

Pulsar	P, s
Tau X-1 (4U0532+21)	0.033
LMC 0540-69	0.0502
MSH 15-52	0.150

The second of them, with the period of 0.0502 s, is almost a complete analog for the pulsar in the Crab Nebula. Observations of its rotation periods actually do not leave / any doubts about that the compact object of X-ray pulsars is a neutron star. Where their masses could be determined, they turned out to be in a range of 0.5 - 1.8 M⊙, with an average value being 1.7 M⊙ (an estimation error is rather great, about 1 M⊙). The masses of optical components are much better determined; they range from 0.1 M⊙ to 30 M⊙. The binary systems considered here fall into two groups. In the first group the optical star is a giant (supergiant) of the 0 - B class with a mass of 10 ÷ 30 M⊙. In the second group there is an optical star with a mass M ~ (0.1 ÷ 2) M⊙.

The first, larger, group comprises 15 objects; only 5 objects belong to the second group. Their absolute stellar values are as great as $4^m.5$ with luminosity up to 10^5 L⊙. The minimal luminosity was determined in 4U 1626-67, L = 0.3 L⊙ and M = 0.1 M⊙. The radii of optical stars in massive pairs range from 10 R⊙ to 30 R⊙. Semimajor axes of the orbits in these systems are known only with an accuracy to the factor sin i. However, in the objects, which are subject to eclipses, the angle i is almost equal to 90° and this gives a fairly reliable picture of the pair. These sources, in the same scale, are shown in Fig. 2 together with the mass of the optical components [9]. The scale 100 l. s. is equal to 0.2 x 1.496 x 10^{11} m or 40 R⊙. The biggest among them is the source GX 301-2 with the star radius of 40 R⊙, the orbital period of $41^d.4$ and semimajor axis of 0.74 x 1.496 x 10^{11} m. As a rule, the eccentricity of the system is small, except for 4U 0115+63 and GX 301-2, where it is equal to 0.4 ÷ 0.5.

A very strong eccentricity is most likely to be found in the system, which is very popular with observers and theorists, viz., A 0535+26, identified with the star Tau V 725: a supergiant Sp = 09.7IIIe, V = $9.^m1$, M = 25 M⊙, and d = 1.8 kpc [10].

The system must have an orbital period of $110.^d5 \pm 0.^d5$, though this interpretation of observational results / might cause some difficulties. The fact is that an X-ray flux from this source can vary from 10^{-3} ÷ 10^{-2} to 3 Crabs within a characteristic time of 10 ÷ 20 days. The period of a neutron star rotation (provided the model is correct) is 104 s and may vary in a rather queer way. It may either increase or decrease by ~ 0.5 s, with the period sharply decreasing, within 10 ÷ 20 days, by 0.1 ÷ 0.2 s in every 111^d cycle observed. The source may well have a 1000-day period as well, with its X-ray flux of 0.1 -: 1 Crabs. The 1000-day period seems to be detected in the optical and IR regions of the spectrum [11]. A 0535 + 26 might be a triple system comprising the component observed. A source, which is considered to be best studied, is Her X-1 (HZ Her) with P = 1.23779226 ± 2 · 10^{-7} s, P_{orb} = 1.700167788. For this source carefully calculated models have been constructed, which account for such fine effects as: (1) X-ray radiation procession by the surface of the A9 ÷ B0 star of the main sequence with M = 2 M⊙ and R = 4 R⊙; (2) changes in the effective temperature of the star observable hemisphere under the influence of an external X-ray flux, which exceeds the radiation of the star itself; and (3) the effects due to the accretion disc, etc. [12]. However, even for this so well studied source, some effects still did not find an adequate explanation. Among them are the 35^d period, which has been observed for 15 years already, and the fading out of the source detected with "Astron" on June 30, 1983, [13]. Moreover, in spite of the expected maximal flux of ~ 150 mCrab, the source luminosity turned out to be 20 times less (~ 5 mCrab), whereas its pulsed radiation (1.24 s) was not detected at all at the ≤ 1% detection level. These two facts, namely, a 35-day period and fading out of the source for several months (the source was "rediscovered" by "Astron" and EXOSAT in 8 months, on March 1, 1984) still await their explanation in terms of the model of binaries and accretion disc.

Now let us proceed to analyze observational data mainly obtained with "Astron". There have been studied 15 X-ray pulsars during 45 contact sessions with "Astron", each about 3 - 5 h. Some pulsars were repeatedly observed: Tau X-1, 6 times; A 0535 + 26, 8 times; Her X-1, 7 times; Vel X-1, 3 times; Cen X-3, 4 times; and A 1627-67, 2 times. Our first aim was to determine (or to isolate) a pulse period. To this end, convolution of the measured counting rate (imp/s) for 0.305 s and 2.27 ms with the unknown period P was made. This required transition from the registration time t on the Earth to the time t' in the coordinate system related to the barycenter of the X-ray system. The procedure was performed in two steps. At the first step the "stationary" (on the Earth) time was recalculated during its registration aboard the satellite by making corrections for the distance between the satellite and the ground-based registration place. At the second step, account was taken of the Doppler effect due to three movements, those of

- the satellite relative to the Earth center,
- the Earth relative to the barycenter of the Solar system,
- a neutron star relative to the barycenter of an X-ray pulsar;

$$t' = \int_0^t \frac{\left(1 - \frac{V_r}{c}\right) dt}{\sqrt{1 - \frac{V^2}{c^2}}}$$

where V^r and V (radial and total velocities) account for all the three movements. The satellite movement was calculated by a well-known program developed by V.I. Prokhorenko (ISR USSR AS) [14]; the movement of the Earth relative to the barycenter, by an American program DE-118 [15]. The last correction was calculated from literature data (except for single pulsars).

$V_r = V_{orb} \sin i \sin \Theta_{orb}$

where

$$\Theta_{orb} = \Theta_0 + \frac{360}{P_{orb}}(t - t_0) - \frac{360}{2}\frac{(t - t_0)^2 \dot{P}_{orb}}{P_{orb}^2}$$

Unfortunately, $\dot{P}_{orb}$ is not known for many pulsars. Its estimates will be dealt with below.

The subintegral function was calculated with a 10^m step, and was then approximated by a ($A t^2 + B t + C$)-type polynomial. After integration, the time on the satellite was recalculated to obtain the sought-for quantity for each telemetric point according to the formula with the known A, B, and C.

$$t' = t\left(C + \frac{B t}{2} + \frac{A t^2}{3}\right)$$

To find the period P, time t' was substituted by a phase according to the formula

$$\Theta = \frac{t'}{P} - INT\left(\frac{t'}{P}\right), \quad 0 \leq \Theta < 1$$

The phase from 0 to 1 was divided into 32, 64, and 128 intervals depending on the source and background luminosities. Then the value χ^2 was calculated

$$\chi^2 = \frac{1}{n}\sum_{i=1}^{n}\frac{(I_i - \bar{I})^2}{\bar{I}}; \quad n = 32, 64, 128$$

where $\bar{I}$ is an average period intensity.

The calculated $\chi^2(P)$ values were used to construct a least- square polynomial of the 2nd and 3rd degree; its maximum corresponded to the sought-for P value. The number of points, for which χ^2 was calculated, was chosen so that the neighboring points were not correlated.

Figure 3 shows brightness curves for the X-ray pulsars Tau X-1, Her X-1, Cen X-3, 4U 1626-67, A 0535 + 26, and Vel X-1.

The thus determined values for the periods of these pulsars allowed tracing the history of the period changes for the last 5 years and sometimes even for the last 10 - 15 years, making use of other observations [16]. Figure 4 demonstrates these results for Her X-1, Cen X-3, Vel X-1, A 0535 + 26, and 4U 1626-673. For a single pulsar in the Crab Nebula a monotonous slowing down of the pulsar allowed calculation of $\dot{P}$ and even $\ddot{P}$: $\dot{P} = -4.214386 \cdot 10^{-13}$ s/s, $\ddot{P} = -2.5 \cdot 10^{-24}$ s^{-1}. It should be remembered that in 1986 the pulsar in the Crab Nebula suffered an abrupt jump in the period (a glitch). Usually, pulsars in binaries are subject to speeding-up (period decrease) due to the substance transfer from a normal star bringing a torque with it. However, an instantaneous value of $\dot{P}$ may be both > 0 and < 0, depending on a delicate balance between accretion and deceleration, which results from emission and other losses of the angular momentum. As to $\dot{P}_{orb}$ the experimental / data are scarce. Its value can be, however, estimated assuming two mechanisms: substance transfer onto a neutron star and the system mass loss into the surrounding interstellar space.
Assuming

$$P_{orb} = \frac{2\pi a^{3/2}}{\sqrt{GM}} \qquad J_{orb} = M_x M_0 \sqrt{\frac{Ga}{M}}$$

and $M = M_x + M_0$,
differentiation gives

$$\dot{P}_{orb} = P_{orb} \cdot \frac{3}{2}\frac{\dot{a}}{a} \qquad \dot{J}_{orb} = J_{orb}\left(\frac{\dot{M}_x}{M_x} + \frac{\dot{M}_0}{M_0} + \frac{1}{2}\frac{\dot{a}}{a}\right) \equiv 0$$

with the system retaining its torque. If the system does not lose any mass into the interstellar medium,

$$\dot{M}_x = -\dot{M}_0, \quad M = const$$

and

$$\dot{P}_{orb} = 3P_{orb}\dot{M}_X\left(\frac{1}{M_0} - \frac{1}{M_X}\right) = -\frac{6\pi a^{3/2}\dot{M}_X(M_0 - M_X)}{M_X M_0\sqrt{G(M_X + M_0)}}$$

$\dot{M}_X$ can be estimated assuming the X-ray luminosity $L_X = 0.15\,\dot{M}_X s^2$. Thus, for Her X-1 $L_X = 10^{37}$ erg/s, for Cen X-3 4 x 10^{37} erg/s, and for SMC X-1 6 x 10^{38} erg/s, which correspond to $\dot{M}_X \approx 10^{-9}$, $4 \cdot 10^{-9}$, and $6 \cdot 10^{-8}$ M⊙/year. Accordingly, $\dot{P}_{orb}$, by the order of the magnitude, will be $1.3 \cdot 10^{-9}$, $2.2 \cdot 10^{-8}$, and $6.7 \cdot 10^{-7}$ day/year ($3.7 \cdot 10^{-12}$, $6.3 \cdot 10^{-11}$, and $2 \cdot 10^{-9}$ s/s). Though small, these values can be measured in some cases. Estimation of the second effect is more complicated because the losses of the system angular momentum into surrounding interstellar space are not known.

Spectra of X-ray pulsars are diverse and imply changeability. Figure 5 shows 6 spectra of pulsars with periods from 1 to 835 s [17]. As a rule in the long-wave region ($E < 10$ keV), the spectra have a power form

$$N(E)\,dE = A\,E^{-\alpha}\exp(-\sigma_E N_H)dE$$

Distant sources with a small galaxy latitude have a "drop" in their spectrum starting with 2 ÷ 4 keV due to absorption in interstellar medium. This was found in the spectra of Cen X-3 and 4U 223-62. The latter has B = 0 at a distance of 2 kpc. Almost all the sources mentioned here have an exponential "drop" in the spectra at E ~ 20 keV, and a line assigned to iron at 6 keV. There is enough evidence for a gyroline with E = 56 keV in the spectrum of Her X-1 [18]. The presence of this line allows the magnetic field strength to be estimated on the surface of a neutron star. The energy of this line is $h\nu = h\nu_B(\nu + \frac{1}{2} + m_s)$, where ν_B is the cyclotron frequency equal to $eB/(2\pi m_e c)$, n = 0, 1, 2 ...; and $m_s = \pm 1/2$ is the electron spin projection along the magnetic field. The first harmonics corresponds to a transition to a level with $m_s = 1/2$ and n = 1; the second harmonics, $m_s = 1/2$, n = 2. A transition with E = 56 keV makes it possible (with due allowance for relativistic effects) to find $B = 6 \cdot 10^{12}$ G for $M_X \approx 1$ M⊙ and $R_X = 10$ km.

Thus, observational results fit the model of an X-ray pulsar, which has a rotating neutron star with a mass ~ 2 M⊙ and a strong (up to 10^{13} G) magnetic field. The neutron star forms a binary with a companion: a blue giant or a supergiant. It should be noted, however, that in binaries of small masses, a pair is possible with a star of the major sequence down to a dwarf of a mass M_o ~ 0.1 M⊙ (4U 1626- 67). In such a system a thin accretion disk is formed near the neutron star. Sometimes, instead of the disc, a spherically symmetrical accretion onto the neutron star (in the mode of the solar wind) occurs. Frequently, the optical component fills its Rochet void and substance is transferred through the internal Lagrange point to form an accretion disc. The star shape in this case may considerably deviate from the spherical, which affects the brightness optical curve. Among important effects, the precession of the accretion disc and, possibly, of the neutron star should be mentioned.

The behavior of substance near a NS resembles that of the Earth magnetosphere stream-lined by the solar wind. There exists an inner zone of a radius 10 - 100 R_X where the magnetosphere behaves in a regular manner due to the dipole magnetic field of the star. Its strength at the surface is $10^{11} \div 10^{13}$ G, and its magnetic moment is 10^{30} - 10^{32} CGS units, which is by 3 ÷ 4 orders of magnitude greater than the magnetic moment of the Earth and is of the order of the Sun's magnetic moment. Like in the solar wind, the magnetic field pressure $H^2/8\pi$ near the magnetosphere boundary (at a distance of $(10 \div 10^2)R_X$) becomes equal to the solar wind pressure ρV^2. A shock wave is formed at about this distance. Because of plasma instability "bunches" of completely ionized plasma freeze into the magnetic field and penetrate into the "cusp" region the near the magnetic poles of the NS. Like on the Earth the area of the polar cusps does not exceed 10^{-3} of the NS surface. Thus, on an area of ~ 1 ÷ 10 km² the substance is deposited at an accretion rate of ~ 1 ÷ 100 ton/cm² · s. Each gram of the substance possesses of about 15% of mc^2 energy, that is, this accretion rate provides 10^{37} - 10^{38} erg/s, which corresponds to a mean luminosity of X-ray pulsars. At a distance of 40 l.s. to the optical component ($\approx 0.1 \cdot 1.496 \cdot 10^{11}$ m) an X-ray emission flux on an illuminated star hemisphere will equal $3 \cdot (10^{36} \div 10^{37})$ erg/s. This is approximately an optical luminosity of the star itself, and in some cases (SMCX - 1, LMC X - 4) even exceeds it by 10 ÷ 30 times. It is to this energy, released near the NS poles, that the X-ray phenomenon of pulsars is due. However, the processes near the Ns surface are not quite well understood. All the energy of the accreting gas is released in a layer of about 10 - 100 meter thick thus providing a very high density of emission. The radiation pressure acting on electron through the Thomson scattering is transmitted to protons by the Coulomb interaction. The substance falling down stops. The problems becomes self-consistent.

In pulsars of enormous X-ray luminosity exceeding 10^{37} - $3 \cdot 10^{38}$ erg/s, an accretion column is formed near the polar cusps, which emits through a side surface and slowly settles down onto the star surface. This model was constructed by R.A. Sunyaev and Yu.E. Lubarskii. According to the model such a column can provide luminosity greater than the

Eddington's, because the main force supporting the accretion column is that of the magnetic field rather than gravitation. With the field strength over 10^{12} G (and even 10^{13} G) an equilibrium temperature may be as high as 10^{10} K, that is, by 3 orders of magnitude higher than in the center of the Sun. The density in the column is lower than in the Sun center by 10^3 - 10^4 times. Under such conditions neutron luminosity, exceeding the X-ray, plays an important role. Such are the sources SMC X- 1 and LMC X- 4 with their luminosity exceeding the Eddington's. If an accretion rate is high enough ($10^{-6}M_\odot$/year and higher) and a neutron star mass is of an order of $2M_\odot$, then at lifetime of about 10^6 years, the star may collapse to a black hole. Whether it is going to be an explosion of a supernova or a "quiet" disappearance of an X-ray pulsar is to be seen. Two bright spots on the NS surface and an accretion disc account for a pulse shape from the sources Her X-1, Cen X-3, 4U1627-67 and other pulsars with a relatively "simple" pulse structure. However, we cannot explain a complicated shape of the pulses from the sources A 0535 + 26 and Vel X-1, which have about ten stable details. One or two of them "wander" all over the spectrum and the rest "sit" in their places all over the 2 $\div$ 25 kev interval.

We are still unable to calculate a directional pattern that would include an emitting plasma temperature, strong magnetic field and cyclotron radiation. X-ray pulsar spectra have been interpreted though there remain some problems waiting for their proper explanation. These are plasma radiation near cusps, shock waves, an accretion column, a cooler accretion disc, etc. A part of a multicomponent radiation formed would pulsate with a frequency of the NS rotation, whereas another part of it would be constant. Their explanation requires further spectral observations both in different rotation phases and in an averaged radiation.

There are X-ray pulsars that generate pulses only in the periods of activity. These are an 111-day periodic source A 0535 + 26, SMC X-1, a 3.6-s pulsar 4U0115 63 and some others. They resemble bursters, whose activity is of an aperiodic transient character.

The mechanism of radiation generation by young single X-ray pulsars with luminosities up to $3 \cdot 10^{36}$ erg/s turned / out to be more complicated in the X-ray range than that of pulsars in binaries.

Presently, some progress has been made in the interpretation of the mechanism [19 - 21]. The mechanism of hard radiation generation by a single rapidly rotating NS with a strong magnetic field consist in the following. When the magnetic field at a NS surface is about 10^{12} G, an electric field of about 10^8 CGS units is generated, which speeds up particles up to the Lorentz-factor $\gamma \sim 10^6$. Accelerated electrons, by the mechanism of bending radiation, generate hard photons with energies up to 100 MeV. In a strong magnetic field these photons give birth to e^- - e^+ pairs. At a high density of these pairs, they screen the electric field and move along its force lines. When this occurs, plasma parameters are far from the Maxwell's: first, a strong spatial inhomogeneity exists, and, second, the distribution function for speeds differs substantially from the Maxwellian. The plasma has an average $\gamma \sim 10$ and a high- speed "tail" up to $\gamma \sim 10^4$. Under these conditions a cyclotron instability develops in the vicinity of a light cylinder, that is, at a distance of $\sim 100\ R_X$. Due to this instability development, Alfv'en waves are generated in the plasma with frequencies up to 10^{13} - 10^{14} Hz.

At these frequencies, e^- and e^+ scattered over pitch angles up to 10^{-3}radians ($\sim 3'$) and generate a spectrum from the optical to gamma-region by a synchrotron mechanism. The observational spectrum is in fair agreement with this model. About 10^{36} erg/s is reemitted during the process with a total pulsar losses 10^{38} erg/s, i.e , emission accounts for 1% of the losses. The rest 99% are used to support the nebula activity, to generate particles, etc.

Our physicists, Drs. D.G. Lominadze, G.Z. Machabeli and V.V. Usov greatly contributed to the development of this model.

Of course, this model is not ideal and it is difficult to confirm experimentally. It would be wonderful, for / example, to record an annihilation line 511 keV from the pulsar in the Crab.

This is our hope that investigations of X-ray pulsars will continue. In the report presented we tried to keep a reasonable balance between optimism ("how much we have learned!") and problems still waiting for their solutions. Pulsars would certainly provide much work for theorists and experimentalists.

The authors would like to express their gratitude to E.K. Sheffer, T.A. Mizyakina, V.M. Zenchenko, A.S. Smirnov, V.V. Usov, G.S. Bisnovaty-Kogan and all the participants of the project "Astron".

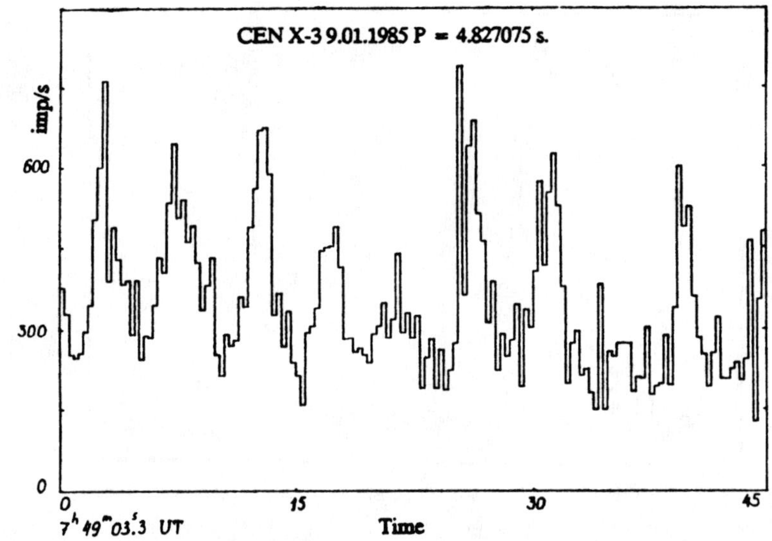

Fig. 1. X-ray Pulsar Cen X-3 (4U 1118-60) observed shortly after leaving an eclipse. Time resolution is 0.305 s.

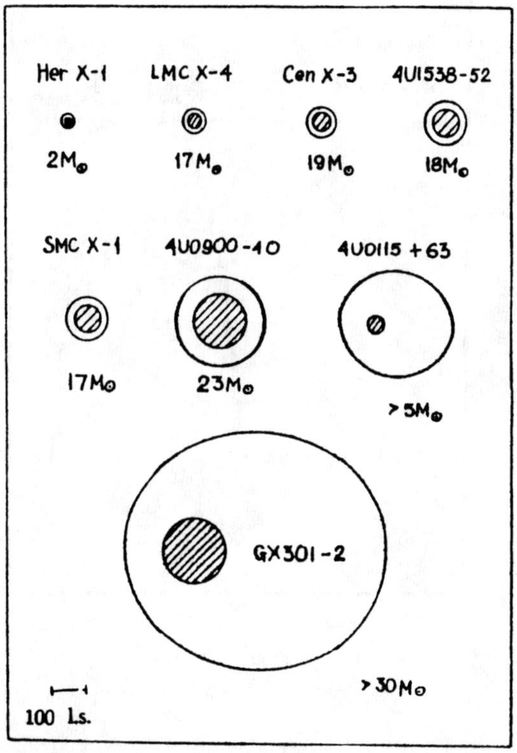

Fig. 2. Parameters of eight X-ray binaries containing orbital sizes and masses of the optical components.

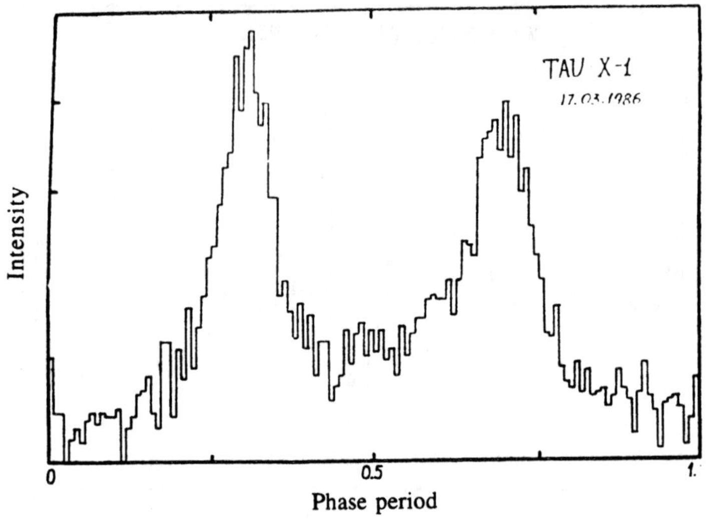

Figure 3a

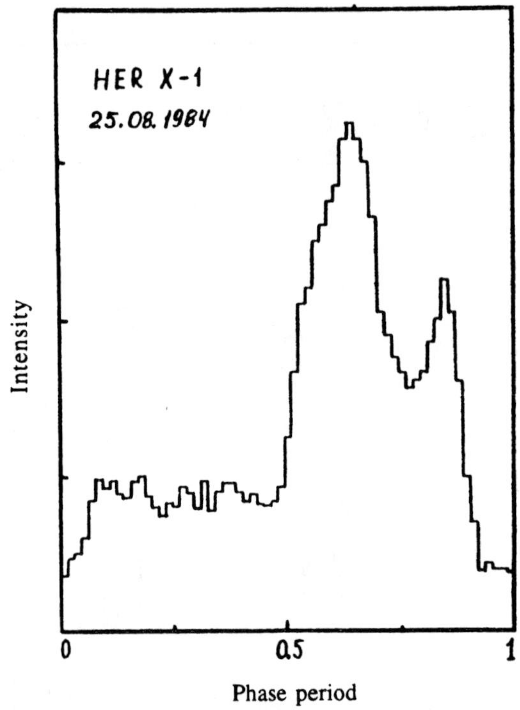

Figure 3b

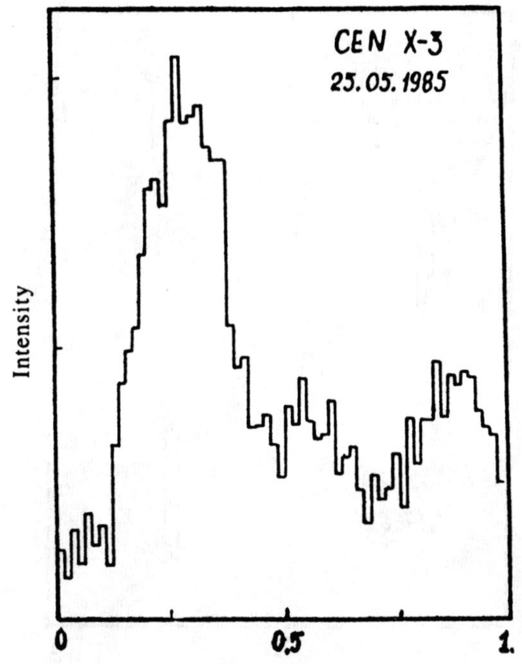

Figure 3c

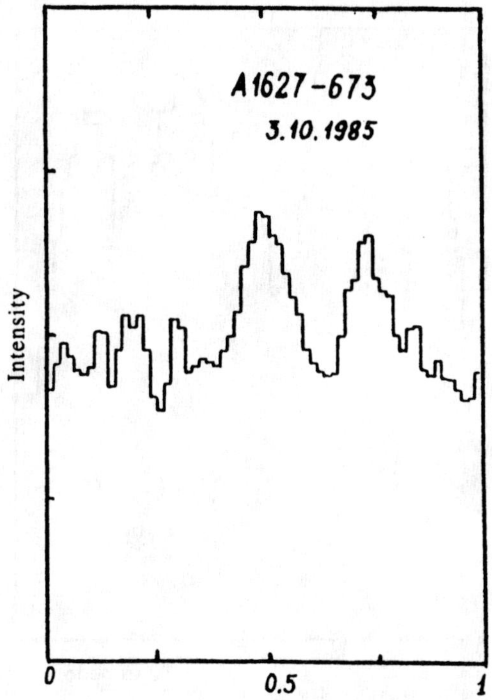

Figure 3d

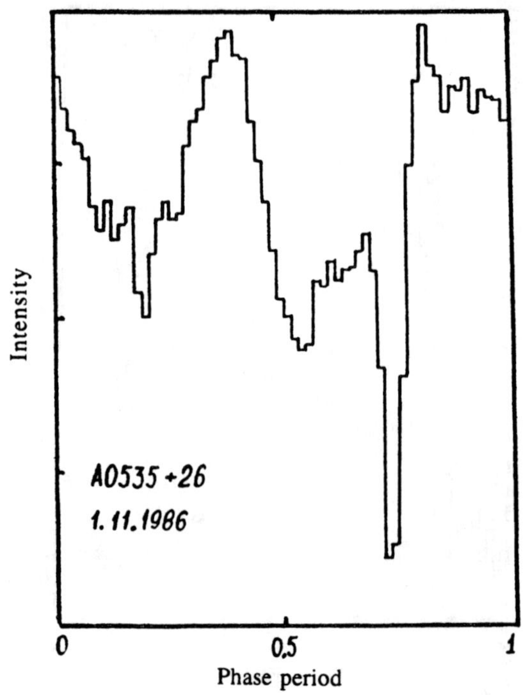

Figure 3e

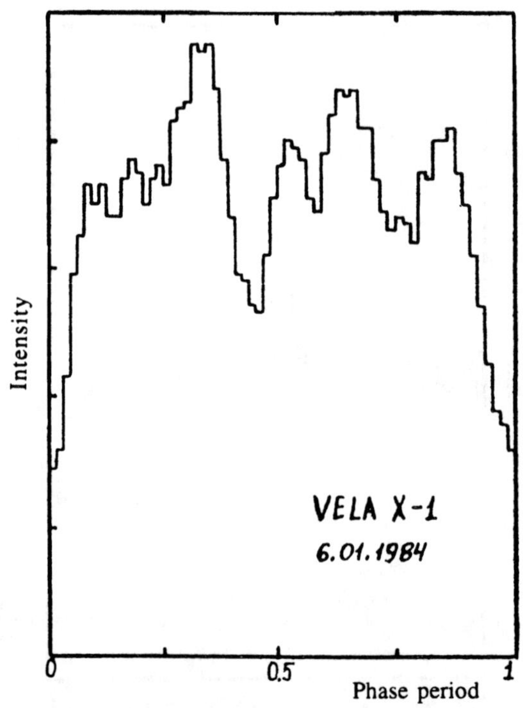

Figure 3f

Fig. 3. Brightness curve for Tau X-1 (a), Her X-1 (b), Cen X-3 (4U1118-60) (c), 4U 1626-67 (d), A 0535 + 26 (e), and Vela X-1 (f) on observational data from "Astron".

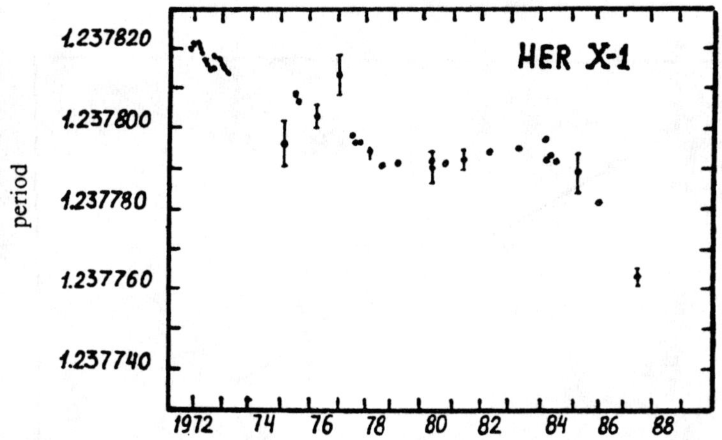

Figure 4a

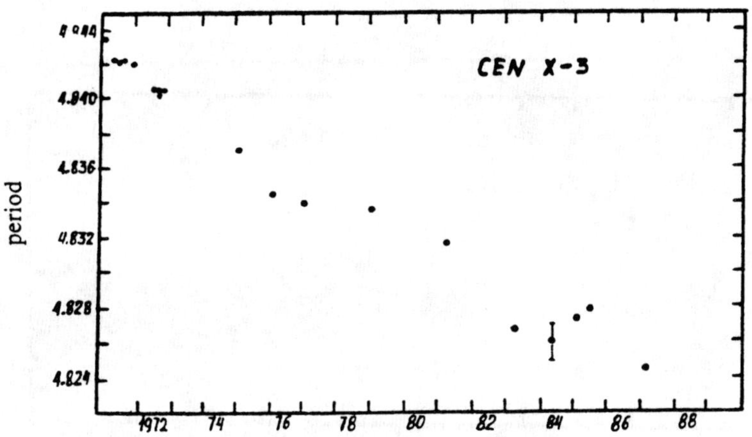

Figure 4b

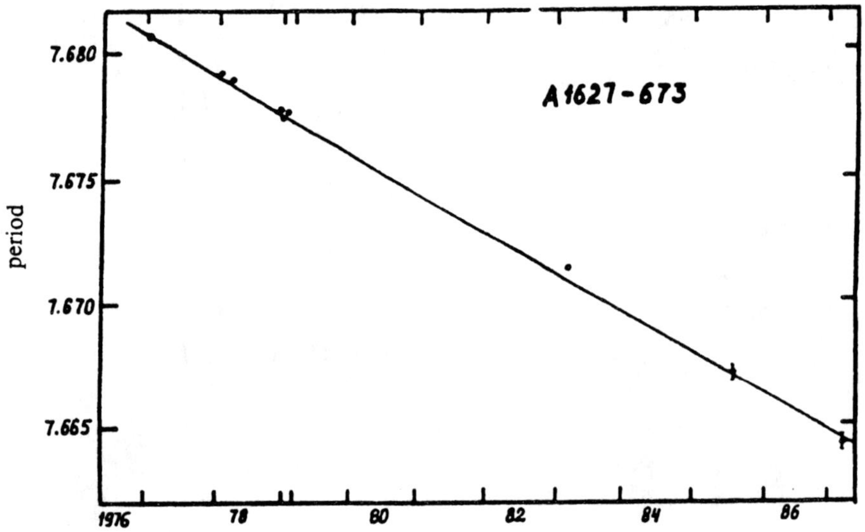

Figure 4c

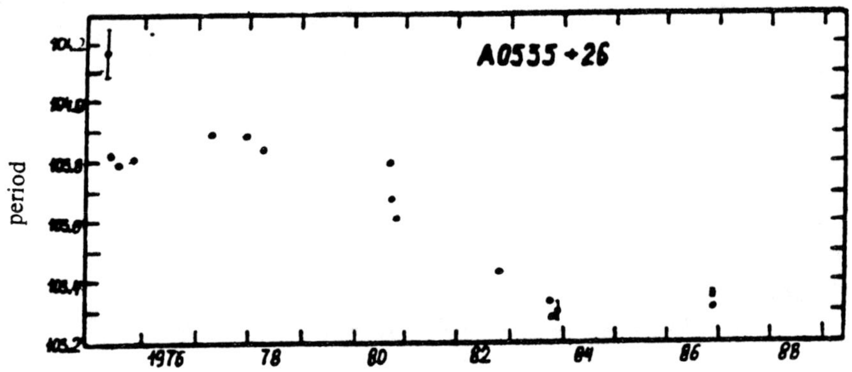

Figure 4d

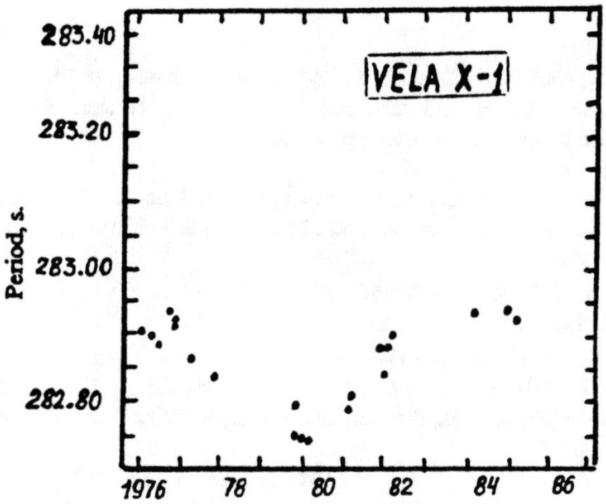

Fig. 4. Period change of pulsar Her X-1 (a), Cen X-3 (b), 4U 1626-67 (c), A 0535 + 26 (d), and Vela X-1 (e).

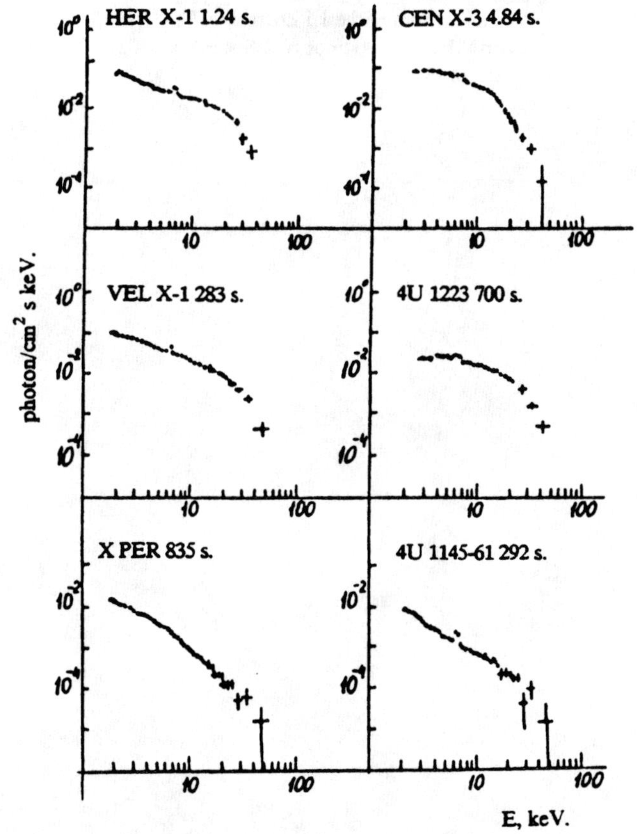

Fig. 5. Spectra of six X-ray pulsars in a range of up to 50 keV. Supplement: photon/cm^2 s keV. E, keV. HER X-1 1.24 s. CEN X-3 4.84 s. VEL X-1 283 s. 4U 1223 700 s. X PER 835 s. 4U 1145-61 292 s.

References

1. H.V.D. Bradt and J.E. McClintock, Ann. Rev. of Astron. and Astrophys., 1983, v. 21, p. 13.

2. S.I. Babichenko, V.A. Silyakin, E.K. Sheffer, and A.V. Shifrin, Problems of Nuclear Science and Technology. Ser. Nuclear Machine Bulding, Atomizdat, Moscow, 1975, v. 28, pp. 8 - 20.

3. V.G. Kurt, Vestnik AN SSSR, 1984, v. 8, p. 6.

4. I.M. Golinskaya, A.Y. Diachkov, V.G. Kurt, et al., Adv. Space Res., 1984, v. 3, no. 10 -12, p. 539.

5. V.G. Kurt, M.S. Burgin, I.M. Golinskaya, et al., Astron. Zh., 1986, v. 63, no. 5, p. 946. 6. H.M. Davis, J.H. Taylor, J.M. Weisberg, and D.C. Backer, Nature, 1985, v. 315, p. 547.

7. L.M. Newton, R.N. Manchester, and D.J. Cooke, MNRAS, 1981, v. 194, p. 841.

8. R.N. Manchester, A.G. Lyne, J.H. Taylor, et al., MNRAS, 1978, v. 185, p. 409.

9. P.C. Joss and S.A. Rappaport, Ann Rev. of Astron & Astrophys., 1984, v. 22, p. 537.

10. I.M. Golinskaya, V.G. Kurt, T.A. Miziakina, et al., Astrophys. & Space Sci., 1985, v. 115, p. 115, p. 393.

11. F. Giovanelli, M. Ferrari-Toniolo, P. Pepsi, et al., Proceed. of the Symposium on X-ray Astronomy'84. Bologna (Italy), 26 - 30 June, 1984, Sp 67-00044, no. 14.

12. E.K. Sheffer, I.F. Kopaeva, M.B. Averintsev, et al., Preprint IKI AN SSSR, no. 1252, 1987.

13. E.K. Sheffer, Pis'ma v Astronomicheskii Zhurnal, 1987, no. 3, p. 1.

14. V.I. Prokhorenko, Preprint IKI AN SSSR, no. 263, 1976.

15. X.X. Newhall, E.M. Standish, and J.G. Williams, Astron. & Astrophys., 1983, v. 125, p. 150.

16. S. Hayakawa and F. Nagase, Proceed. of a Workshop Held at the Max-Planck Institute, July 19 - 29, 1982, p. 14.

17. N.E. White, Proceed. of a Workshop Held at the Max- Planck Institute, July 19 - 29, 1982, p. 29.

18. J. Trumper, Astrophys. J., 1978, v. 219, L. 105.

19. G.Z. Machabeli and V.V. Usov, Pis'ma v Astronomicheskii Zhurnal, 1979, v. 5, no. 9, p. 445.

20. D.G. Lominadze, G.Z. Machabeli, and V.V. Usov, Astrophys. & Space Sci., 1983, v. 90, p. 19.

21. P.E. Hardee, Astrophys. J., 1979, v. 227, p. 958.

Physics of Neutron Stars
Edited by A. M. Kaminker

ISBN 1-56072-164-2

HER X-1/HZ HER: 35-day CYCLE, FREELY PRECESSING NEUTRON STAR AND ACCOMPANYING EFFECTS

N.I. Shakura
Shternberg State Astronomical Institute, 117234, Moscow

Abstract - Effects caused by a neutron star free precession with a period of≈ 35 days in the binary system Her X-1/HZ Her are discussed. The precession modulates an induced star wind off normal component surface and, as a result, the position of the outer boundary of an accretion disk. When in its maximum size the disk shields the X-ray radiation flux from the observer on the Earth.

Since the discovery of a 35-day cycle of modulation of Her X-1 X-ray emission more than 15 have passed [1]. However, the nature of this regular change remains a mystery. There are several models to account for this phenomenon: (a) a freely precessing neutron star [2], (b) precessing of an accretion disk, whose plane is inclined to the system orbital plane [3 - 6], (c) existence of a third body in the system [7], (d) excited nonlinear oscillations at an outer disk edge [8]. Though the model of a precessing accretion disk, whose plane periodically shields the X-ray flux from the Earth observer, can explain some details of an X-ray and an optical brightness curves of the system [9], some doubts about its feasibility still remain [10, 11].

X-ray observations of Her X-1 from the EXOSAT satellite during a 35-day cycle [12] favor the mechanism of free precession of a neutron star.

Free precession of a symmetric vortex is well known to occur when the axis of one of its leading inertia momenta (the vortex symmetry axis) differs from an angular momentum vector $\vec{M}$. Let $I_{||}$ be the inertia momentum parallel to the symmetry axis, and $I_{\perp}$ be two other momenta equal to each other. Let b be the angle between $I_{||}$ and vector $\vec{M}$. Then, together with the main rotational motion with an angular velocity $\dot{\varphi} = M / I_{\perp}$ (the pulsar frequency) directed along vector $\vec{M}$, an additional vortex rotation occurs as a whole entity around the symmetry axis with an angular velocity

$$\dot{\Psi} = \dot{\Psi} \cos b \cdot \left(\frac{I_{\perp}}{I_{||}} - 1\right)$$

In the case of a symmetrical vortex, angle b is constant. As a consequence of the precession motion, both the polar distance δ of some point N and its longitude λ counted from a fixed meridian are variable (see Fig. 1). In the text to follow, the point will be treated as a luminous spot on the neutron star surface. Variability of coordinate δ will lead to periodic (with a period $P_{pr} = 2\pi / \dot{\Psi}$) disappearance and appearance of the luminous spot N, whereas variability of the spot longitude λ can change an observed pulsar frequency [14]:

$$\omega = \dot{\varphi} + d\lambda \,/ dt$$

Let us consider the effect in more detail. If the rate of the λ change were constant and monotonous ($d\lambda \,/ dt$ = Const), the observed periodicity of pulses from the pulsar would remain unaltered. In fact, the longitude of a luminous spot on the star surface varies in a more complicated manner because of precession. Let spot N be at an angular distance a from the axis of the inertia momentum I (Fig. 1). The angle is unchanged during the spot precession motion. Assume that the arc of the large circle passing through points M and I is the prime meridian (λ = 0) and let λ increase in the pulsar rotation direction. Formulas of spherical trigonometry [15] are used for the triangle MIN on the star surface:

$$\cos\delta = \cos a \cdot \cos b + \sin a \cdot \sin b \cdot \cos D \qquad (1)$$

$$ctg\, b \cdot \sin a = \cos a \cdot \cos D + \sin D \cdot ctg\, A \qquad (2)$$

$\lambda = A$ at $0 < \lambda < \pi$ and $\lambda = 2\pi - A$ at $\pi < \lambda < 2\pi$. For the precession of a constant speed Ψ, the velocity of the angle change $\dot{D} = -\Psi$ at $0 < \lambda < \pi$ and $\dot{D} = \Psi$ at $\pi < \lambda < 2\pi$. Then, after differentiation of formula (2) with respect to time, we obtain

$$\frac{d\lambda}{dt} = \Psi \frac{\cos A \cdot (\cos b \cdot \sin D \cdot \sin A - \cos D \cdot \cos A)}{ctg\, a \cdot \sin b - \cos b \cdot \cos D} \qquad (3)$$

Two fundamentally different paths of the spot on the star surface exist depending on the ratio between angles a and b (Fig. 2a, b). When $a > b$ the spot longitude changes monotonically, and at $D = \pi$ and $A = 0$ in the lower connection we have

$$\frac{d\lambda'}{dt} = \Psi \frac{\sin a}{\sin(a + b)}$$

whereas at $D = 0$ and $A = \pi$ in the upper connection we obtain

$$\frac{d\lambda^{||}}{dt} = \Psi \frac{\sin a}{\sin(a - b)}$$

Thus, in this case an observed period will be minimal in the upper connection and maximal in the lower if the direction of the pulsar rotation around its axis $\dot{\varphi}$ coincides with the direction of its precession.

When $a < b$ in the lower connection ($D = \pi$, $A = 0$)

$$\frac{d\lambda'}{dt} = \Psi \frac{\sin a}{\sin(a + b)}$$

and in the upper connection ($D = 0$, $A = 0$)

$$\frac{d\lambda^{||}}{dt} = -\Psi \frac{\sin a}{\sin(b - a)}$$

that is, in this case the spot longitude decreases in the upper connection. Let us now find rates of change in the observed period of impulse appearance averaged over the precession period. For $a > b$

$$<\frac{dP}{dt}> = \pm 4 \left(\frac{P_p}{P_{pr}}\right)^2 \frac{\sin a \cdot \cos a \cdot \sin b}{\sin(a - b) \cdot \sin(a + b)} \qquad (4)$$

and for $b > a$

$$<\frac{dP}{dt}> = \pm 4 \left(\frac{P_p}{P_{pr}}\right)^2 \frac{\sin^2 b \cdot \cos b}{\sin(b - a) \cdot \sin(a + b)} \qquad (5)$$

where $P_p = 2\pi / \dot{\varphi}$ is the pulsar rotation period, $P_{pr} = 2\pi / \Psi$ is the precession period. If the pulsar were observable during all the precession cycle, the proposed model could be unequivocally identified from periodic variations of the observed pulsar frequency. In reality the pulsar in Her X-1 is observed only during a small "window" of the precession cycle when no sign reversal occurs in dP/dt. The precession in this system is likely to proceed in the direction opposite to that of the pulsar rotation ($\Psi < 0$) when $a > b$. Then, restrictions on angles a and b can be found, assuming that the observed changes in the pulsar period $dP/dt = (-2 \pm 1) \cdot 10^{-13}$ s/s detected during an observation set by EXOSAT are due to precession. They are

$$\frac{\sin a \cdot \cos a \cdot \sin b}{\sin(a - b) \cdot \sin(a + b)} = 0.3^{+0.45}_{-0.15} \qquad (6)$$

For example, at $a = 60°$, $b = 30°$

$$\frac{\sin a \cdot \cos a \cdot \sin b}{\sin(a - b) \cdot \sin(a + b)} = 0.43$$

that is, the changes in the pulsar period in Her X-1 observed at that instant might be accounted for by precession effects.

The character of the X-ray spectrum change in Her X-1 with a 35-day period indicates that when the directional pattern approaches the orbital plane because of precession, the warming-up of the adjacent component increases. The increase causes the material to flow at a higher rate to simultaneously increase of the outer radius and thickness of the accretion

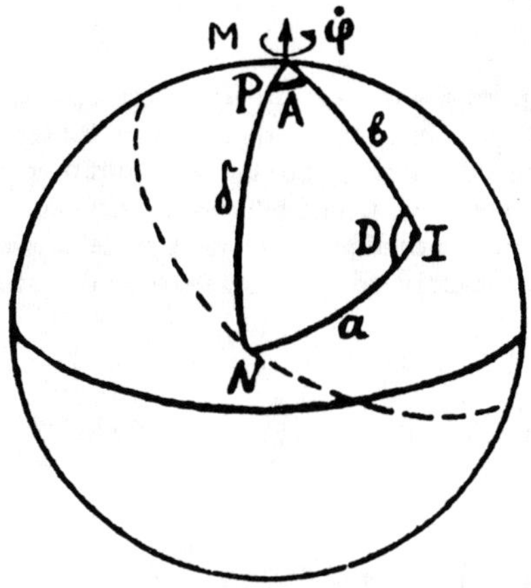

Figure 1

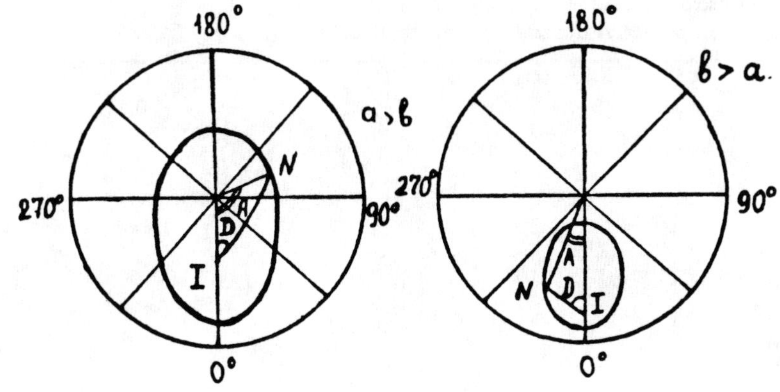

(a) (b)

Figure 2

disk. As the material evaporates from the accretion disk surface in greater amounts, the impulse component of the central source starts to be strongly scattered in the matter, so that radiation in the direction of the observer on the Earth decreases without essential spectral changes.

On the other hand, when the directional pattern moves away from the orbital plane, the rate of an induced material flow, the outer radius and thickness of the disk decrease, so that at some moment the central source becomes visible, the outer edge of the disk unveiling it. However, a very strong absorption on the outer disk edge at the moment should cause drastic changes in the low-energy radiation region observed from the Earth. It should be noted that phase changes of the X-ray "drops" in the brightness curve with time seem to confirm the proposed model provided these "drops" are identified with the point, where the radiation flux meets the material inflowing onto the outer disk boundary.

References

1. H. Tananbaum, H. Gursky, E.M. Kellog, et al., Astrophys. J., 1972, v. 174, p. L143.
2. K. Brecher, Nature, 1972, v. 239, p. 325.
3. J.I. Katz, Nature Phys. Sci., 1973, v. 246, p. 87.
4. J.W. Roberts, Astrophys. J., 1974, v. 187, p. 575.
5. J.A. Petterson, Astrophys. J., 1975, v. 201, p. L61.
6. J.A. Petterson, Astrophys. J., 1977, v. 218, p. 783.
7. T. Mzeh and J. Shaham, Astrophys. J., 1977, v. 213, p. L17.
8. F. Meyer and E. Meyer-Hofmeister, Astron. & Astrophys., 1984, v. 140, p. L35.
9. D. Gerend and P.E. Boynton, Astrophys. J., 1980, v. 209, p. 562.
10. J. Papaloizou and J.E. Pringle, MNRAS, 1982, v. 200, p. 49.
11. Y. Kondo, van Flandren, and C.L. Wolf, Astrophys. J., 1983, v. 273, p. 716.
12. J. Trümper, P. Kahabka, H. Ogelman, et al., Preprint MPI, n. 41, 1985.
13. L.D. Landau and E.M. Lifshits, Mechanics, 1973, Nauka, Moscow.
14. E.K. Sheffer, Pis'ma Astron. Zh., 1987, v. 13, p. 204.
15. K.A. Kulikov, Course in Spherical Astronomy, 1969, Nauka, Moscow.

Physics of Neutron Stars
Edited by A. M. Kaminker

ISBN 1-56072-164-2

X-RAY BURSTERS

E.V. Ergma
The Astronomy Board of the USSR Academy of Sciences, 109017, Moscow

New observational data on burster obtained by "EXOSAT" are presented. Different models of bursters and their evolution are discussed.

The majority of recent observational data on small mass X- ray binaries have been obtained by "EXOSAT". "EXOSAT" was the first high orbital satellite ($P \approx 4$ days) which could carry continuous observations of sources.

The results of continuous observations (for about 80h) of X-ray bursters are of great interest for the development of theoretical models. Four sources were mainly observed: EXO 0748 - 676 [1], MXB 1636 - 53 [2, 3], MXB 1735 - 44 [4], 4 U 1705 - 44 [5, 6].These observations permit seeking mutual correlations between such parameters as the value of direct current which determines the rate of accretion, an interval between outbursts and energy released in explosions.

For two sources, 4U 1705 - 44 and EXO 0748 - 676 an anticorrelation was established between the value of direct current and outburst periodicity. For 4U 1705 - 44 20 outbursts (flashes) were detected with $\nu_R \sim 0.5\ h^{-1}$ at $F_x \sim (1.3 - 1.8) \cdot 10^{-9}$ erg/cm^2 - s; when F_x increases to $2.3 \cdot 10^{-9}$ erg/cm$^2 \cdot$ s the number of outbursts decreased to $\nu_R \sim (0.5 - 0.1)\ h^{-1}$. As the constant background to increase $F_x \sim 1.07 \cdot 10^{-8}$ erg/cm$^2 \cdot$ s, the outbursts stopped taking place [5, 6]. For EXO 0748 - 676 [1] the periodicity of outbursts is also in anticorrelation with the F_x value: $\nu_R \sim 0.75\ h^{-1}$, $F_x \sim 3 \cdot 10^{-10}$ erg/cm$^2 \cdot$ s; $\nu_R \sim 0.12\ h^{-1}$ $F_x \sim 10^{-9}$ erg/cm $\cdot$ s. A correlation between the flux E_b, integrated over the outburst time t_R, and the interval between outbursts t_R was established for EXO 0748 - 676, MXB 1636 - 53 and MXB 1735 - 44. At small t_R the ratio of E_b and t_R is almost linear, and at t_R 6h the E_b dependence on t_R becomes more flat (Fig.1).

Analysis of the data for the above three sources led Van Paradijs and Lewin [7 - 9] to the following conclusion: these sources release almost equal amounts of energy in their outbursts irrespective of the accretion rate at a given time of the outburst expectation t_R.

Another important conclusion made by those authors concerns a drastic of α (α is the ratio of constant flux, integrated over the time t_R, and the outbursts flux, integrated over the outburst time).This change cannot possibly be due only to substitution of thermonuclear fuel in the shell (hydrogen by helium), which would ultimately change α by a factor of 7. This implies that permanent combustion of thermonuclear fuel may take place not only in very bright sources but in less bright as well. Estimations of the α value require, however its dependence on a model of a binary, i. e. on the angle of the orbital inclination i, should be taken into consideration. Lapidus and Sunyaev showed [10] that at $\mu \sim 1$ ($\mu = \cos i$) the α value is underestimated by about 1,5 -3.5 times whereas at $\mu \sim 0$ it is overestimated by 1.5 -2.5 times as compared to the generally accepted simplest estimate. Moreover, α for the same source can vary with the accretion rate.

Analysis of the flashes from MXB 1636 - 53 (data on more than 100 flashes are now available for it [2, 3, 8]) revealed considerable changes in the integral and maximal fluxes of the outbursts (by a factor of 7) and in the intervals between the outbursts (by a factor of 20) whereas the constant flux did not change very much ($\sim$ 2.4 times). "Dwarf" outbursts with $E \approx 10^{-7}$ erg/cm^2 and $t_R \approx 100$ min (Fig. 1) recorded from this source also demand a new approach to the theoretical models for bursters(see below).

Complicated brightness curves.

The flash profiles in some bursters have been recently found to be of a double structure, especially in a harder region [11, 12]. Latest observations by "EXOSAT"detected such a structure in hard channels $E > 6$ keV in the flashes of MXB 1820 - 30 [13].

The brightness curve from MXB 1636 - 53 behaved quite differently [2, 14]. In May 1984 eight flashes were recorded from this source, four of them gave an an usual brightness curve (see Fig. 2a). The source is unusual in that the bolometric brightness curve had a double structure. And, secondly, the energy released in these flashes is of the same order that in ordinary flashes though their brightness curves do not behave in a similar manner. Even a triple flash was recorded from this source [Fig. 2b].

Determination of orbital periods in low-mass X-ray systems (LMXRS).

Determination of orbital periods in LMXRS remained a difficult problem for observers. The probability of an X-ray eclipse to be detected is very small because of X-ray emission anisotropy especially in the systems with large inclination angles $60^{\circ} < i < 90^{\circ}$ [10]. However, orbital periods have been recently determined for more than 20 systems owing mainly to the "EXOSAT" observations. Some systems showed "gaps" against the constant X-ray background; these "gaps" appear with a certain periodicity. They are most likely formed in a more dense area near the accretion disc periphery, where a gas flux from the second component strikes against the disc. The disc becomes thicker and part of x-ray radiation is lost for the observer. Both complete eclipses and "gaps" were observed in EXO 0748 - 67 (P = 3.8 h) and MXB 1659 - 29 (P = 7.2 h). However deep and complete eclipses of X-ray sources are very rare to observe.

A most short period system was discovered not long ago, being unique not only among LMXRS but cataclysmic variables as well. This system, MXB 1820 - 30, is the first X - burster discovered by Grinday and Heise in 1975 [16]. The source period determined from the X-ray beam modulation (within 2 - 3%) is 685s [17]. The period was confirmed by the data from SAS - 3 [18]. These two sets of data, from SAS - 3 and "EXOSAT" (with an interval of 8 years between them), allowed determination of $P = 685.0114 \pm 0.0003$s and at the 3σ level $\dot{P} < 0.11$ ms/yr. Fig. 3 shows distribution of MXRS and cataclysmic variables with $P < 10$h according to periods. The data for cataclysmic variables were borrowed from Ritter catalogue [19] and for LMXRS from Refs [15, 19].

Theoretical analysis.

The effects of the accretion rate M., interior thermal structure and magnetic field of a neutron star on the outburst character were analyzed in Refs [20, 21]. Works [22 - 24] developed a single zone model, which allows investigation of the effect of different parameters (M, P, Z_{CNO}) on the character of nuclear energy release. The model has some advantages as compared to the multiple zone model. If an explosive combustion of hydrogen is taken into consideration, then a system of nuclear kinetics equations should be used [25 - 27] to include nuclei of not only group elements but more heavier nuclei as well. These calculations take too much computer time, therefore a simplified system is used in the multizone model, which includes only nuclei of major elements. The scheme works well when the initial concentrations of hydrogen and helium remain unchanged before the explosion. If, However, the helium concentration in the accreting material is great or a major portion of hydrogen has burnt before the outburst, then the multizone scheme give erroneous results.

Comparison of observational data requires theoretical calculations of outburst cycles. Woosley and Weaver [28] were the first to make such calculations. For the major series with $\dot{M} = 2 \cdot 10^{-9}$ $M_{\odot}$/yr and $Z_{CNO} = 0.02$ twelve outbursts were calculated at $t_R = 2.4$h. It should be noted that the concentration of the CNO group elements strongly affects the brightness curve shape, energy released in the outburst and time interval between the outbursts.

"Dwarf" outbursts of MXB - 1636 - 53.

First simplified estimates in terms of the spherically symmetrical thermonuclear model [29, 30] showed that an explosive instability arises after some critical mass of material is accumulated. Thus, there exists some lower limit of the energy liberated in the explosion. the critical mass is determined by the formula [24]:

$$\Delta M_{en} \approx 7.6 \cdot 10^{-13} \left(R / 10\,km\right)^2 \left(0.02 / Z_{CNO}\right)^{0.5} M_{\odot}$$

and, hence, the total energy released in the outburst (for an observer at some distance *) is

$$\varepsilon_b^* = q \Delta M_{en} (1 + Z_s)^{-1} \approx 3.3 \cdot 10^{39} \left(R / 10km\right)^2 \left(1.31 / (1 + Z_s)\right) \left(0.02 / Z_{CNO}\right) erg$$

$$(1 + Z_s) = \left(1 - 2GM / RG^{2}\right)^{-0.5}; \; q = \left(1.5 \cdot 10^{18}\, erg/g\,(He); 4.5 \cdot 10^{18}\, erg/g\,(H)\right)$$

Luminosity is expected the lowest when hydrogen concentration in the shell does not practically change. In this case

$L^{*}_{max} \approx 0.6 L^{*H}_{Edd} = 0.6\ 4\pi cGM\ /\ (1 + Z_s)\ae$

the factor o.6 shows that about 35% of the energy is lost with neutrinos during β^+ decay.

According to [31] the lower limits for the maximal flux in dwarf outbursts and energies in them can be written down as $F_{max} \geq 0.6 F^{H}_{Edd} = 2.1 \cdot 10^{-8}\ erg/cm^2 \cdot s$,

$$E_b = \varepsilon_b^{*} \frac{F^{H}_{Edd}}{L^{*H}_{Edd}} \geq 7.6 \cdot 10^{-7} \left(R\ /\ 10km\right)^2 \left(\frac{1.4M_\odot}{M}\right) \left(\frac{0.02}{Z_{CNO}}\right)^{0.5} erg\ /cm^2$$

Thus, theoretical values are appreciably greater than observational data. Fudjimoto et al [31] suggested that an additional intermixing of some fresh material with the material of deeper layers should be included to resolve the discrepancy. This intermixing may result from hydrodynamics instabilities. A similar mechanism was proposed for new stars to explain an excessive concentration of CNO group elements in the ejected material.

We think that a two stage model of nuclear energy release is worthwhile considering to account for a complicated curve of MXB 1636 - 53 brightness. Preliminary calculations (Kudryashov and Ergma, unpublished results) show that the brightness curve does have a double humped shape at a certain dependence between the shell mass and concentrations of helium and hydrogen (see also [21 -23]). Simultaneous combustion of hydrogen and helium gives a curve with two maxima for the nuclear energy release. The first maximum is due to intensive helium combustion, the second results from the residual combustion of more heavy elements. Whether this structure remains depends on the time of photon diffusion and the sum of times of positron decays of the most neutron- deficient isotopes with 56 > A > 20. To establish line details, a scheme of nuclear kinetics should be chosen with great care.

Determination of masses and radii of neutron stars.

Shank et al. [33] were the first to show a burster spectrum may be described with the help of a black body spectrum with a temperature (kT) varying from 0.9 - 2.3 eV. For the black body emission we have

$$L^{*}_{b} = 4\pi R^{*2} \sigma T^{*4} = 4\pi D^2 F^{*}_{b}$$

Where T is the effective temperature. The formula is valid for isotrope emission. With anisotropy [10] and bolometric correction ε 1(energy is radiated not only in the X-ray region) we have

$$L^{*}_{b} = 4\pi R^{*2} \sigma T_e^{*4} = 4\pi D^2 \xi_b\, \varepsilon\, F^{*}_{b}$$

The black-body interpretation encounter some difficulties. First, the main contribution into nontransparence of the burster photosphere comes from Thomson scattering, which distorts the Planck spectrum. Second, the black-body interpretation involving the effects due to general relativity theory gives very small upper limits for the mass and radius of a neutron star excluding the case when its luminosity considerably exceeds the Eddington's in the flash maximum [34].

Following Van Paradijs and Lewin [7], two methods may be used to find R^{*}. The first states that at the stage of Eddington luminosity,when R » R• (R• is the radius of a neutron star) we have

$$L^{*}_{Edd} = 4\pi D^2 \xi_b\, \varepsilon\, F^{*}_{Edd} \approx \frac{4\pi c\ GM}{\ae}$$

When the radius of the photosphere becomes R ≈ R• and the luminosity is still at the Eddington level, we obtain

$$L^{**}_{Edd} = 4\pi D^2 \xi_b\, \varepsilon\, F^{**}_{Edd} = \frac{4\pi c\ GM}{\ae} \left(1 - \frac{2\,GM}{R_* c^2}\right)^{0.5}$$

These two expressions can give the value of M/R* if F_{Edd}^{*} (R » R•) and F_{Edd}^{**}(R ~ R•) are known. The precision of this method is not great.

The second method uses observations of different phases in the outburst (expansion stage = cooling stage). According to [7] we have

$$F_b^* / F_{Edd}^* = R_*^2 \left(1 - \frac{2GM}{R_* c^2}\right)^{-1} \cdot \frac{æ}{c\,GM} \sigma T_e^{*4}$$

Where F_b^*, F_{Edd}^* and T_c^* are observed values. Most difficult to determine in this method is T_e^*. We have mentioned above that the emission spectrum during the outburst is strongly different from the Planck spectrum because of the effect of Thomson scattering. Hence, the observed temperature T_c is not equal to the effective temperature T_c [35, 36]. At present the relationship between T_c and T_e is not well understood [35, 36] (see, however, Titarchuk in this volume).

Kaminker et al. [37] calculated a possible mass and radius of the neutron star in the burster MXB 1728 - 34, assuming that at maximum luminosity the spectrum is formed in the isothermally scattering atmosphere of the neutron star, in which radiation pressure and the OTO effect play a certain role.

A third possibility to determine R and M for a neutron star is make use of the absorption lines which were found in the spectra of some bursters. If line E = 4.1 keV for MXB 1636 - 536 [38] could be interpreted as a helium like line of iron (6.7 keV) but displaced due to the gravitational red shift, then the radius of the neutron star would small ($\sim 1.5\ R_g$) [39]. However, the reverse Doppler effect makes the equation for R and M look like

$$R = (2.4 \pm 0.1)\,R_g = (9.3 \pm 0.3)\,(M \,/\, 1.4 M_\odot)\ km$$

Evolution of LMXRS and formation of ultrashort-period systems.

Detection of orbital periods in more than 20 systems raises the problem of evolution status of such objects.

The most important aspect in the evolution of these systems is the mechanism of mass transfer [4]. Analysis of the evolution of a binary system with masses M_1 and M_2 (the star losing its mass, $M_2 < M_1$) shows that the Rocha radius and distances between the components increase with time if the system orbital angular moment and its total mass remain unchanged (i. e. under conservative evolution conditions) Therefore an additional mechanism of this moment loss by the system should be sought for. One of this mechanism may be the loss of the angular moment during the emission of gravitational waves, first suggested by Paczynski [42]. The characteristic evolution time, if this mechanism works, is shorter than the cosmological provided $P < 10$ h. The rate of the orbital angular moment loss is

$$\dot{J} = -3.46 \cdot 10^{-67} M_1^2 M_2^2 \,/\, (M_T + M_2^1)^{2/3} P^{-7/3}$$

Nondegenerate hydrogen-helium stars have a minimal period about 80 min [43, 44]. A characteristic value of the accretion rate for such stars is about 10^{-10} $M_\odot$/yr ($P < 10$ h) owing to the emission of gravitational waves. Observations of LMXRS show that $\dot{M}$ must be, at least, an order of magnitude greater.

One more mechanism of the angular moment loss is that of stellar wind.

Verbunt and Zwaan [45] proposed an approximate formula for the angular moment loss via the magnetic stellar wind

$$\dot{J} = -0.5 \cdot 10^{-28} f^{-2} k^2 M_2 R_2^4 \omega^3$$

where $k^2 = 0.1$, $f = 0.78$ or 1.78 and ω is the orbital frequency.

If the loss of the orbital angular moment is caused by the magnetic stellar wind, then $\dot{M}_2$ may be from 10 to 100 times greater than in the case of the gravitation wave emission.

Let us consider the evolution a tight binary system in which one the components is a compact object (a white dwarf or a neutron star [46 - 48]).

Figure 4. shows evolution tracks for different models.

Stars that fill the Rocha cavity because they are homogeneous, become completely convective at $M < 0.3$ M. In the work [49] it was shown that at complete intermixing the magnetic stellar wind disappears, and then the star radius decreases,the material exchange temporarily stops and would renew after the second component fills the Rocha cavity again when the components come close to each other due to the emission of gravitational waves. This process sometimes explain the deficit of cataclysmic variables with $P \sim 2^h - 3^h$.

The evolution track in Fig. 4 presents the evolution of a homogeneous-helium star. Its rate of mass loss was artificially increased by an order of magnitude as compared to case 1. Under these conditions the star becomes completely convective already at $P \sim 4^h$ and the material accretion stops. This may be the cause of periods $2^h - 4^h$ deficit in LMXRS.

We showed earlier [46, 48] that ultrashort periods in binary sisters can appear if hydrogen in the second component core is completely consumed in combustion by the moment the Rocha cavity becomes filled: $10^{-6} \leq X_c \leq 10^{-2}$. The gradient of the chemical compositions prevents the convection from penetrating inside the star, and complete intermixing occurs only if the star mass is $\sim 0.02 - 0.03M_\odot$, and hydrogen concentration $X \sim 0.1$. Greater helium abundance may cause the orbital period to decrease down to 10^m. Fig. 4 shows the evolution track for a system with the following parameters: $1.25 + 1M_\odot$, $X_c \sim 10^{-6}$. The rate of the material accretion for such systems ($p < 1$ h) must be high, which is observed.

Let us now consider other possibilities of ultrashort periodic LMXRS formation. If a neutron star companion is a giant, then a tight system may form comprising the neutron star and a helium or a carbon-oxygen dwarf, the giant core, following the stage of a common shell after the giant fills the rocha cavity.

On the other hand, if in a tight binary system the component with a mass $M < 2.5M_\odot$ fills the Rocha cavity at the stage of hydrogen combustion in a layered source, it wold transform into a nondegenerate helium star as a result of material loss. Savonije et al. [50] noted that such helium may be a neutron star companion during helium combustion in the core. A characteristic time for the components of a binary system to approach each other due to the emission of gravitational waves is

$$\tau_{GR} \approx 5.4 \cdot 10^6 \,(M_{He} \,/\, M_\odot)^{1.27} \,(M_t \,/\, M_\odot)^{0.33} \text{ years}$$

$M_t = M_{He} + M_1$. The helium star filling the Rocha cavity will remain in thermal equilibrium if
$\tau_{kh} \approx 6 \cdot 10^5 (M_{He} \,/\, M_\odot)^{-2.5} \textit{years} < \tau_{GR} < \tau_{He} \approx 10^{7.2} \,(M_{He} \,/\, M_\odot)^{-3.1}$
at $0.4 \leq M_{He} / M_\odot \leq 1$ or $22^m < P < 47^m$. Reaching the period of 20^m, the helium star loses it thermal equilibrium. A track for a star in thermal equilibrium is shown in Fig. 4. Periods in the systems with equilibrium helium nondegenerate components are liable only to decreasing. Therefore the sign of $\dot{P} / P$ indicates the chemical composition of the optical component.

Evolution status of MXB 1916 - 05 and MXB 1820 - 30.

Both of systems are X-ray bursters and a joint analysis of their X-ray outbursts (from the viewpoint of their chemical composition) and and evolution status of the optical component seems appropriate. Excluded from the analysis should be the systems in which the optical component is a homogeneous hydrogen-helium star ($P \geq 80$ min) by the time of the Rocha caving filling.

Swank et al. found that a candidate for the optical component of MXB 1916 - 05 should be a star with a helium core, a hydrogen shell and out of thermal equilibrium. Such a star would account for the intervals between the outbursts and the observed $\dot{M}$.

However, the idea of a normal chemical composition of the material before an outbursts ($X = 0.7$, $Z = 0.004$, because hydrogen cannot be consumed completely in combustion for a time of 10^4) seems incompatible with outbursts lasting only for $t < 10$ s [52]. If the evolution of MXB 1916 - 05 proceeds according to variant 2 (Fig. 4), When the Rocha cavity is filled after hydrogen is exhausted in the second component core, then the surface of the neutron star becomes substantially enriched with helium ($Y \sim 0.73$ at $P \approx +$‘ 50 min). Because of a high helium concentration in the material before an outburst, a characteristic time of it may be short (< 10 s).

According to this scenario MXB 1820 - 30 is a system at a more advanced stage of evolution as compared to MXB 1916 - 05. Our calculation allow-explanation of a high X-ray luminosity of the system, the value of $| \dot{P} / P |$ and an extremely short period. The parameters of double outbursts of MXB 1820 - 30 require, according to ref. [53], that hydrogen concentration in the material following onto the neutron star be low ($X \sim 0.1$).

"Blanks" in orbital periods in LMXRS.

Distribution of LMXRS and cataclysmic variable stars in orbital periods reveals "blanks" (Fig. 3). "Blanks" for cataclysmic variables are now considered to result from "switching off" of the magnetic stellar wind in completely convective stars ($M < 0.3M_\odot$) at $P \sim 3^h$ [49]. when the stellar wind drops, the loss of the orbital angular moment decreases, the secondary star contracts and the material exchange stops. the system becomes unobservable. The exchange renews only after the components come closer to each other due to the emission of gravitational waves and the Rocha cavity is filled with the second component. Further evolution proceeds in the time scale of gravitation wave emission.

Comparison of distributions in periods for LMXRS and cataclysmic variables shows that the former also lack periods of 1 to 2 hours, i.e. their "blank" is wider than for variable stars.

We have made an attempt to account for the "blank" in the periods for LMXRS, Ref [48]. Let us now consider the accretion onto a neutron star with a weak magnetic field ($H \sim 10^8 - 10^9$ G). After the secondary, chemically homogeneous, component fills the Rocha cavity, it continues to lose its material for about $(6 - 8) \cdot 10^8$ years at a rate $10^{-7} - 10^{-9}\ M_\odot$/year [46, 47]. Accretion onto a magnetized neutron star speeds it up the so-called equilibrium period P_{eq} in which the co-rotation radius $R_c = (G M P^2 / 4\pi^2)^{1/3}$ is equal to the Alfvien radius R_A. Following Lipunov and Shakura [54], we can write down

$$R_A \approx 3.1 \cdot 10^8 \dot{M}_{-10}^{-2/7} B_8^{4/7} R_6^{12/7} m^{-1/7}\ cm \qquad (1)$$

$$P_{eg} \approx 2.9 \cdot 10^{-3} \dot{M}_{-10}^{-3/7} B_8^{6/7} R_6^{18/7} m^{-5/7} c \qquad (2)$$

where $\dot{M}_{-10}$ is the accretion rate $10^{-10} M_\odot$/yr, B_8 is the magnetic field strength 10^8 G, R_6 is the radius of the neutron star 10^6 cm and m is the mass of the neutron star in $M_\odot$.

A change in the velocity of the neutron star rotation can be described with the equation

$$\frac{d(I\omega)}{dt} = \dot{M}(GMR)^{1/2} \qquad (3)$$

$I = 0{,}4MR^4$ is the inertia moment of the neutron star, $\omega = 2\pi / P$. A characteristic time of the rotation acceleration up to P_{eq} is

$$\tau_{acs} \approx 4.4 \cdot 10^8 \dot{M}_{-10}^{-3/7} B_8^{-8/7} R_6^{-10/7} m^{9/7} \qquad (4)$$

less than the evolution characteristic time until the magnetic stellar wind stops. As soon as P_{eq} is reached the "propeller" mechanism switches on [55], and the material is pushed away from the neutron star. Accretion and "propeller" phases would evidently alternate in the further evolution, whereas the revolution period would remain close to its present value $P_{eq}(M)$.

After the magnetic stellar wind drops and the material exchange stops, the neutron star continues to rotate at a high speed ($P \approx P_{eq} \approx 10^{-3}$ s). When the exchange is resumed the star becomes a "propeller", because the characteristic $\dot{M} \approx 5 \cdot 10^{-11} M_\odot$/yr is more than an order of magnitude smaller than $\dot{M}$, which determines P_{eq} and, hence, $R_A > R_C$. From expression (1 - 4) we can derive the time for the neutron star slowing down to a period where accretion becomes possible: $\tau_{slow} \approx \tau_{accel} (\dot{M}_1 / \dot{M}_2)^{6/7}$. Here $\dot{M}_1$ and $\dot{M}_2$ are the ratios of the material flow at the stages of acceleration and slowing down, respectively. The value τ_{slow} is comparable with the time of the system evolution to the minimal orbital period about $(1 - 4) \cdot 10^9$ yr when the rate of the material flow falls drastically. Thus, when an X-ray source "switches on" it turns out to be weak ($L_x < 3 \cdot 10^{34}$ erg/s) because of a low-accretion rate.

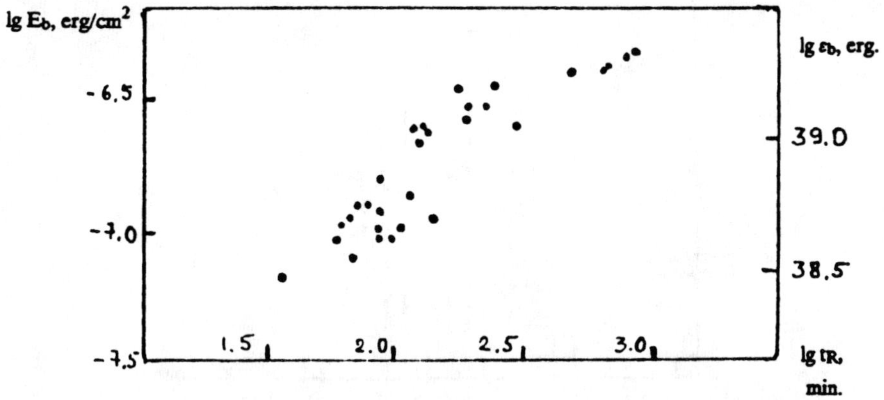

Fig. 1. The dependence of Eb on tR for MXB 1636 - 53.

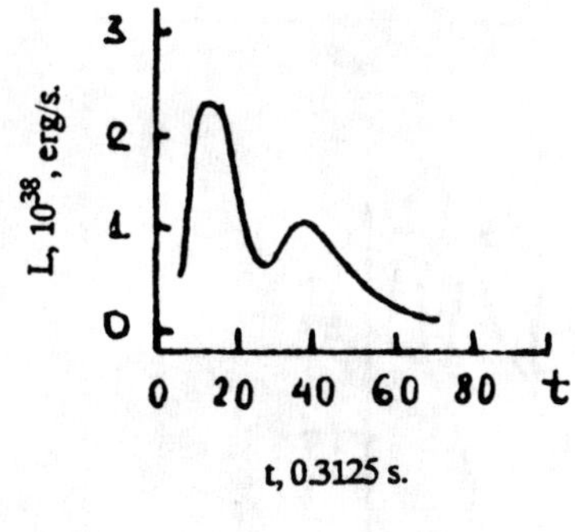

Fig. 2a. A double flash of MXB 1636 - 53 [2].
Supplement: L, 10^{38}, erg/s. t, 0.3125 s.

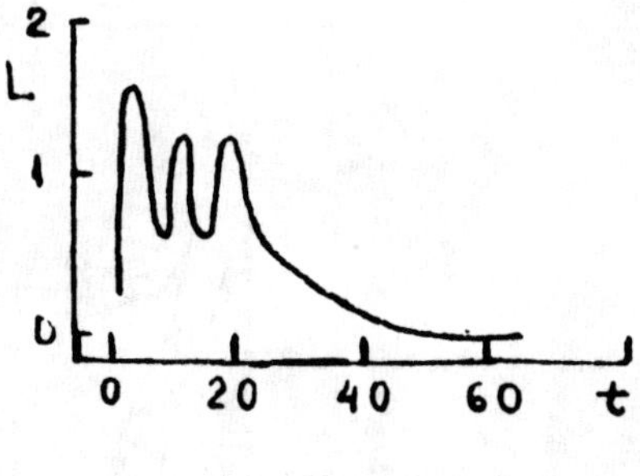

Fig. 2b. A triple flash of MXB 1636 - 53 [14].

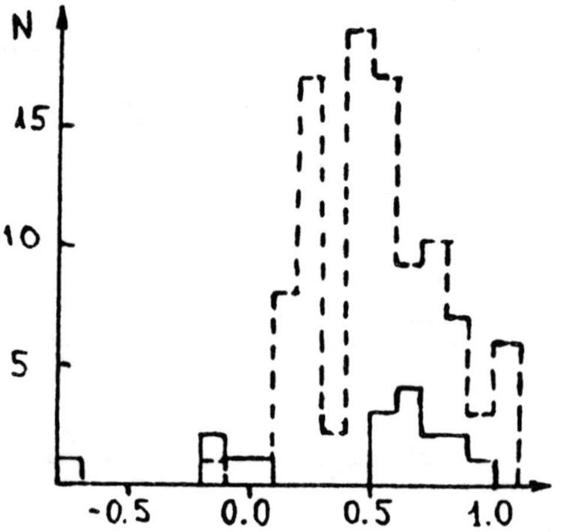

Fig. 3. Distribution of low mass X-ray sources (solid lines) and cataclysmic variables (dotted lines) in orbital periods. Supplement: lg P, hours.

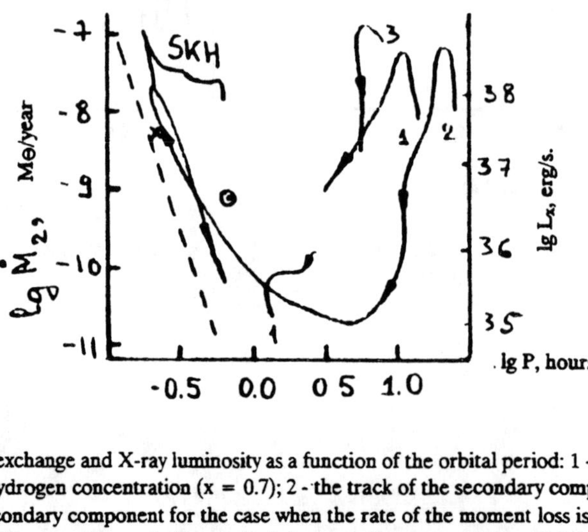

Fig. 4. The rate of material exchange and X-ray luminosity as a function of the orbital period: 1 - the track of the secondary component with a normal hydrogen concentration (x = 0.7); 2 - the track of the secondary component after its evolution is over; 3 - the track of the secondary component for the case when the rate of the moment loss is artificially increased by an order of magnitude. Dotted lines are the evolution track of the secondary component which is a homogeneous thermally equilibrium helium star or a helium dwarf. Line SKH presents the calculated results for a helium component, x - MXB 1820 - 30, ϴ - MXB 1916 - 05

REFERENCES

1. M. Gottwald, F. Haberl, A.N. Parmar, N.E. White, Astrophys. J. 1987, v. 308, p. 213.

2. M. Szatajino, J. Van Paradijs, W.H.G. Lewin et al., Astrophys. J., 1985, v. 299, p. 487. 3. W.H.G. Lewin, W. Penninx, J. Van Paradijs et al., Astrophys. J., 1987, v. 319, p. 893.

4. J. Van Paradijs, W. Pennix, W.H.G. Lewin et al., Astron. and Astrophys., 1987.

5. A. Langmeier, M Sztajno, G. Hasinger, J. Trumper, MPE Preprint N. 77, 1986.

6. M. Gottald, F. Haberl, A. Langmeier et al., EXOSAT Preprint N. 55, 1987.

7. J. Van Paradijs, W. Penninx, W.H.G. Lewin, Preprint CSR - HEA - 87 - 16, 1987.

8. T. Ohashi, Preprint ISASR, N. 141, 1981.

9. J. Van Paradijs, W.H.G. Lewin, Preprint CSR - HEA - 87 - 15, 1987.

10. I.I. Lapidus, R.A Sunyaev, NNRAS, 1985, v. 217, p. 291.

11. W. H. G. Lewin , J.A. Hoffman, J. Doty et al., NNRAS, 1976, v. 177, p. 95P.

12. J.A. Hoffman, L. Cominski, W.H.G. Lewin, Astrophys. J.,1980, v. 240, p. L27.

13. F. Haberl, L. Stella, N.E. White et al., Astrophys. J., 1987, v. 314, p. 266.

14. J. Van Paradijs, M. Sztajino, W.H.G. Lewin et al., NNRAS 1986, v. 221, p. 617.

15. N.E. White, in :"Evolution of Galactic X-ray Binaries", ed. J Trumper, 1986, p. 227.

16. J. Grinday and J Heise, IAUC N. 28779, 1975.

17. L. Stella, W. Priedhorsky, and N.E. White, Astrophys. J , 1987, v. 312, p. L17.

18. E. Morgan, IAUC N. 4261, 1986.

19. H. Ritter, MPA Preprint N. 285, 1987.

20. J.P. Joss, F. Li, Astrophys. J., 1980, v. 228, p. 287.

21. S. Aysly, P.C. Loss, Astrophys. J., 1982, v. 256, p. 637.

22. A.D. Kudryashov and E.V. Ergma, AZh, 1983, n. 60, p. 923. 23. A.D. Kudryashov and E.V. Ergma. AZh, 1983, n. 60, p. 1148.

24. T. Hanawa and M.Y. Fujimoto, PASJ, 1982, v. 34, p. 495.

25. A.D. Kudryashov and E.V. Ergma, PAZh, 1980, n. 6, p. 712.

26. R.E. Wallace and S.E. Woosley, Astrophys. J. Suppl. Ser., 1981, v. 45, p. 389.

27. T. Hanawa, D. Sugimoto, M. Hashimoto, PASJ, 1983, v. 35, p. 491.

28. S.E. Woosley and T.A. Weaver, In: AIP Conference Proceeding N. 115 "High Energy Transient in Astrophysics", ed. S.E. Woosley, 1984, p. 273.

29. E.V. Ergma and A.V. Tutukov, Astron. Astrophys., 1980, v. 84, p. 123.

30. M.Y. Fujinoto, T. Hanawa, S. Miyaji, Astrophys. J., 1981, v. 246, p. 269.

31. M.Y. Fujimoto, M. Sztajno, W.H.G. Lewin, J. Van Paradijs, Astrophys. J. , 1987, v. 319, p. 902.

32. R. Kipenhahn, H.C. Thomas, Astron. Astrophys., 1978, v. 63, p. 625.

33. J.H. Swank, R.H. Becker, E.A. Boldt et al., Astrophys. J.,1977, v. 212, p. L73.

34. H.L. Marshall, Astrophys. J., 1982, v. 260, p. 815.

35. I.I. Lapiduc, R.A. Sunyaev, L.G. Titarchuk, PAG, 1986, v. 12, p. 918.

36. R.A. London, R.E. Taam, W.M. Howard, Astrophys. J., 1986, v. 306, p. 170.

37. A.D. Kaminker, G.G. Pavlov, U.A. Shibanov et al., PAJ, 1988, v. 14, p. 918.

38.I. Waki et al., PASJ, 1984, v. 36, p. 819.

39. M.Y. Fujimoto, Astrophys. J., 1985, v. 293, p. L19.

40. M.YY. Fujimoto, R.E. Tamm, Astrophys. J. 11986, v. 305, p. 246.

41. R.E. Taam, In: "The Evolution of Galactic X-ray Binaries", ed. J. Trumper, 1986, p. 205.

42. B. Paczynski, Acta Astrono., 1967, v. 17, p. 287.

43. B. Paczynsi, R. Sienkiewich, Astron. J., 1981, v.1981, v. 248, p. L27.

44. S. Rappoport, P.C. Joss, R.E. Webbinks. Astron. J., 1982, v. 254, p. 616.

45.F. Verbunt, C. Zwaan, Astron. Astrophys. , 1981, v. 1000, p. L7.

46. A.F. Tutukov, A.V. Fedorova, E.V. Ergma, L.R. Yungel'son, PAZh, 1985, v. 11, p. 123.

Physics of Neutron Stars
Edited by A. M. Kaminker

ISBN 1-56072-164-2

SPECTRA OF BURSTERS IN FLASHES AND A CONTINUOUS EMMISSION MODE

L.G. Titarchuk
Institute of Space Research, USSR Academy of Sciences, 117810, Moscow

An analytical model of the spectrum formation for exploding neutron stars is proposed. Different radiation conditions in a burster atmosphere are analyzed as a function of luminosity proximity to the Eddington limit. Formulas are derived to estimate the color temperatures of the Planck law, which adequately approximates a short-wave range of the burster spectrum. Some problems of burster spectrum formation in a continuous emission mode are discussed. Theoretical spectra are calculated in terms of the double-temperature accretion model as a function of luminosity and for different temperatures of the emitting region. The calculated spectra are compared with observations.

Introduction.

Outbursts of many bursters are considered to be due to thermonuclear explosions in the subsurface layers of accreting neutron stars, whereas the state of continuous emission is a result of matter accretion onto a neutron star [1, 2].

The report presents the results of our work on the burster spectrum formation both in outbursts and a continuous emission mode.

The first part of the report deals with the spectrum formation in outbursts. Theoretical models for bursters have been discussed in a number of papers [3 - 8]. All the authors agree that the main feature of the burster atmosphere is the predominance of the Thomson scattering in nontransparency, which results in color temperatures of X-ray bursters greatly exceeding the Eddington limit.

It should also be noted [6, 8] that determination of an X-ray burster emission spectrum is a typical stationary problem in the sense that X-ray photons are emitted almost instantaneously. A burst of X-ray radiation lasts a few seconds, which is much longer than the time the radiation ploughs through the atmosphere upper layers ($t_f \ll 10^{-5}$ s) and the time for hydrostatic equilibrium ($t_s < 10^{-5}$ s) [6] to be reached.

Numerical calculations show [4, 6, 7] that a burster atmosphere can be described by a rather simple physical model. At luminosities far from the critical ($l = q/q_{crit} < 0.5$), the rate of temperature changes $T \sim \tau^{1/4}$ down to the surface of a neutron star is governed by the retarding processes ($\tau \geq 1$). And only at the very boundary of the atmosphere, Compton processes increase the temperature up to the color one [7]. The Compton processes do not affect the emission spectrum in a wide energy range $h\nu \leq 10$ keV. When luminosity reaches the critical $l \to 1$, the isothermal zone near the boundary penetrates into the atmosphere, and, in the long run, Compton processes start to determine the emission spectrum in a wider energy range up to $hn \ll kT_0$ (T_0 is the effective emission temperature).

Model of the atmosphere.

A model of the burster atmosphere was considered in great details in [6, 7]. Here we shall give its main parameters and their functional relations. The hydrostatics equation relates electron density and temperature with the Thomson optical depth $\tau = \int_r \sigma_T N_e \, dr$, counted from the upper edge of the atmosphere:

$$N_e k T = \frac{4(1-L)}{8-5Y} \frac{g m_p}{\sigma_T} \tau \qquad (1)$$

where Y is the abundance of helium and heavy elements (by weight), $l = q/q_{crit}$,

$$q_{crit} = \frac{g m_p c}{\sigma_T} \frac{2}{2-Y} \qquad (2)$$

The hydrostatics equation should be supplemented by the equation of radiation equilibrium, which guarantees the stability of a total flux H(t) at all the depths:

$$H(\tau) = \int_0^\infty H_\nu(\tau)\, dx_0 = const \qquad (3)$$

$X_o = h\nu / kT_o$ is the dimensionless frequency related to the effective temperature T_o.

Basic equations.

The emission field of a burster atmosphere can be described with a diffusion equation [9]:

$$\frac{kT}{m_e c^2} \frac{1}{x_0^2} \frac{\partial}{\partial x_0} x_0^4 \left(\frac{\partial n}{\partial x_0} + \frac{T_0}{T} n \right) + \frac{1}{3} \frac{\partial n}{\partial \tau^2} = \frac{\alpha_{ff}}{\alpha_T} \left(n - \frac{B_\nu}{x_0^3} \right) \qquad (4)$$

Here $n = J\nu/x_o^3$ is the density number of the photon phase space,

$$\frac{\alpha_{ff}}{\alpha_T} = D\tau \frac{\tilde{g}(x)}{x^3} \left(1 - e^{-x}\right) = D_0 \frac{\tau \tilde{g}(x_0 T_0 / T)}{x_0^3} \left(1 - e^{-\frac{x_0 T_0}{T}}\right) \qquad (5)$$

α_{ff} and α_T are the coefficients of retarding absorption and Thomson scattering, respectively, $D_o = D\,(T/T_o)^3$, $\tilde{g}(x) = \frac{\sqrt{3}}{\pi} e^{x/2} k_0(x/2)$ is the Gaunt factor, $K_o(x)$ is the Macdonald function,

$$D = \left(\frac{T}{T_0}\right)^{-4.5} \frac{0.18\, g_{14}^{-1/8}}{\left(1 - \frac{Y}{2}\right)\left(1 - \frac{5}{8} Y\right)} \frac{1 - l}{l^{9/8}} \qquad (6)$$

g_{14} is the free-fall acceleration onto the surface of a neutron star in the units 10^{14} cm/s^2. The equation should be supplemented with a boundary equation

$$\frac{\partial n}{\partial \tau} - \frac{3}{2} n \,/_{t=0} = 0 \qquad (7)$$

An equilibrium blackbody spectrum can be presented in a dimensionless form as:

$$B_\nu = \frac{x_0^3}{\exp(x_0 T_0 / T) - 1} \qquad (8)$$

It follows that with such $B\nu$ a dimensionless flux H will be equal to $\pi^5/15$.

Integrating equation (4), preliminarily multiplied by X_o^3, with respect to the frequency X^o and taking the stability in depth into account $\left(\frac{dH}{d\tau} = \frac{4\pi}{3} \frac{d^2 J}{d\tau^2} = 0 \right)$, an equation for the temperature can be obtained

$$\frac{kT_e}{m_e c^2} \left(4 \int_0^\infty J_\nu\, dx_0 - \frac{T_0}{T} \int_0^\infty x_0 J_\nu\, dx_0 \right) - \qquad (9)$$

$$- D_0 \tau \left(\int_0^\infty \frac{J_\nu \tilde{g}(x_0 T_0 / T)}{x_0^3} \left(1 - e^{-x_0 T_0 / T}\right) dx_0 - \frac{2\sqrt{3}}{\pi} \frac{T}{T_0} \right) = 0$$

which involves energy exchange between photons and electrons during scattering.

Burster spectra as a function of luminosity. Review of the results.

At low luminosities $l \ll 0.02$ the spectrum of outgoing emission will be influenced by the blanket effect, which results from a strong frequency ($a\nu^{-3}$) dependence of the nontransparency coefficient due to retarding processes. This spectrum will be different from the blackbody one and can be presented by the expression

$$F_\nu \sim x_0^3 \exp(-2.81 \sqrt{x_0}) \qquad (10)$$

For effective temperatures, greater than 1 keV ($l \geq 0.1$), electron scattering plays a dominating role in the spectrum formation.

In [6] a spectrum of outgoing radiation was obtained that does not include frequency changes due to the Compton effect:

$$F_\nu \sim x_0^3 \exp\left(-1.577 A_2^{1/13} x_0^\beta\right) \quad (11)$$

$\alpha = 599/286 \approx 2.1$ and $\beta = 15/26 \approx 0.577$

$$A_2 = \frac{3}{(0.75)^{1/4}} \sqrt{\frac{3}{\pi}} \frac{0.18 g_{14}^{-1/8}}{\left(1-\frac{Y}{2}\right)\left(1-\frac{5}{8}Y\right)} \frac{1-l}{l^{9/8}} \quad (12)$$

From the solution of equation (11) it follows, in part, that intensity at each frequency is determined by the region of an optical depth of the order

$$\tau_0 \cong 0.44 A_2^{-4/13} x_0^{22/13} \quad (13)$$

The effect of the Compton processes may be evaluated in the following way [6]. The outgoing radiation spectrum (11) has a maximum at

$$x_0^{max} = \left(\frac{2.1}{1.577 A_2^{1/13}\, 0.577}\right)^{\frac{1}{0.577}} \gg 1 \quad (14)$$

Radiation with $h\nu > kT_o$ passes through a layer of a given optical thickness relative to scattering τ_o. An average number of scattering in this layer is close to $3\tau_o^2/\pi^2$. Hence, when the parameter $P = (h\nu / m_e c^2)(3\tau_0^2/\pi^2)$ reaches unity, the recoil effect should lead to drastic changes in the emission spectrum and, vice versa, at $P \ll 1$ Compton effects are negligible.

At luminosities $l \sim 0.1$ the Compton effects are operative only at $X_o \gg 1$ ($X_o > X_o^{max}$). At these frequencies a Wien spectrum (11) is formed with a temperature

$$T_c = \frac{T}{T_0} = \frac{1}{4} \frac{\int_0^\infty x_0 J_\nu\, dx_0}{\int_0^\infty J_\nu\, dx_0} = 1.97\, \varphi(l, Y, g_{14}) \quad (15)$$

$$\varphi(l, Y, g_{14}) = \frac{l^{3/20}}{(1-l)^{2/15}} \left[\left(1-\frac{Y}{2}\right)\left(1-\frac{5}{8}Y\right)\right] g_{14}^{1/60}$$

At greater $l \sim 0.4 \div 0.5$, the value of X_o^{max} increases up to 1/4 and the Compton processes become important for all frequencies right of the maximum. It can be seen that the position of the spectral distribution maximum remains unchanged. It follows inequivocally that the temperature of the Wien "tail" should be determined from the position of the spectral distribution maximum:

$$T_c = \frac{T}{T_0} = \frac{x_0^{max}}{2.7} = 1.7\, \varphi(l, Y, g_{14}) \quad (16)$$

The analysis made shows that for all $P \le 1$ ($l \le 0.5$) the outgoing radiation spectrum can be described as

$$J_\nu = \begin{cases} c_1 x_0^{2.1} \exp\left(-1.577 A_2^{1/13} x_0^{0.577}\right) & at\ x_0 \le x_0^{max} \\ c_2 x_0^3 / \left(\exp(x_0/T_c) - 1\right) & at\ x_0 \ge x_0^{max} \end{cases} \quad (17)$$

The coefficients C_1 and C_2 are found from joining these asymptotics in the point X_o^{max} and normalizing J_ν to the flux:

$$H = 2\pi \int_0^\infty J_\nu\, dx_0 = \pi^5/15$$

$$c_1 = A_2^{0.28}\, 2.56 \cdot T_c^3 c_2 \quad (18)$$

$$c_2 = 0.474\, T_c^{-4}$$

Figure 1a shows a good agreement between the theoretical spectrum (17) and the results of numerical calculations [4] for a model of a burster atmosphere with T_o = 2.541 keV, g_{14} = 10, Y = 0.3, and l = 0.48. T_c = 1.65 was calculated from equation (16).

A low-frequency asymptotics $J_\nu \sim X_o^{26/16}$ at $X_o \ll 1$ [6] is also in a fairly good agreement with numerical calculations. At l > 0.5 the value of P increases and the Compton processes come to play a more important role for the frequencies left of X_o^{max} as well.

At luminosities close to the critical, the so-called saturated Compton process takes place when $4kT_0\tau_0^2/m_ec^2 \gg 1$. In this case in a wide range of frequencies up to $X_\bullet$, where the rates of the Compton and retarding processes become equal, the Compton process has time to bring about Bose-Einstein photon distribution, and the emission spectrum can be presented as

$$J_\nu = \frac{1}{2} T_c^{-4} \frac{x_0^3}{\exp\left(x_0 / T_c\right) - 1} \qquad (19)$$

where

$$T_c = \frac{T}{T_0} = \frac{1.32\,\varphi\,(l, Y, g_{14})}{\ln^{4/15}(2.35 / x_\bullet)} \qquad (20)$$

and
$$x_\bullet = \sqrt{\frac{\sqrt{3}}{2\pi}\frac{m_e c^2}{k T} D T_c^4 \ln\frac{2.35}{x_\bullet}} \qquad (21)$$

To compare the theoretical spectra (17) and (19) (normalized to the flux $\pi^5/15$) with observations (in erg $cm^{-2}s^{-1}$ keV^{-1} units) the emission intensity should be multiplied by the coefficient :

$$C_N = 2\pi \cdot 3.7 \cdot 10^{23} \left(\frac{k T_0}{1.646\,keV}\right)^4 \frac{1}{k T_0}$$

$$F_\nu = C_N J_\nu \; (erg\,cm^{-2} s^{-1} keV^{-1}) \qquad (22)$$

Figure 1b shows the results of comparing the theoretical spectrum (17) with numerical calculations [8] for a model with T_o = 2.84 keV, l = 0.99, and Y = 0 (Y = 1).

The spectra in Figs. 1c, 1d correspond to the model with T_o = 1.42 keV, l = 0.9 or T_o = 1.615 keV, l = 0.79. The results of numerical calculations from [8] and [4], respectively, are also given for comparison.

These two cases are intermediate between a weak ($l \le 0.5$, $P < 1$) and a saturated ($1 - l \ll 1$, $P \gg 1$) Compton process. Therefore, the T_c coefficient is determined from equation (20), and the spectrum is constructed using the distribution according to (17).

In conclusion consider the formation of burster spectra in a continuous emission mode. According to the present-day accretion models [10], the region of the main energy release can be regarded as a double-temperature medium with a relatively cool optically thick deeper layer and a hot upper layer. A flux radiated by the cold layer undergoes Compton scattering by hot electrons, and some of short-wave photons of the hot region would return back to the dense cold layers of a disc or a neutron star, warming them up. The atmosphere would become an additional source, emitting in low-frequency region.

Mathematical formulation of the problem.

Let atmosphere consist of two layers: an upper one with a Thomson optical thickness τ_o and temperature T_e^h, and a deeper layer, semi-infinite with a temperature T_e^C, with $T_e^C \ll T_e^h = T_e$. Let then assume that radiation in the deeper layer is subject to Compton recoil and photoionization of heavy elements. As a result of these processes, energy photons, ploughing from the upper layer, would lose their energy in it. It is this loss of energy that determines the energy flux from the deeper to the upper layer, provided no other sources of energy exist in the deeper layer. It must be clear that a total emission flux πF coming from the depth τ_o is related to the temperature of low-frequency emission:

$$\pi F = \sigma T_r^4 / \frac{3}{4}\left(\tau_0 + \frac{2}{3}\right) \qquad (23)$$

A flux of low-frequency photons determines the original sources inside the hot layer:

$$f(x, \tau) = \frac{F_\nu}{2}\int_0^{-1} \exp\left(-(\tau_0 - \tau)/\mu\right) d\mu = \frac{F_\nu}{2} E_1(\tau_0 - \tau) \qquad (24)$$

In order to find a radiation field inside the atmosphere, equation (4) with a relevant right-hand part - $f(x,\tau)/x^3$ (see [24]) should be solved, imposing the boundary condition (7) at $\tau = 0$ and the condition

$$\frac{\partial n}{\partial \tau} + \frac{3}{2}\frac{1-A_\nu}{1+A_\nu} n = 0 \quad at \quad \tau = \tau_0 \qquad (25)$$

The last condition is that of reflection, which allows for photons scattered in the deeper semiinfinite layer.

All these equations should be supplemented by the condition of self-consistency so that a total integrated flux be equal to the amount of energy absorbed in the deeper layer

$$\pi F = \int_0^\infty \pi F_\nu dx = 2\pi \int_0^\infty \frac{1-A_\nu}{1+A_\nu} J(x, \tau_0)\, dx \qquad (26)$$

The reflectivity of the deeper semiinfinite layer can be found by solving a boundary problem for equation (4) at zero temperature and taking into account photon annihilation during photoionization of heavy elements [11]. Condition (7) can be considered as a boundary one in this case. Then, τ is an optical depth counted from the upper boundary of the deeper semiinfinite layer. A distribution similar to that described by expression (24), $f^c = J^h(x, \tau_0) E_1(\tau)$, can be taken as original emission sources.

Astrophysical supplement.

Comparison of theoretical spectra, obtained in terms of the double-temperature model, with observations may help determine physical conditions of the region where an X-ray emission spectrum is formed. More than thirty models were used for calculations with different luminosities $l = 10^{-3} \div 1$ and hot layer temperatures $T_e^h = 4 \div 26$ keV. All the spectra obtained are of typically Compton nature. At sufficiently low energies ($h\nu < k T_e^h$), there is a power region, which turns into a Wien "tail" with increasing energy. The higher T_e^h, the longer and flatter the power region is. At low temperatures $T = 4 \div 6$ keV, spectra of Wien type correspond to a high luminosity ($l > 0.1$).

A short-wave region ($h\nu \gg k T^h$) in the majority of spectra can be described with the help of a spectrum due to Compton scattering [12]

$$F_\nu \sim x^3 e^{-x} \int_0^\infty t^{\alpha-1} e^{-t} \left(1+\frac{t}{x}\right)^{\alpha+3} dt \qquad (27)$$

where α is the index of the power region in the spectrum. In Fig. 2 theoretical spectra are compared with the spectra of two bursters MXB-1728-34 [13] and 4U-1705-44 [14]. It is seen that observations of MXB-1728-34 (curve d) are in agreement with the theoretical curve 2 at $T_e^h = 4.09$ keV, $l = 1.7 \cdot 10^{-2}$ and $\tau_0 = 6.9$. The results for the second source (curves a, b, and c) coincide with the theoretical curve 1 at $T_e^h = 4.09$ keV, $l = 8.3 \cdot 10^{-3}$, and $\tau_0 = 6.1$. It is evident that experimental curves b and c correspond to theoretical models with still lower luminosities ($l < 8.3 \cdot 10^{-3}$) and τ_0 values ($\tau_0 < 6.1$). In addition, a similar comparison was made for the source EXO 0748-676 [15]. Its experimental spectrum, approximated by

$$F_\nu \sim (h\nu)^{-\gamma} \exp(-h\nu / kT) \exp(-\sigma N_H)$$

where $\gamma = 0.92 \pm 0.12$, $N_H = (7.8 \pm 1) \cdot 10^{21}$, $kT = 11.1^{3.5}_{-2}$ keV, $\sigma = \sigma_T (8\,\text{keV} / h\nu)^3$ resembles, by its shape, the spectra for the 4U 1705-44. In Fig. 2 curves a, b, and c are also approximated by the spectra (27) - the dashed lines. Here, the experimental curves correlate best with the theoretical spectra at $\alpha = 0.9 \div 1$ and $kT \cong 2.5$ keV.

In conclusion, it is worth noting that all observations of bursters in a continuous emission mode are of the same form and correspond to theoretical curves 1, 2 at low luminosity $l \leq 10^{-2}$. On the other hand, fluxes received in a continuous emission mode are, on the average, only ten times smaller than the corresponding fluxes in outbursts. It leads to a conclusion that in a quiet state the area of an emission region must be, at least, ten times that of a neutron star and, hence, it must be at a distance of about three radii of the neutron star.

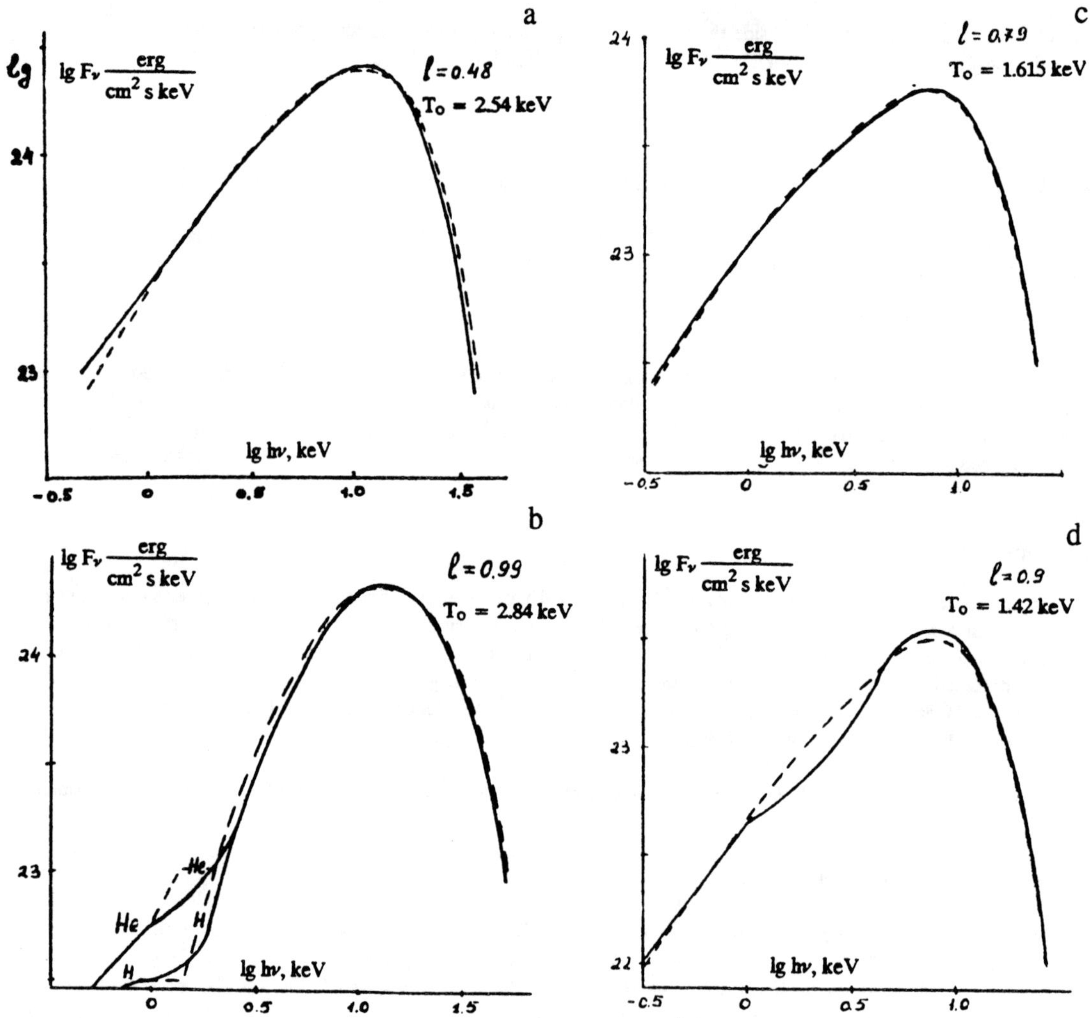

Fig. 1. Burster spectra in outbursts. Solid lines represent numerical calculations [4] for a, c, and d; and [8] for b. Dashed lines for a, c, and d correspond to the theoretical spectra (17), and for b, to the spectrum (19). a, at T_o = 2.541 kev, l = 0.483; b, T_o = 2.84 keV, l = 0.99; c, T_o = 1.615 keV, l = 0.788; d, T_o = 1.42 keV, l = 0.9.
lg hν, keV.

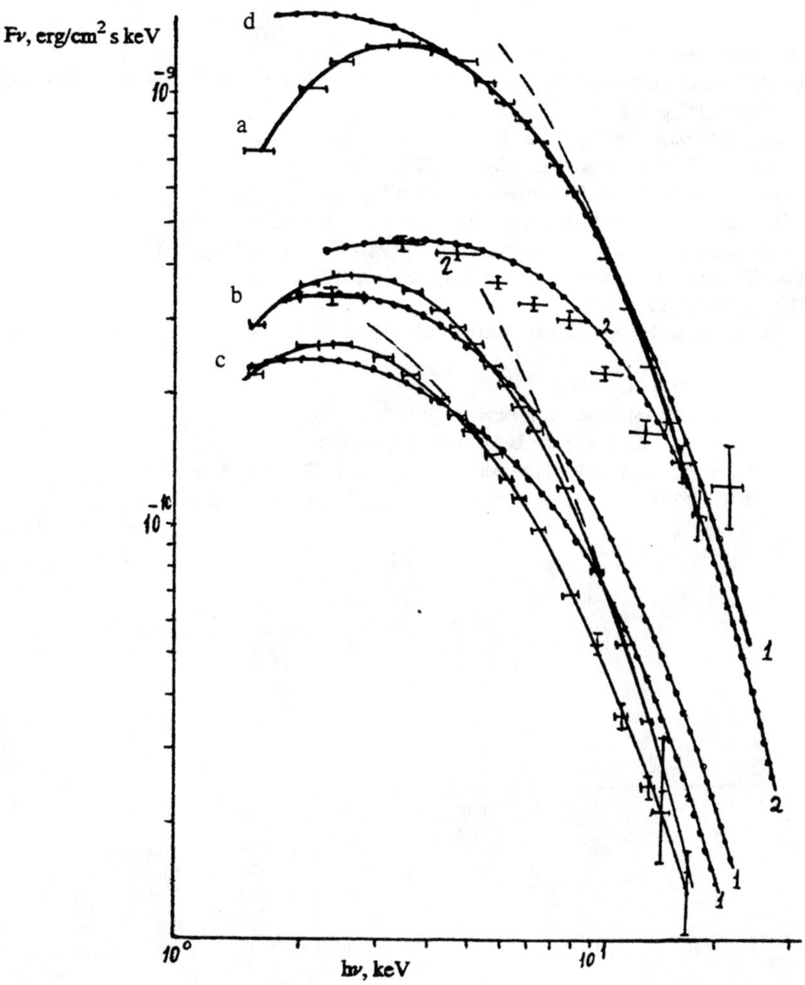

Fig. 2. Comparison of theoretical spectra with observations of bursters in a continuous emission mode. Dotted solia lines correspond to theoretical spectra of the following models:curve 1, at T = 4.09 keV, $l = 8.3 \cdot 10^{-2}$, and $\tau_o = 6.1$; curve 2, at T = 4.09 keV, $l = 1.7 \cdot 10^{-2}$, and $\tau_o = 6.9$. Curves a, b, and c represent observations of the source 4U1705-44 [14], and curve d, of the source MXB-1728-34 [13]. Dashed lines correspond to the Compton spectra (27) at $\alpha = 1$ (for 1a) and $\alpha = 0.9$ (for 1c) and kT = 2.5 keV.

References

1. W.H.G. Lewin and P.C. Joss, Space Sci. Rev., 1981, v. 28, p. 3.
2. M. Oda, In: Gamma-ray Transients and Related astrophysical Phenomena, Ed. R.E. Lingenfelter, H.S. Hudson, and D.M. Worral, AIP, New York, 1982, p. 319.
3. J. van Paradijs, Astron. & Astrophys, 1982, v. 107, p. 51.
4. R.A. London, R.E. Taam, and W.M. Howard, Astrophys. J., 1986, v. 306, p. 170.
5. T. Ebisuzaki and K. Nomoto, Astrophysical J. (Letters), 1986, v. 305, p. 67.
6. R.Q. Sunyaev and L.G. Titarchuk, Letters to Astrophysical J., 1986, v. 12, p. 857.
7. I.I. Lapidus, R.Q. Sunyaev, and L.G. Titarchuk, Letters to Astrophysical J., 1986, v. 12, p. 918.
8. A. Babul and B. Paczynski, Preprint of Princeton Observatory, no. 209, 1987.
9. Q.S. Kompaneets, JETPh, 1956, v. 31, p. 876.
10. L.A. Pozdnyakov, I.M. Sobol', and R.A. Sunyaev, Astrophys. & Space Science Rev., Ed. R.A. Sunyaev, 1983, v. 2, p. 189.
11. L.G. Titarchuk, Astrofizika, 1987, v. 26, p. 97.
12. R.A. Sunyaev and L.G. Titarchuk, Astron. & Astrophys., 1980, v. 86, p. 121.
13. E.K. Sheffer, L.G. Titarchuk, et al., Kosmicheskie Issledovaniya, 1988, v. 26.
14. A. Langmeier, M. Sztajno, G. Hasinger, and I. Trumper, Astrophys. J., 1987, v. 323, p. 288.
15. A.N. Parmar, N.E. White, P. Giommi, W. Gottwald, et al., Astrophys. J., 1986, v. 308, p. 199.

Physics of Neutron Stars
Edited by A. M. Kaminker

ISBN 1-56072-164-2

GENERATION OF GAMMA-RADIATION BY NEUTRON STARS

I.G. Mitrofanov
Institute of Space Research, USSR Academy of Sciences, 117810, Moscow

Basic physical processes of g-radiation generation by neutron stars are discussed. Some models to describe sources of g-bursts and g-pulsars are considered. New experimental tasks are suggested.

Introduction.

At present two main types of Neutron Stars (NS), sources of cosmic gamma-radiation, are known. The first type are radiopulsars. According to the present day theoretical knowledge gamma-rays play an important role in the generation of radiation by radiopulsars. Two young pulsars, Crab and Vela, were unequivocally identified with pulsing point sources of gamma-rays. In many cases identification of gamma-sources with pulsars is a strong probability. Generation of gamma-rays by radiopulsars is believed to be due to the electromagnetic cascade in the NS magnetosphere. However, a complete model of pulsar gamma-radiation is still unavailable [1, 2].

The second type of NS is associated with sources of gamma-bursts, though identification in this case cannot be considered unambiguous. The cause of gamma-bursts has not yet been established. Quite a number of models have been proposed, which either supplement or contradict each other [3, 4].

Investigations of gamma-radiation from NS require, on the one hand, a thorough theoretical examination of new physical processes observed and, one the other hand, application of the theoretical ideas derived to interpret observational results and to suggest further experiments. The importance of proper understanding of the new physical processes is due to a unique nature of physical conditions near NS surfaces: sources of gamma-rays. First, there may exist superstrong magnetic and electric fields there. Second, the density of beam energy may be enormous, with one quantum of it being sufficient to generate an electron- positron pair. Third, the mechanism of gamma-radiation is, most probably, non-thermal. And lastly, the characteristic time of a beam energy density change in the source is comparable with its light sizes.

The physical processes, which strongly affect the observed properties of gamma-rays are associated with a mechanism (or mechanisms) of radiation generation. In section 2 the basic processes of e^{+} interaction with photons in a strong magnetic field are enumerated. Section 3 deals with possible generation mechanisms of gamma-rays by NS. In section 4 in order to verify alternative models of NS, observations are suggested, which might help elucidate the nature of these objects.

2. Basic processes of $e^{\pm}$ interaction with gamma -photons in magnetospheres of neutron stars.

2.1 Generation of synchrophotons in a $e^{\pm}$ quantum transitions in magnetic field.

The conditions, under which quantum consideration of the synchrotron radiation $e^{\pm}$ in the magnetic field is of necessity, have been discussed elsewhere [5 - 7]. First, quantization of the $e^{\pm}$ motion should be considered if the magnetic field B ≥ E(E + 1) (the transverse energy of the particle E and the magnetic field B are in the units mc^2 and $B_c = m^2c^3/e\hbar = 4.4 \cdot 10^{13}$ G). Second, the quantum recoil and spin reversals are of importance if the energies of ejected photons are close to E. It follows from the conditions that a quantum approach should be appropriate for consideration of synchrotron gamma-radiation generated by NS.

The works [7, 8] deal with the probabilities of synchrophoton emission during spontaneous $e^{\pm}$ transitions between Landau levels, their angular distribution and polarization. These studies gave values of the characteristic times: the life time of an excited state τ_l, the time of an excited state transition into the ground one τ_{dec} and the cooling time τ_c (Fig. 1). Synchrophoton emission depends on the relationship between the above times and the characteristic excitation time τ_o [7]. In Fig. 1 the isochrones of τ_l, τ_{dec}, and τ_o at different E values are given for the values of 10^{-16} (curve a), 10^{-17} (curve b), 4 $\cdot$ 10^{-18} (curve c), and 10^{-18} s (curve d). The dashed and solid lines in the Fig.1 correspond to two spin projections relative to the field direction: along and opposite the field, respectively.

If $\tau_l \gg \tau_0$, particle distribution in the Landau levels can be described by a stationary distribution function. In this case a stationary synchrotron radiation occurs (SSR). If particles are thermally excited, their distribution function in energies is thermal and a thermal synchrotron radiation (TSR) is generated.

On the contrary, at $\tau_l \ll \tau_0$ a particle, after being excited, passes into the ground state. In this case, of physical sense is the distribution function in the initial excited states. If excitation occurs permanently, the ensemble generates synchrotron cooling radiation (SCR); its total intensity is equal to $t\tau_l / \tau_0$ of that of SSR [7]. Comparison of TSR and SCR, resulting from thermal collisions [7], showed that their spectra do not differ substantially, though the SCR spectrum has a smaller short-wave region (Fig. 2). The dashed line in the Fig.2 represents a degree of linear polarization (ω_B is a cyclotron frequency).Because of recoil effects, synchrotron radiation spectra are strongly dependent on the direction. Therefore, at $E \geq 1$ it is more convenient to present the spectra not as a function of energy but of the harmonics number [8]. The difference of the spectra of photons, ejecte\d from a given level and during their transition onto a ground level, is demonstrated by Fig.2 from Ref. [8].

An important feature of synchrotron radiation is its polarization. In the nonrelativistic region $E \ll 1$, photon emission is mainly due to electric transitions and it is linearly polarized perpendicular to the plane with a wave vector $\vec{k}$ and magnetic field $\vec{B}$. At $E \geq 1$ magnetic-type transitions resulting from spin reversals greatly contribute to the emission. The radiation corresponding to them is linearly polarized in the above-mentioned plane, and; therefore, the quantum effect of the spin reversals causes [7] depolarization of synchrotron radiation with high harmonics (Fig. 2).

2.2 One- and two-photon birth of $e^{\pm}$ in magnetic field

When considering this process, two situations should be distinguished: the case when an individual photon cannot form a pair, and the case when a photon can form a pair.

In the first case a two-photon birth of a pair causes radiation self-absorption in the NS magnetosphere. Though self-absorption is possible at $B = 0$ [9], at $B \sim B_c$ it acquires some important features [10, 11]. First, the birth threshold is lowered in many directions of photon propagation, because in the magnetic field the pulse conservation law is substituted by the conservation low of its projection along $\vec{B}$. Second, the magnetic field substantially increases self-absorption in the fluxes, propagating perpendicular to $\vec{B}$. Two fluxes of this kind suffer almost the same self-absorption irrespective of that whether they propagate toward each other or at a small angle (to each other). Third, the cross-section of the process has maximums at the energies, which correspond to the thresholds of $e^{\pm}$ pair generations, with certain quantum states. Fig.3 shows a cross-section of a two-photon birth as a function of two similar photon energies, propagating transverse the magnetic field $B = 1$. The solid line represents polarization ($||$, $||$), the dashed line polarization ($||$, $\perp$), and the dash-dotted line polarization ($\perp$, $\perp$)[10]. Figure 3 also shows a cross-section for $B = 0$, where solid and open circles correspond to parallel and and perpendicular photon polarizations. The peaks in it arein complete agreement with a one-photon birth [12]. And lastly, self-absorption is strongly dependent on photon polarization and may result in an integrated polarization of radiation passing through the NS magnetosphere.

As to the second situation, both a one-photon (see, for example Ref. [12]) and a two-photon processes would lead to gamma-ray absorption and; thus, to an $e^{\pm}$ pair generation. At given photon energies, a predominance of one of the processes depends on the relationship between the magnetic field and photon density [11]. If gamma-rays propagate in thermal radiation, the relationship will depend on the magnetic field and temperature [13].

2.3 Annihilation of $e^{\pm}$ pair in the magnetic field

In the magnetic field two types of annihilation are possible: one-photon and two-photon (at $B = 0$ only two- photon annihilation works). The cross-sections of these processes must depend on the quantum states of annihilating particles. One-photon annihilation from the ground and excited levels was considered in [14]. Two-photon annihilation in the magnetic field has been analyzed only for particles at the ground levels [15]. An important feature of annihilation in the magnetic field is a linear polarization of photons.

The prevalence of one-photon or two-photon annihilation depends on the quantum states of annihilating particles and on the magnetic field strength. At the annihilation from the ground level, cross-sections of these processes become similar in a magnetic field of $\sim 10^{13}$ G [16], whereas in a stronger field one-photon annihilation prevails. Estimation of the limiting field for excited particle annihilation requires further investigations [17].

3. Mechanisms of gamma-ray generation by neutron stars

3.1 Gamma-ray generation in an electromagnetic cascade in the NS magnetosphere

Rotation or vibration of a NS generates an electric field in its magnetosphere and, hence, accelerates charged particles along the lines of magnetic force. The bending radiation of these particles initiate an electromagnetic cascade, in which true gamma-quanta transform into electron- positron pairs and photons (see, for example, [18]). Photons in the cascade are generated both due to the bending radiation of the secondary particles and as a result of SCR. The maximal energy emitted from the photon magnetosphere can be determined from its transparency with respect to the $e^{\pm}$ -pair birth. This mechanism is generally accepted for explanation of relativistic particle flux generation in pulsars [18]. A similar model was proposed in [19] to account for annihilation bursts of gamma-fluxes.

The ratio of the number of bending photons to the number of synchrotron photons in a cascade depends on the energy of nucleating bending photons and the magnetic field strength. This ratio determines the coefficient of particle breeding in a cascade, the coefficient being an important parameter of radiopulsar models [18]. The ratio also determines whether the model of an inner gap [1] can be applicable to explain gamma-emission from pulsars.

Photons of synchrotron and bending radiations have a different linear polarization. The polarization direction of the synchrotron radiation is normal to the plane passing through the magnetic field and wave vector. In a magnetic field of > 10^{13} G a strong synchrotron depolarization occurs due to spin-reversal-induced transitions (see Sec. 1, 2).

The bending radiation can be easily shown by Lienar- Vichert formula to be linearly polarized along the above plane. The degree of polarization does not depend on the magnetic field strength.

3.2 Gamma-ray generation by radiative relativistic ejection of photoshere

It is known that radiation pressure on electron-proton plasma in a flux of thermal radiation above the NS photosphere can, under certain conditions, exceed gravitational attraction. In the absence of the magnetic field, this depends only on the emission temperature: radiational ejection of the photoshere occurs at

$$T > T_{cr} = \left(g_{NS}\; m_p c \;/\; \sigma_T \,\sigma_{S-B}\right)^{1/4} \sim 2 \cdot 10^7\; K \qquad (1)$$

If radiation comes uniformly from all the star surface, the temperature T_{crit} corresponds to the Eddington luminosity limit of 10^{38} erg/s.

In a strong magnetic field the electron cyclotron resonance in the cross-section of the radiation interaction with substance substantially increases radiation pressure on plasma [20, 21]. Radiational ejection of the photoshere may take place at T < T_{crit}.

Ejection in this case greatly depends not only on the directional pattern but on the spectrum of out-going radiation [22]. The maximum gamma-factor of an ejected plasma is determined by a kinematic constraint, which corresponds to the condition that total pulse from the radiation to substance is equal to 0 in an accompanying reference system. The value of the kinematic constraint does not depend on the radiation power. When optically thin plasma is accelerated over an emitting surface with a thermal spectrum $\gamma^{(o)}{}_{kin} \sim 2$, and over a hot spot

$$\gamma_{kin} \sim \gamma^{(0)}_{kin} \left(R \;/\; R_0\right)^2$$

, where R_o and R is the spot radius and distance to the spot, respectively.

Dissipation of the photosphere thermal radiation in an ejected plasma causes gamma-ray generation. On the basis of this mechanism a model was proposed for radiation generation of gamma-bursts over a hot NS photosphere [22, 23]. The flux of gamma-rays may change with a characteristic time of the thermal photoshere inertia equal to ≤ 1s. A maximum energy of the photons formed in a transparent ejecting layer of the photosphere is equal to

$$E_{max} \sim 6\,k\,T \left(\gamma^{(0)}_{kin}\right)^2 \leq 300\; T_8\; keV \qquad (2)$$

Account of the cyclotron absorption in an ejected plasma may increase the estimates of a limiting gamma-factor [22]. However, it is difficult to explain the origin of gamma-rays with energies » 1 MeV in this case. At the same time photons with energies up to 30 MeV [24] are known to be observed in the spectra of gamma-bursts. Therefore, attempts to construct a complete model of gamma-ray generation assuming a radiative relativistic ejection of a NS photosphere have not been a success yet.

3.3 Generation of gamma-rays by electrically accelerated particles over the photosphere

Acceleration of charged particles by an electric field in an active NS magnetosphere over its photosphere is accompanied by radiative friction [25]. The particle acceleration is reduced to zero when the power of the electric force completely transforms into a gamma-radiation flux (the mechanism of a gamma-gun). In a strong magnetic field, where the main contribution to the radiative force is due to a cyclotron resonance, the radiation friction is maximally high for particles with gamma-factors of $\sim \hbar\omega_B / kT$.

Relativistic particles, accelerated by an electric field over the photosphere along lines ofmagnetic force, generate gamma-photons with energies

$$E \geq \left(\hbar\omega_B\right)^2 / kT \sim 10^3 B_{13}^2 T_8^{-1}\ keV \qquad (3)$$

These energies correspond to the first harmonics of an electron cyclotron resonance. On the whole, all the harmonics provide a power SCR spectrum in the region of high energies.The mechanism of a gamma-gun accounts for the basic spectral features of gamma-bursts. Due to relativistic radiation anisotropy in every point, the mechanism lifts up rather strict constraints on the distance to a source and on the strength of a magnetic field, which are usually derived from observations of hard spectra for models neglecting anisotropy [4]. This mechanism also suggests a special class of pulsars, in which radiative friction prevents a pulsed avalanche in the magnetosphere [25].

Excitation of a NS magnetosphere, as a result of star quakes and interchange of the force line connections, may be of different nature. On the one hand, active regions of a size $\leq 10^5$ cm with a noncompensated electric field may arise in the magnetosphere. On the other hand, beams of accelerated particles in the active regions may break into separate bundles, due to different instabilities. These two processes can generate photons by short microsecond bursts.

3.4 Generation of annihilational and nuclear gamma- radiation

Generation of gamma-rays in a NS magnetosphere would inevitably lead to $e^{\pm}$ pair birth. On the other hand, the $e^{\pm}$ pairs must undergo annihilation. Situations are usually distinguished when the region of gamma-ray generation is optically thick or optically thin with respect to the process of pair birth.

In the absence of a magnetic field, a self-consistent photon spectrum is formed in the optically thick source because of the emission self-absorption (Sec. 2.3) and pair annihilation (Sec. 2.4). Also forms the distribution function of $e^{\pm}$ pairs [26], and the average number of particles and photons remains unchanged. However, at $B = 0$ such a source could not but expand with a relativistic speed due to inner pressure (see, for example, [27]). Under these conditions annihilation radiation should be observed as a burst from a relativistic fireball.

If the magnetic field $B > 2 \cdot 10^{10} \left(\frac{n}{10^{26}\ cm^{-3}}\right)$ G,

the annihilation of relativistic $e^{\pm}$ pairs contributes but slightly to the gamma-ray flux as compared with synchrotron radiation [3]. Therefore, in the presence of a magnetic field, contributions from the annihilation continuum of both optically thin sources over a NS surface and fireballs to the radiation flux can be neglected. This conclusion is in disagreement with the annihilation interpretation of hard g- burst spectra proposed in [28].

However, relativistic $e^{\pm}$ pairs with energies γmc^2 can pass from a NS magnetosphere to its surface layer, cool down in it and generate an annihilation line. The time of the layer thermal inertia is ≤ 1 s. If the surface is bombarded with short bursts « 1 s, annihilation bursts can be observed from time to time [19]. Naturally, the power oi the thermal radiation from the NS surface should be γ times an average power in the annihilation line [4].

If protons and nuclei are accelerated in a NS magnetosphere, nuclear lines can be radiated because of proton and nucleus interactions with the surface. The number of nuclei in the surface layer transparent for gamma-rays is $\sim 10^{23}$ cm^{-2}. If the radiation is caused by multiple excitations of the same nuclei, the life time of excited states of $\sim \sim 10^{-11}$ s would restrict the maximum photon flux of nuclear radiation from a NS surface to $\leq 10^{47}$ s^{-1}. If generation of gamma-photons in the nuclear line is associated with nuclear decay, a stationary nuclear radiation would be impossible without substance withdrawal from the surface. When the withdrawal does occur, the rate of photon generation would be set by its pace.

The generation of a nuclear iron line Fe^{56} 847 keV by a NS discussed in work [29] on the basis of observational data. It follows from the Table that no nuclear transformations occur at the excitation of Fe^{56} when the energies of rushing protons are pcE_p < 7 MeV [30].

This should be taken into account when interpreting a stationary flux of Fe nuclear radiation (similar restrictions are imposed on other nuclei).

Table

Line energies (MeV) (given in brackets are relative intensities)		Reaction
0.847	(1)	
1.238	(0.5)	$^{56}Fe(p, p)^{56}Fe^*$, $E_p \geq 3$ MeV
1.772 + 1.811	(0.64)	
2.094 + 2.113	(0.22)	
0.812	(1)	
0.158	(1)	$^{56}Fe(p, n)^{56}Co^*$, 7 MeV < E_p < 20 MeV
1.434	(1)	
0.931	(0.65)	$^{56}Fe(p, \alpha)^{52}Cr^*$ E_p > 20 MeV
1.408	(0.60)	

4. Observations necessary to continue investigations of NS gamma-radiation

4.1 Interrelation between gamma-pulsars and their radiation polarization

Two physical models were proposed to explain gamma- radiation from the pulsars Crab PSR 0531 + 21 and Vela PSR 0833 - 45. The first model relates gamma-radiation to the inner gap in a NS magnetosphere over its pole [1], assuming that SCR synchrophotons basically contribute to radiation fluxes with E < 1000 MeV. According to the second model, gamma-rays are generated due to the bending radiation of relativistic $e^{\pm}$ in an outer gap of the magnetosphere [2]. Gamma-ray generation in the outer gap is supported by a simple geometrical model [31] accounting for a pulsar radiation in all frequency ranges.

In terms of the inner gap model SCR is generated inside a narrow cone with its axis along the polar axis. The radiation correlates with small peaks in the pulsar phase curve. Relativistic particles in the pulsar outer gap radiate in a wide fan-like directional range and can be regarded as a flux during a major fraction of a period.

It is known that the linear polarization of gamma- photons during their registration in a spark chamber can be measured from the distribution of electron-positron pair planes with respect to the direction of the radiation arrival [32]. Such an analysis was made using the data obtained by the Cos-B satellite [33]. The observations imply a strong linear polarization of gamma-radiation from Vela for the phases corresponding to an interpulse and the second peak. This leads to the conclusion that the model of bending radiation in the outer gap is more appropriate for explaining gamma-radiation in this phase range.

Future gamma-telescopes should be developed so that they could afford gamma-polarimetry (see [32, 33]). Observations with such instruments would help elucidate the mechanism of gamma-ray generation by pulsars and develop a physical model of these objects (see 3.1).

4.2 Variability of gamma-bursts

All the models proposed to explain gamma-bursts can be arbitrarily divided into: (a) models with a second variability time (thermonuclear combustion, substance accretion from a disc-accumulator), (b) models with a millisecond variability time (thermonuclear combustion generating Alfve'n and other magnetohydrodynamic waves, star quakes), and (c) models with microsecond and shorter variability times ("chromosphere" bursts, beam instabilities in the models of a "gamma-gun" type).

Each time the resolution time for gamma-bursts was increased, some rapid variations of the flux were detected (see, for example,[34, 35]). Of paramount importance for further investigations of gamma-burst variability with time is the question whether the readings on the detector are due to the Poison flux with a characteristic variability time much greater than the minimum time between counts ≥ 1 ms(microsecond), or they are due to the interaction of microsecond (or even shorter) bursts on the detector, i.e., the variability time is comparable or shorter than an average recording time [36]. An answer to this question is crucial for the choice of a gamma-burst model.

Unfortunately, the dead time during photon recording in scintillation counters ~ 1 ms(microsecond) does not allow detection of microsecond variations without special sophisticated procedures. It is deemed necessary to design telescopes with special detectors to search for microsecond flux variations in gamma-bursts.

4.3 Investigation of "instantaneous" gamma-burst spectra

Each physical mechanism of gamma-ray generation predicts a certain spectral photon distribution (see Sec. 2.1 - 2.3). Therefore, a generation mechanism can be derived from spectral observations and an astrophysical model of its sources can be constructed. Moreover, if the radiation spectrum is known, a distance scale to the sources can be derived from self-absorption estimates, that is, their astronomical nature can be identified. However, this is possible only if "instantaneous" spectra have been measured, which reflect photon spectral distribution in the source in every moment of the real time.

This necessitates determination of a minimum time of spectral variability to carry on gamma-burst spectrometry taking this time into account. Assuming that the spectral variability is physically interrelated with the flux variability and has a comparable characteristic time, it should be admitted that in all the experiments made up to now, the time of spectral measurements was too long as compared to the time of their variation. Therefore, "instantaneous" spectra have not been measured. Moreover, if the contribution of microsecond bursts to the flux turns out to be substantial, the spectrometry with scintillation counters would prove impossible (energy release in the counter in this case will correspond to a total contribution of several photons, see [36]).

If the characteristic time of spectral variability is taken as > 100 ms, instantaneous spectra of gamma-bursts can be considered measured [4, 37, 38]. The power laws detected in many cases point to an athermal mechanism of generation. To explain the continuum, the mechanism of a "gamma-gun" seems most preferable (see Sec. 3.3). There are observational data on annihilational radiation bursts of ≤ 500 ms duration [39]. To rationalize these observations, positrons in burst sources should be assumed to be generated as primary particles. If so, an electromagnetic cascade in the active magnetosphere of a cold NS (see Sec. 2.1) may be practically the only mechanism possible [4, 19]. Recording of burst spectra at a < 100 ms resolution would allow the spectral evolution to be followed and physical conditions of photon and $e^{\pm}$ pair generations to be estimated, which would help to construct a self-consistent model of such sources.

If nuclear lines were detected in the spectra of gamma- bursts, this would lead to practically inequivocal conclusions about their astrophysical nature, because the chemical composition of a NS surface layer would have been established. In nuclear NS spectrometry measurements should be made with the help of semiconductor counters of a maximally high spectral resolution. But it be remembered that nuclear lines can be generated as short bursts; therefore, the time resolution in these measurements should also be high enough.

4.4 Polarimetry of gamma-bursts

The synchrotron and bending mechanisms of gamma-burst generation discussed above correlate with different linear polarization of photons. Moreover, polarization may arise during photon propagation through a magnetosphere if polarization dependent processes of photon absorption and fission play an appreciable role [40].

In a magnetic field annihilational radiation is polarized [15]. Recording of the polarization of annihilation lines would enable measurements of a NS magnetic field strength, if the polarization of nuclear lines helps to exclude the polarizing effect of a magnetosphere.

Most suitable instruments for gamma-burst polarimetry are Compton gamma-telescopes. The photon polarization dependence of Compton scattering geometry allows its measurements. Unfortunately, photons coming from sources as microsecond bursts make the work of a Compton telescope impossible.

5. CONCLUSION

For further investigations of the nature of NS, sources of gamma-radiation, elucidation of the following processes and problems seems topical:

(1) Generation of synchrotron and annihilation photons in a NS magnetosphere.
(2) Formation of emitting particles and possible processes of their heating up.
(3) Measurement (or estimation of the upper level) of pulsar gamma-ray polarization and its dependence on the period phase.
(4) Measurement of a minimum time of flux variability for gamma-bursts.
(5) Measurements and detailed examination of instantaneous spectra and gamma-ray polarization.

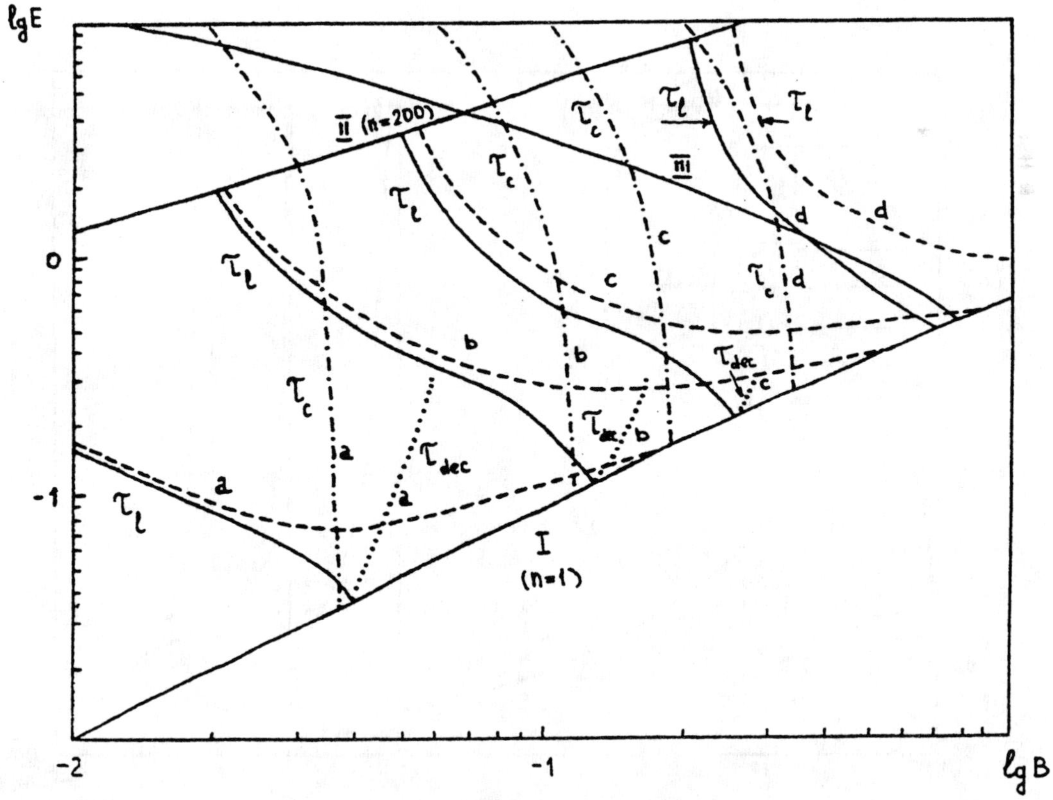

Figure 1

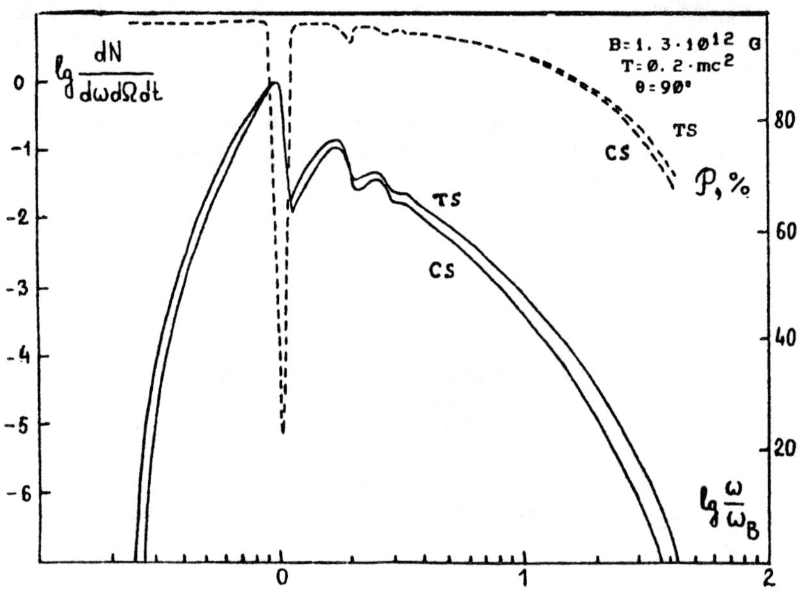

Figure 2

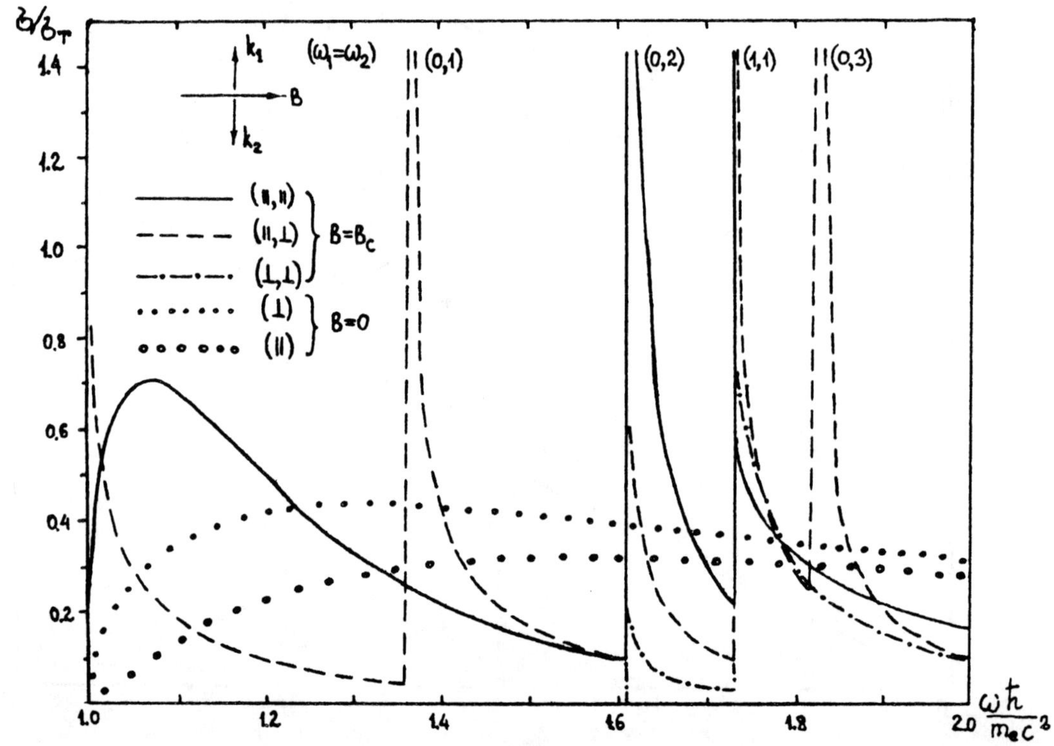

Figure 3

REFERENCES

1. A. Cheng, M. Ruderman, and P. Sutherland, Astrophys. J., 1976, v. 212, p. 808.
2. J.K. Daugtherty and A.K. Harding, Astrophys. J., 1982, v. 252, p. 337.
3. D.Q. Lamb, Proc. of 11th Texas Symp. on Relativistic Astrophysics, ed. D. Evans, 1983.
4. I.G. Mitrofanov and C. Barat, Proc. of Workshop "Nuclear Spectroscopy of the Astrophys. Sources," ed. R. Lingenfelter, 1987.
5. J.J. rainerd and D.Q. Lamb, Astrophys. J., 1987, v. 313, p. 231.
6. G.G. Pavlov and S.V. Golenetskii, Astrophys. Sr. Sci., 1986, v. 212, p. 341.
7. I.G. Mitrofanov and A.S. Pozanenko, ZhETF, 1987, v. 93, no. 6, p. 1951.
8. I.G. Mitrofanov and A.S. Pozanenko, Fizika Neitronnykh Zvezd. Obrazovanie, Stroenie i Evolyutsiya (Physics of Neutron Stars. Formation, Structure, and Evolution), Leningrad, 1988, p. 205.
9. A.I. Akhiezer and V.B. Berestetskii, Kvantovaya Elektrodinamika, 1981, Nauka, Moscow.
10. A.A. Kozlenkov and I.G. Mitrofanov, ZhETF, v. 91, no. 12, p. 1978.
11. A.A. Kozlenkov and I.G. Mitrofanov, Fizika Neitronnykh Zvezd. Obrazovanie, Stroenie i Evolyutsiza (Physics of Neutron Stars. Formation, Structure, and Evolution), Leningrad, 1988, p. 99.
12. V.S. Beskin, Astrofizika, 1982, v. 18, p. 437.
13. A.A. Kozlenkov and I.G. Mitrofanov, Astron. Zh, 1988, v. 65.
14. J.K. Daugherty and R.W. Bussard, Astrophys. J., 1980, v. 238, p. 296.
15. A.D. Kaminker, G.G. Ravlov, and P.G. Mamradze, Astrophys. Sp. Sci., 1987, v. 138, p. 1.
16. J.K. Daugherty and A.K. Harding, Positron-Electron Pairs in Astrophysics, eds. M.L. Burns, A.K. Harding, and R. Ramaty, AIP Conf. Proc., no. 101, 1983, p. 163.
17. A.A. Kozlenkov, I.G. Mitrofanov, and A.S. Pozanenko, Preprint IKI AN SSSR, 1988.
18. J. Arons, Positron-Electron Pairs in Astrophysics, eds. M.L. Burns, A.K. Harding, and R. Ramaty, AIP Conf. Proc., no. 101, 1983, p. 163.
19. I.G. Mitrofanov, Astrophys. Sp. Sci., 1984, v. 105, p. 245.
20. I.G. Mitrofanov and G.G. Pavlov, MNRAS, 1982, v. 200, p. 1033.
21. I.G. Mitrofanov, Astrophys. Sp. Sci., 1987, v. 84, p. 155.
22. I.G. Mitrofanov and A.I. Tsygan, Astrophys. Sp. Sci., 1982, v. 84, p. 35.
23. A.I. Tsygan, Astrophys. Sp. Sci. v. 77, p. 187.
24. P.L. Nolan, G.H. Share, S. Matz, et al., High Energy Transients in Astrophysics, ed. S.E. Woosley, AIP Conf. Proc., no. 115, 1984, p. 399.
25. N.S. Kardashev, I.G. Mitrofanov, and I.D. Novikov, Astron. Zh., 1984, v. 61, no. 5, p. 1113.
26. J.I. Katz, Astrophys. J., 1982, v. 260, p. 371.
27. G.J. Heuter and R.E. Lingenfelter, Positron-Electron Pairs in Astrophysics, eds. M.L. Burns, A.K. Harding, and R. Ramaty., AIP Conf. Proc., no. 101, 1983, p. 163.
28. S.V. Golenetskii, E.P. Mazets, R.L. Aptekar, et al., Astrophys. Sp. Sci., 1986, v. 124, p. 243.
29. B.J. Teegarden and T.L. Cline, Astrophys. J., 1980, v. 236, p. L67.
30. R. Ramaty, B. Kozlovsky, and R.E. Lingenfelter, Astrophys. J. Suppl., 1979, v. 40, p. 487.
31. F.G. Smith, MNRAS, 1986, v. 219, p. 729.
32. A.A. Kozlenkov and I.G. Mitrofanov, Astron. Zh, 1985, v. 62, no. 3, p. 1021.
33. P.A. Caraveo, G.F. Bignami, I.G. Mitrofanov, and G. Vacanti, Astrophys. J., 1988.
34. J.G. Laros et al., Preprint LANL, 1986.
35. E. Eveno, C. Barat, and G. Shambon, Adv. Sp. Res., 1986, v. 6, no. 4, p. 27.
36. I.G. Mitrofanov, Preprint IKI AN SSSR, 1988.
37. E.P. Mazets, S.V. Golenetskii, Yu.A. Guryan, et al., Positron-Electron Pairs in Astrophys., eds. M.L. Burns, A.K. Harding, and R. Ramaty, AIP Conf. Proc., no. 101, 1983, p. 36.
38. V. Sh. Dolidze, A.A. Kozlenkov, R.N. Kucherova, et al., this issue, p. 112.
39. C. Barat, K. Hurley, M. Niel, et al., Astrophys. J., 1984, v. 286, p. L11.
40. I.G. Mitrofanov, A.S. Pozanenko, V.Sh. Dolidze, et al., Astron. Zh., 1986, v. 63, no. 5, p. 1118.

Physics of Neutron Stars
Edited by A. M. Kaminker

ISBN 1-56072-164-2

DEVELOPMENT OF ELECTROMAGNETIC CASCADES IN A NEUTRON STAR CRUST

S.V. Bogovalov and Yu.D. Kotov
Moscow Engineering and Physics Institute, 115 409, Moscow

Monte Carlo calculations were made of the parameters of the cascade induced by 10^{12} - 10^{13} eV primary particle rushing into the material 10^4 g/cm^3 dense along the ~ $5 \cdot 10^{12}$ G magnetic field. Unlike cascades in ordinary material, elementary processes under the conditions of a neutron star are strongly modified by the Landau-Pomeranchuck effect at great energies, and to by synchrotron radiation at small energies. As a result, the cascade maximum shown to be several times deeper than in ordinary material. The coefficient of positron knocking out from the surface has been estimated and turned out to be about unity at a primary particle energy of 10^{13}.

Rotation of a neutron star with the magnetic field of about 10^{12} G generates, near its surface, electric fields which can accelerate particles up to energies of about 10^{13} eV. Moving in the magnetic field a curvature radius ~ 10^7 cm, these electrons (positrons) emit bending photons [1], whose energy is sufficient for the birth of e^+e^- pairs in the magnetic field. These pair then generate hard synchrotron radiation which, in its turn, converts into e^+e^- pairs too. As a result, an electromagnetic cascade develops in the pulsar magnetosphere. Its properties determine the characteristics of radiation observed.

The cascade develops, partly, in the area whose electric field component is not zero along the magnetic field. Therefore a portion of particles will be revolved and accelerated towards the surface. The electrical charge of these particles would be, evidently, opposite to the charge of the main particle flux from the pulsar. This inverse current plays an important role in all present day models of pulsars. All pulsar can be divided into two types depending on the charge sign of particles accelerated from a pulsar. If the value of $(\vec{\Omega}\vec{\mu}) > 0$, where $\vec{\Omega}$ is the vector of the angular velocity of the star rotation, $\vec{\mu}$ is the vector of the magnetic moment, then electrons will be accelerated from the star. According to the model proposed by Arons [2] the inverse current of positrons will warm up the star surface up to 10^6 K°, which if sufficient for practically free emission of electrons from the surface. In the case when $(\vec{\Omega}\vec{\mu})) < 0$, the pulsar ejects positively charged particles. Ion ejection will not generate e^+e^- plasma, because a small Lorentz factor of the ions prevents them from generating photons sufficiently hard for e^+e^- pairs birth. It also seems most probable ions would not be able to escape from the surface at all because of a great binding energy with the crystal lattice [3. 4].Therefore this model requires a mechanism of permanent positron birth over the star surface.

Ruderman and Sutherland [5] proposed a model for a pulsar with $(\vec{\Omega}\vec{\mu}) < 0$. At present this model has been developed by Gurevich, Beskin and Istomin [6 - 8]. In their latest works they analyzed a self-consistent model of a stationary vacuum break-down in a relatively narrow vacuum gap. They showed that positrons over the pulsar surface would be generated by an inverse beam of ~ 10^{13} eV ultra relativistic electrons by two mechanisms. First, positrons will form over the surface due to the conversion of bending photons. This mechanism, however, is not efficient enough. A more efficient mechanism of positron births will be the interaction of the beam with the pulsar surface.

Zones [9, 10] proposed a concrete mechanism of positron formation over the star surface. It consists in the following; electrons, accelerated towards the surface by electric field , fall onto the dense material of the neutron star crust. An electromagnetic cascade must develop in the dense material which is most likely iron [11]. The cascade photons, via photonuclear absorption, excite iron nuclei. The excitation is relaxed by the emission of neutrons. The neutrons show down, diffuse and are caught by a ^{56}Fe nucleus. As a result, an excited ^{57}Fe nucleus is formed which transforms into a ground state by emitting a linear shaped spectrum of 1 - 7 MeV photons. If a neutron is caught near the surface, photons may escape from the material. This mechanism actually provides a high energy photon albedo when an ultrarelativistic electron falls on the material. The processes of repeated particle scattering and photon Compton scattering cannot provide a substantial albedo of photons with energies greater than 1 MeV. Escaping photons will form e^+e^- pairs provided the condition $E_\gamma \sin \Theta \geq 2\, m\, c^2$ is fulfilled Θ is the angle between the photon impulse and the magnetic field direction). This will guarantee a stationary mode of high energy particle generation in the pulsar.

Thus construction of a self-consistent quantitative model of a pulsar requires knowledge of the cascade characteristics in the neutron star material. Conventional calculations of the electromagnetic shower development are inadequate under the conditions of a neutron star. First, because polarization of the medium and Landau-Pomeranchuck effect influence the development of the cascade due to the material high density. Second, strong magnetic field generates synchrotron radiation which may exceed the collision radiation under certain conditions [12]. The cascade in this medium and subsequent neutron transfer were first investigated in works [9, 10] which assumed that the cascade develops in the usual fashion. The magnetic field was included through decreasing the radiation length by a factor of 2. The effects of the medium on the collision radiation and e^+e^- pair formation were not taken, into consideration. Similar investigations were partly made in Ref. [13] too. Its authors obtained numerical estimates which implied that maintaining of stationary particle generation is impossible. Our analysis however, shows that the above effects were overestimated in Ref [13].

In the present work we modelled electromagnetic showers induced by ultrarelativistic electrons in the crust of a neutron star with $4.4 \cdot 10^{12}$ G magnetic field, by the Monte-Carlo method. The albedo of photons and the number of positrons formed by these photons near the star surface have been calculated.

Generation of photons and e^+e^- pairs in the dense material in strong magnetic field.

1. Collision radiation.

The material density in the surface layer of a neutron star is determined by magnetic field. According to [4] the dependence of nuclei concentration on the magnetic field strength can be described by the expression $n = 4.74 \cdot 10^{25} (z/26)^{-3/5} (B/10^{12}\,G)^{6/5}$ (1/cm^3). A some what different value for the concentration was obtained in [14]: $1.5 \cdot 10^{25} (z/26)^{-1} (B/10^{12}\,G)^{3/2}$ (1/cm^3).. Both of the formulas give approximately equal values for the density which are about (1 - 3) $\cdot 10^4$ g/cm^3 at $B = 0.1 \cdot B_c$ ($B_c = 4.41 \cdot 10^{13}$ G). Modelling the cascade, we assumed the material density to be $2 \cdot 10^4$ g/cm^3, magnetic field strength B = 0.1 Bc, and Z = 26.

At electron energies $E = 10^{12}$ - 10^{13} eV and the material density $2 \cdot 10^4$ g/cm^3, collision radiation undergoes crucial modification. The modification is to due two factors. The first is polarization of the medium. Photon energy E_γ and its wave vector $\vec{k}$ in the medium with a dielectric constant $\varepsilon(E_\gamma)$ are related through the expression $\sqrt{\varepsilon(E_\gamma)}\, E_\gamma = \hbar k c$. For photon energies much higher than the excitation energy of atomic electrons, $\varepsilon(E_\gamma)$ is determined by the relation $\varepsilon = 1 - (\hbar\omega_p)^2 / E_\gamma^2$ [15], where $\omega_p{}^2 = 4\pi n e^2 z/m$, e and m are the electron charge and mass, respectively. At z = 26 and $\rho \sim 10^4$ g/cm^3, $\hbar\omega_p \sim 2$ - 3 keV. As the light velocity changes, radiation is suppressed in the medium in the region where $E_\gamma < E \cdot (\hbar\omega_p)/m c^2$ [16]. The second factor is radiation suppression by repeated electron scattering (the Landau-Pomeranchuck effect) The probability of collision radiation in the medium, where repeated scattering and medium polarization are taken into account, can be written down as [17]:

$$\frac{dW}{dx} = \frac{\Psi(s)}{3L}\,\frac{dE_\gamma}{E_\gamma}\left\{\frac{2(E^2 + (E - E_\gamma)^2)\,\Phi(s)}{E^2(1 + (\hbar\omega_p E / E_\gamma mc^2)^2)} = \left(\frac{E_\gamma}{E}\right) G(s)\right\} \tag{1}$$

Here Φ(s) and G(s) are the Migdal functions

$$\Phi(s) = 12 s^2 \int_0^\infty cth\left(\frac{x}{2}\right) e^{-sx} \sin(sx)\, dx - 6\pi s^2$$

$$G(s) = 48 s^2 \left[\frac{\pi}{4} - \frac{1}{2}\int_0^\infty e^{-sx}\, \frac{\sin sx}{shxs\, dx}\right] \tag{2}$$

and the S parameter is equal to

$$S = (1 + (\hbar\omega_p E / E_\gamma mc^2)^2)\sqrt{\frac{(mc^2)^3 E_\gamma L}{8\lambda\varepsilon_s^2 E (E - E_\gamma)\,\Psi(s)}} \tag{3}$$

The Φ(s) function has the form

$$\Psi(s) = \begin{cases} 2 & , s \le s_1 = (190 z^{-1})^{-2} \\ 1 + (\ln s / \ln s_1), & s_1 < s < 1 \\ 1 & , s \ge 1 \end{cases} \tag{4}$$

In all the formulas L is the radiation length equal to , $1/\ (4\alpha n (z r_0)^2 \ln (190 z^{-1/s}))$, $\alpha = 1/137$, $r_0 = e^2/mc^2$ is the classical electron radius,

$\varepsilon_s = \sqrt{4\pi/\alpha}\ mc^2 \approx 21$ MeV, $\lambda\!\!\!^{-} = \hbar/mc$.
The function G(s) and Φ(s) at S ≥ 1 approach unity. At Φ(s) = G(s) = 1 , cross- section (1) transforms into the Bethe-Heitler cross-section of collision radiation within the limits of complete screening. At s « 1 , G(s) ~ S^2 and Φ(s) ~ 6 s. At the chosen parameters

$$S = (1 + (E/200E_\gamma)^2)\sqrt{1.4 \cdot 10^3 \frac{E_\gamma}{E - E_\gamma}\frac{mc^2}{E}} \qquad (5)$$

Therefore, at all Eγ, except a narrow-region with (E − Eγ) / Eγ ~ 10^{-3}, the cross-section will be radically different from the Bete-Heitler one. Fig.1 shows the spectral distribution (solid line) of the photon number for collision radiation calculated from formulas (1) - (3) for z = 26, density 2.6 · 10^4 g/cm^3 and E = 10^6 MeV. The dotted line shows radiation distribution in ordinary material in the complete screening approximation. The obvious difference is due to Landau-Pomeranchuck effect at high energies and polarization effect at energies below 10^3 MeV.

2. Synchrotron radiation.

Work [12] showed that in the medium with a density of about 10^4 g/cm^3 synchrotron electron radiation is more intensive than collision radiation, provided the magnetic field is 10^{12} G electron energies ≤ 10^4 MeV, due to repeated Coulomb scattering. The transverse impulse the particle gains is about 10mc i.e. the electron transverse motion is relativistic. Therefore, calculations were made using the formula for synchrotron radiation in the quasiclassic approximation including the medium polarization [18]

$$\frac{dW}{dx} = \frac{\alpha}{\sqrt{3}\pi}\frac{1}{\lambda\!\!\!^{-}}\frac{mc^2}{E}\frac{du}{(1+u)^2}\left\{\int_{\frac{2}{3}\frac{u}{\chi}q^2}^{\infty} K_{5/3}(y)\,dy + \frac{u^2}{(1+u)^2}K_{2/3}\left(\frac{2}{3}\frac{u}{\chi}q^2\right)\right\} \qquad (6)$$

where $u = E_\gamma/(E - E_\gamma)$, $q^2 = 1 + (\hbar\omega_p E/E_\gamma mc^2)^2$, $\chi = \frac{B}{B_c}\frac{E\sin\Theta}{mc^2}$

Θ is the angle etween the electron motion direction and the vector $\vec{B}$. $K_{5/3}$ and $K_{2/3}$ are the McDonalds functions of the order 5/3 2/3.

Figure 1 shows the distribution of the synchrotron radiation (dot-dash line) generated by a 10^6 MeV particle and a 10 MeV/s transverse impulse in a dense material . It can be seen that the photon spectrum in the E ~ 10^3 MeV region markedly differs from the photon spectrum of synchrotron radiation in vacuum (in vacuum the spectrum in this energy region behaves as $E^{-2/3}$). This difference is wholly due to the polarization effect.

When modeling the cascade, we presented a complete particle radiation as a sum of collision and synchrotron radiation probabilities . Strictly speaking, this is not correct. The particle radiation must be regarded as a single process. A detailed analysis, however, showed that the expression used in our calculations causes differences in a narrow-energy region, ~ 10^4 MeV, where both processes interfere. Therefore, we neglected these differences.

3. Birth of e^+e^- pairs on nuclei in a dense medium.

The probability of a pair birth with an electron energy E induced by a Eγ photon has the following form [17] if the Landau-Pomeranchuck effect is included

$$\frac{dW}{dx} = \Psi(s)\frac{dE_-}{3E_\gamma L}\left\{G(s) + 2(E_-^2/E_\gamma^2 + (1 - E_-^2/E_\gamma^2)\Phi(s)\right\} \qquad (7)$$

Here S is determined by the formula

$$S = \sqrt{(mc^2)^3 E_\gamma L / 8\lambda\!\!\!^{-}\varepsilon_s^2 E_- (E_\gamma - E_-)\Psi(s)} \qquad (8)$$

and Φ, G and Ψ functions are given by expressions (2) and (4). The probability (7) was obtained in the approximation of complete screening. When calculating the photon path, we calculated the cross-section of a pair birth with an adequate account of the atom screening and Landau- Pomeranchuck effect. The deformation of atomic shells by magnetic field was neglected [4]. According to our estimates the latter effect does not considerably change the radiation length; the change was found to be ~ 25%.

Shown in Fig. 2 is the probability of a pair formation per a length unit in iron with $\rho = 2.6 \cdot 10^4$ g/cm^3 as a function of photon energy. At low energies the cross-section behaves like it does in ordinary material: it increases logarithmically to reach a plateau, which corresponds to the atom complete screening. At energies $\geq 10^3$ MeV the pair birth experiences the effect of repeated scattering. As a result, the cross section decreases. At energies ~ 10^6 MeV it becomes almost three times smaller than the maximal.

The Landau-Pomeranchuck effect influences the distribution of energy between the pair component as well. The insertion in Fig. 2 shows the distribution, in E_+/E_γ. value, of positrons formed by 10^6 gamma-quanta. Also shown the distribution in iron in the case of complete screening. Both distribution are normalized to unity. Because of the Landau- Pomeranchuck effect the probability of all the photon initial energy transfer to one component of the pair only becomes greater. However, it is not so great as was assumed in [13], where the probability of a pair birth with energies satisfying the condition

$$\frac{E_-}{E_\gamma}\left(1-\frac{E_-}{E_\gamma}\right) > \frac{(mc^2)^3 L}{8\hbar\varepsilon_s^2 E_\gamma} \qquad (9)$$

was considered negligibly small.

4. Repeated electron scattering.

It was noted above that synchrotron radiation, in our problem, arises because electrons have a non-zero pitch angle. Estimates show that repeated scattering will be the main mechanism of the angle enlarging. changes in the particle movement direction during radiation or e^+e^- pairs birth would hardly affect the angular distribution of particles in the cascade. To model repeated scattering, the Molier distribution [19] was used including three terms of expansion

$$f(\tilde{\Theta}) = \left(\frac{\sin\tilde{\Theta}}{\tilde{\Theta}}\right)^{1/2} \tilde{\Theta}\left(f^{(0)}_{(\tilde{\Theta})} + \frac{1}{\tilde{B}} f^{(1)}(\tilde{\Theta}) + \frac{1}{\tilde{B}^2} f^{(2)}(\tilde{\Theta})\right) \qquad (10)$$

where $f^{(1)} \sim 2e^{-\tilde{\Theta}^2}$, and $f^{(1)}(\tilde{\Theta}) \sim f^{(2)}(\tilde{\Theta}) \sim 1/\tilde{\Theta}^4$ at 1. The angles Θ and $\tilde{\Theta}$ are related by the expression

$\Theta = \tilde{\Theta}/\sqrt{\chi c^2 \tilde{B}}$, $\chi_c = 4\pi n e^4 z(z+1)t/Ap^2(v/c)^2$ and $\tilde{B}$ value is the solution of the equation

$$\tilde{B} - \ln\tilde{B} = \ln 6680 \pm (z+1)z^{1/3}/(v/c)^2 A(1 + 3.34 z^2 d^2) \qquad (11)$$

Here A is the atomic number, t is the path in d/cm^2 and p is the particle impulse.

5. Nuclei excitation by gamma-quanta.

Excitation of nuclei mainly proceeds during photonuclear absorption of photons in the region of a gigantic resonance. We have approximated the cross-section of this process by the function [20]

$$\sigma = \sigma_0 E^2\Gamma^2/\left((E^2 - E_0{}^2)^2 + E^2\Gamma^2\right) \qquad (12)$$

where, according to [21]

$$\sigma_0 = 60z(A-z)(1 + 0.1A^2/((A-z)z)) \cdot A\pi\Gamma \ (\mu Barn) \qquad (13)$$

The resonance width is 6.1 MeV for Fe^{56} and $E_o = 70_{-1/3}$ [22].

6. Diffusion of neutrons and radiation albedo.

An excited Fe^{56} nucleus emits a neutron in 80% of the acts. The bending energy of a neutron in iron is 11MeV. After the neutron is emitted, the excitation energy of the nucleus in the excited state is $\approx$ 5.5 MeV [10]. At first, the neutron has an energy of about 1.5 MeV. At this energy the main source of energy loss is inelastic collisions. After several interactions the neutron energy falls below 850 keV, which is the energy of the first excitation state of Fe^{56}. After that the neutron loses its energy via elastic collisions only. In fact, the neutron passes, all the way to absorption interacting elastically with nuclei.

The ratio of the neutron mass to the scatter mass (Fe^{56} nucleus) is 1/56. Because of a great difference in the masses, the neutron loses only a small fraction of its energy per one scatter. The rate of energy loss (to an accuracy of 1/56) is determined by the expression

$$\dot{\varepsilon}\,(E_{\mathrm{H}}) = 2\frac{E_{\mathrm{H}}}{A}\sigma_{tr}\, n\, v_{\mathrm{H}} \qquad (14)$$

where σ_{tr} is the transport cross-section of neutron scattering, v_N and E_N is the neutron velocity and energy. The equation for neutron transfer in the diffusion and continuous energy loss approximations has the form [23]

$$\frac{\partial N}{\partial t} = \Delta D\,(E_{\mathrm{H}})\,N - \alpha\,(E_{\mathrm{H}})\,N - \frac{\partial}{\partial E_{\mathrm{H}}}\dot{\varepsilon}\,(E_{\mathrm{H}})\,N + Q\,(\vec{r},t) \qquad (15)$$

where $D = \frac{1}{3}\frac{v_{\mathrm{H}}}{\sigma_{tr}\,n}$ is the diffusion coefficient, @a(E@H) is the coefficient of neutron

absorption, $Q\,(\vec{r}, t)$ is the source function. The values for σ_{tr} and $\alpha\,(E_{\mathrm{H}})$ are borrowed from Ref. [24].

We assumed the neutron flux at the material boundary to be zero. Physically speaking, it means that all the neutrons, escaping from the material, are revolved by the gravitational field and return into the material, with the gravitational field influencing the process of diffusion itself but slightly.

For the physics of pulsars the temperature dependence of the positron knock-out coefficient is of great significance.The observed surface temperature for some pulsars falls in the interval 10^6-10^7 K° [25]. Under the conditions considered here the cascade characteristics do not dependent on temperature, because the density of thermal photons is small enough for the process $\gamma + \gamma \rightarrow e^+ e^-$ and inverse Compton scattering to contribute substantially. However, the neutron transfer in the medium does depend on temperature because of energy dependence of the diffusion coefficient and probability of a neutron capture. To estimate the temperature dependence of neutron transfer, we assumed that the neutron first slows down to ≈ 3/2kT (k is the Baltzman constant) and then continues to move with this energy.

Neutron capture results in excitation of Fe^{57} nucleus. The excitation is relaxed through electromagnetic transitions emitting a line spectrum in the region from 1 to 7 MeV. Calculating the probability of these photons escaping from the material we took into consideration photoabsorption, Compton scattering and conversion processes in e^+e^- pairs both in the nucleus and magnetic fields. To calculate the probability of photon conversion in magnetic field, we used the expression [27]

$$\alpha = \frac{3\sqrt{3}}{16\sqrt{2}}\frac{\alpha}{\lambda\!\!\!^{-}}\frac{B}{B_c}\sin\Theta\,\exp\left(-\frac{8}{3}\frac{B_c}{B}\frac{mc^2}{E_\gamma\sin\Theta}f\,(E_\gamma\,,B\,)\right) \qquad (16)$$

which includes the vicinity of the reaction threshold. Here Θ is the angle between $\vec{B}$ and the direction of photon movement, and

$$f\,(E_\gamma\,,B\,) = 1 + 0.42\cdot(E_\gamma\sin\Theta\,/\,2\,mc^2\,)^{-2.7}B^{\,-0.0038} \qquad (17)$$

Some processes were not taken into account in our calculations. These are:
1. bending emission;
2. photon conversion in magnetic field;
3. pairs birth in $\gamma - \gamma$ interactions with thermal photons;
4. inverse Compton scattering on thermal photons;
5. formation of δ-electrons;
6. a single scattering of electrons at large angles;
7. particle losses due to ionization;
8. channeling of charged particles along atom chains aligned by magnetic field;
9. Messbauer effect during neutron diffusion in solid bodies [28].

Our estimates have shown that these processes do not appreciably influence the distribution of electromagnetic showers and neutrons in the crust of a neutron star at energies of interest for us.

Discussion of the results. Astrophysical implications.

We have noted above that collision radiation and e^+e^- pairs birth on nuclei during the development of a cascade in the crust of a neutron star with strong magnetic field turn out to be appreciably suppressed as compared a cascade in ordinary material. This substantially increases the depth of the shower penetration. Fig. 3 shows the distribution of 10 MeV gamma-quanta as a function of the penetration depth in radiation length units (solid line). The radiation length for the parameters chosen here is equal to $0.7 \cdot 10^{-3}$cm. The dashed line presents a similar distribution in ordinary material, calculated in the A approximation [29]. It is evident that the cascade maximum is at a much deeper level, though this depth is 300 radiation units smaller than the value obtained in Ref.[13] at the same parameters. This is due to the overestimation in Ref. [13] of the energy fraction which can be transferred by a secondary particle.

Figure 3 shows the histogram for the number of excited Fe^{56} nuclei as a function of depth. The total number of nuclei in the interval of primary particle energy variations E_o from 10^3 to 10^7 MeV can be described by the low $N_{nucl} = 214 (E_o/10^6 Mev)$. Our estimate is close to that obtained by zones for a 10^6 MeV primary electron [9]. However, his estimate of the cascade penetration depth seems to be erroneous. Assuming that the maximum of the photon number is at 5 radiation units, zones obtained the positron knock-out coefficient K_k equal to $2 \div 3$ at $E_o = 10^6$ Mev, which is an order of magnitude greater than our estimate, $K_k \sim 0.2$.

The dot-dash line in Fig. 3 shows the distribution of exited nuclei Fe^{57} formed in the reaction (n, γ) after neutron diffusion at $T = 10^6$ K. The distribution maximum of Fe^{57} is near the material boundary. This is a result of the boundary condition chosen by us and corresponding to the zero neutron flux through the boundary.

Shown in Fig. 4 is the temperature dependence of the knocking-out coefficient K_k for two values of a primary particle energy. It can be seen that at $E_o = 10^6$ MeV $K_k \sim 0.2$ and at $E_o = 10^7$ MeV $K_k \sim 1$. This value changes with temperature by about 20%. At different energies K_k has a different temperature dependence. This is, obviously, a result of different cascade depths in the situations considered.

Calculations of the cascade at $E_o = 10^8 \div 10^9$ MeV take too much computer time, and we could not follow the development of showers to the end. We stopped their simulation at a particle energy of 10^6 MeV. Analysis of the cascade curves showed that the shower penetration depth at $E_o = 10^8$ MeV would be greater than 100 and at $E_o = 10^9$ MeV greater than 350 radiation units. It means that at a characteristic length of a neutron capture $\sim$ 27 radiation units, neutrons would not be able to reach the surface i.e. at energies $E_o > 10^7$ MeV a decrease of K_k value should be expected.

The K_k value determine at least two of the observed pulsar characteristics. In work [7, 30] an attempt was made to estimate K_k on the basis of astrophysical information available on these two characteristics. The authors of Ref. [7] used the K_k dependence of the pulsar boundary damping. They obtained $K_k \sim 10 \div 100$. This estimate, however, can be considered crude and not quite reliable, because the boundary is dependent on a number of other parameters, including the angle between $\vec{\mu}$ and $\vec{\Omega}$, magnetic field and so on.

A more reliable value can be obtained from the comparison of the energy L_+ in a beam of positrons ejected from the pulsar and the energy L_- in the reverse beam of electrons. The first value can be determined in terms of the Ruderman and Sutherland model, the second can be derived from the measurements of a soft X-ray flux from the polar cap warmed up by an electron beam. It stands to reason that the relation $L_+ = K_k L_-$ should hold between these values. Such an analysis was made in [30] for pulsar PSR 1509 - 58, for which the X-ray flux the polar cap is well documented [25]. A fair agreement was found for $K_k = 2$ at particle energies 10^7 MeV. This value is close to ours ($K_k \sim 1$). The discrepancy ($K_k = 2$ and $K_k \sim 1$) can hardly be regarded as fundamental in view of the approximations made in our calculations and by the authors of work [7, 30].

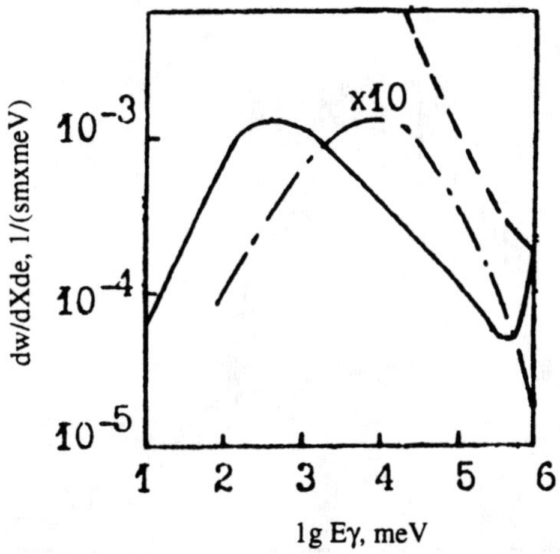

Fig. 1. Spectral distribution of photons emitted by a 10^6 MeV electron in iron with the density of $2.6 \cdot 10^4$ g/cm^3. The solid line represent the collision radiation, the dashed line shows the collision radiation without the Landau- Pomeranchuck effect. The dot-dash line is for the particle synchrotron radiation with a 10 MeV/s transverse impulse.

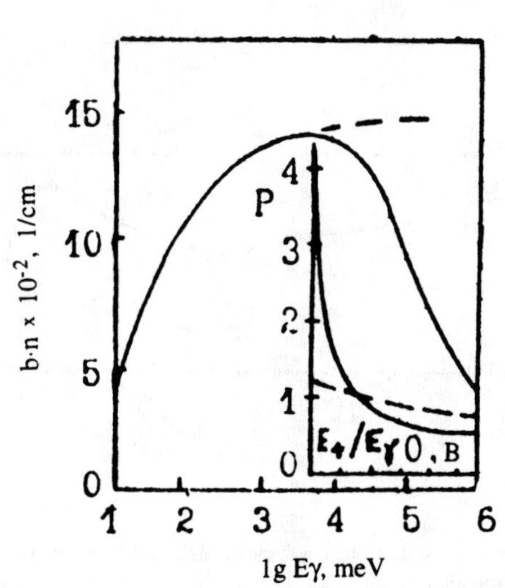

Fig. 2. The probability of an e^+e^-pair birth per a length unit as a function of the photon energy in iron $2.6 \cdot 10^4$ g/cm^3
The dashed line shows the dependence for ordinary iron. The insertion shows the distribution of pair components in energy for a 10^6 MeV photon.
The dashes show the distribution for ordinary material in the complete screening approximation. The areas are normalized to unity.

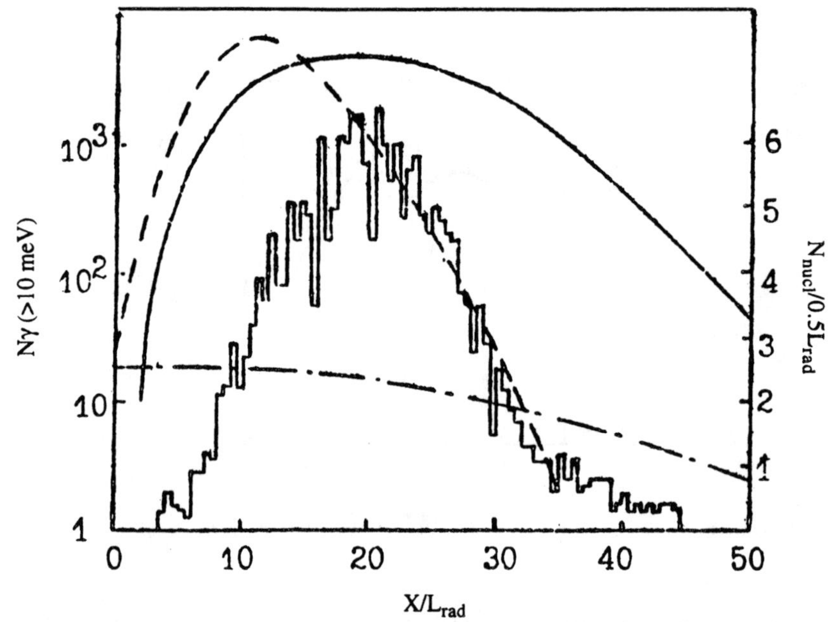

Fig. 3. The solid line represents the number of 10 MeV photons in the shower initiaded by a 10^6 Mev electron. The histogram shows the distribution of F^{56} exited nuclei. The dot-dash line is for the Fe^{57} nuclei distribution.

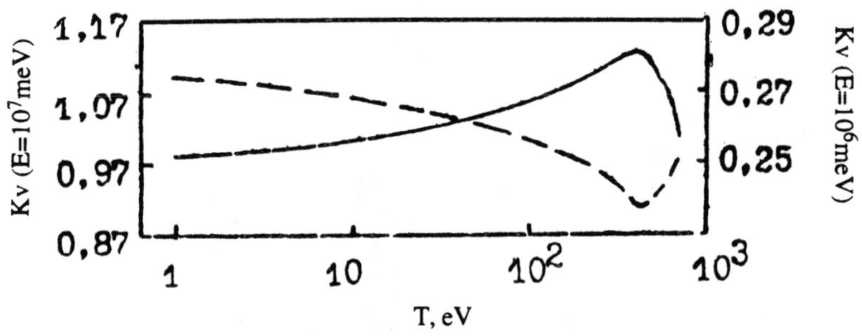

Fig. 4. The temperature dependence of knocking out coefficient. The solid line represents the initial electron energy, 10^7 Mev, the dashed line is for 10^6 MeV energy.

Reference

1. V.V. Zheleznyakov, Electromagnetic waves in cosmic plasma. M.: Nauka, 1977, p. 432.
2. J. Arons, Astrophys. J 1983, v. 266, p. 215.
3. V.L. Ginzburg, V.V. Usov, Pis'ma ZhETF, 1972, v. 15, p. 280.
4. E.G. Flowers, J.F. Lee, M.A. Ruderman et. al., Astrophys. J. 1977, v. 215, p. 291.
5. M.A. Ruderman, P.G. Sutherland, Astrophys. J., 1975, v. 196, p. 291.
6. B.C. Beskin, A.B. Gurevich, Ya.N. Istomin, ZhETF, 1983, v. 85, p. 401.
7. A.V. Gurevich, Ya.H. Istomin, ZhETF, 1985, v. 89, p. 3.
8. V.S. Beskin, A.B. Gurevich, Ya.N. Istomin, UFN, 1986, v. 150, p. 257.
9. P.V. Jones, M. N. R. A. S., 1978, v. 184, p. 807.
10. P.V. Jones, Astrophys. J., 1979, v. 228, p. 536.
11. Ya.B. Zel'dovich, I.D. Novikov, Relativistic Astrophysics. 1967, M., Nauka, p. 654.
12. Yu.D. Kotov, C.P. Kel'nor, C.V. Bogovalov, Pis'ma AZh, 1986, v. 12, p. 168.
13. I.L. Rosental, V.V. Usov, Astrophys. and Space Sci., 1985, v. 109, p. 865.
14. P.B. Jones, Phys. Rev. Letters, 1985, v. 55, p. 1338.
15. V.L. Ginzburg, Theoretical Physics and Astrophysics 1981, M., Nauka, 1981, p. 503.
16. V.L. Ter-Michaelyan, The medium Effect on electromagnetic processes at high energies, 1969. Erevan, the Armenian SSR Publishers, p. 147.
17. A.B. Migdal, ZhETF, 1957, v. 32, p. 633.
18. V.N. Baiyer, V.N. Katkov, V.S. Fadin, Emission of Relativistic Electrons, 1973, M., Atomizdat.
19. H.A. Bethe, Phys. Rev., 1953, v. 89, p. 1256.
20. J.I. Hubbel, Photon cross sections, attenuation coefficients and energy absorption coefficients from 10 keV to 100 GeV, NSRDS - NBS, 2969, 1976.
21. J.S. Levinger, Nuclear photo-disintegration, 1960, Oxford University press.
22. K. Okowoto, Phys. Rev., 1958, v. 110, p. 143.
23. K. Bekurtz, K. Virtz, Neutron Physics, 1968, M., Atomizdat, p. 456.
24. I.V. Gordeev, D.A. Kardashev and A.V. Malyshev, Nuclearphysical constants,1963, M. Gosatomizdat.
25. F.D. Seward, F.R. Harnden, Astrophys. J., 1982, v. 265, p. L45.
26. L.D. Groshev, A.M. Demidov, V.N. Lutzenko, Album of gamma-ray spectra from radiation capture of thermal neutrons, 1958, M., Gosatomizdat.
27. J.K. Daugherty, A.K. Harding, Astrophys. J., 1983, v. 273, p. 761.
28. I.I. Gurevich, L.V. Tarasov, Physics of low-energy neutrons, 1965, M., Nauka.
29.I.P. Ivanenko, Electromagnetic cascade processes. 1972, M.,Izd. MGU.
30. B.S. Beskin, A.V. Gurevich, Yu.N. Istomin, Proceedings of the Joint Varenna-Abastumant International School, 1986, p. 361.

Physics of Neutron Stars
Edited by A. M. Kaminker

ISBN 1-56072-164-2

ELEMENTARY PROCESSES OF GENERATION OF HIGH-ENERGY RADIATION IN A NEUTRON STAR CRUST

S.V. Bogovalov, I.O. Zhukovskii, S.R. Kelner, and Yu.D. Kotov
Moscow Engineering and Physics Institute, Moscow

Processes of collision, synchrotron and magneto-coulomb radiations in the medium of density n and strong magnetic field, h, have been studied for ultrarelativistic electrons. Ranges of the n and H parameters have been determined in which one of the processes is dominating.

1. Introduction

A stationary generation of an electron--positron plasma in the pulsar magnetosphere may be maintained by particles being knocked out from the neutron star surface. A concrete mechanism of the knocking out may be as follows [1]. An $\sim 10^{12}$ eV electron, rushing into the dense material of the neutron star, generates an electromagnetic cascade. The cascade photons with energies typical of the giant resonance excite ^{56}Fe nuclei which mainly make up the star crust. The excitation is relaxed primarily through the escape of neutrons. The latter diffuse towards the surface and are caught by ^{56}Fe: this process produces excited ^{57}Fe nuclei located close to the surface. When ^{57}Fe transforms into the ground state a (2 - 8) MeV gamma-quantum is emitted; it escapes from the material and converts into an electron- positron pair in the strong magnetic field.

According to works [1, 2] a stable plasma generation in the magnetosphere occurs when the coefficient of the particle knocking out $K \sim 10^{2}$. The K value primarily depends on the characteristics of the electromagnetic cascade developing in a dense medium and strong magnetic field. Estimates obtained in Ref. [3] suggest that high density of the material brings about, due to the Landau-Pomeranchuck effect, radical depression of collision radiation and formation of pairs on nuclei. Hence, the depth of the cascade propagation turns out to be so great that K will be close to zero ($K \sim 10^{2}$). However, we do not think that this estimate is correct because in Ref. [3] the fraction of the energy received by a secondary particle is strongly overestimated and as a result overestimated is the path length of the cascade development.

If an electron, in the moment when it rushes into the dense medium, moves along the force lines of magnetic field, that is, it is at the zero Landau level, then repeated scatterings would excite the transverse levels. In the case when only the ground and the first excitation levels are taken into account (the two-level approximation), magnetocoulomb radiation arises [4] that substantially exceeds the intensity of oscillation radiation. If levels with great quantum numbers are excited, the transverse electron movement becomes quasiclassical to cause synchrotron radiation which restricts the growth of a transverse impulse. The aim of this work is to analyze the emission by electrons moving at low angles towards a uniform magnetic field in a dense medium.

2. Emission of an ultraralativistic particle in a dense material and strong magnetic field.

Figure 1 shows areas in which different approximations of an electron movement in the neutron star crust are applicable ($Z = 26$) [5]. In the area below line 1 a two-level approximation applies. The main contribution in this area into intensity is from the magnetocoulomb emission. In the area between lines 1 and 2 the electron transverse motion is nonrelativistic but excitation onto higher levels can occur. Between lines 2 and 3 the transverse motion is ultrarelativistic, however classical electrodynamics is still applicable. In the area below line 3 the classical approximation is invalid and results of the quantum theory of synchrotron radiation should be used. Line 4 determines the relationship between concentration and the initial electron energy, at which synchrotron radiation becomes equal to collision. In the area above line 4 energy losses through collision radiation exceed synchrotron losses. The strength of the magnetic field was taken equal to $4 \cdot 10^{12}$ G.

Let $W_{nn'}(P_z,\Theta)$ be the probability of an electron transition per a unit of time from the state (n, P_z) into the state ($n'P'_z$) accompanied by a ω photon emission at the angle Θ to the field (n is the number of a Landau level, P_z is the electron transverse impulse). The explicit form of $W_{nn'}(P_z,\Theta)$ is given in Ref. [6]. From the concentration laws it follows

$$\omega(n',\Theta) = \frac{\varepsilon - p_z\cos\Theta}{\sin^2\Theta}\left(1-\sqrt{1-\frac{2m\,\omega_H\,(n-n')\sin^2\Theta}{(\varepsilon - p_z\cos\Theta)^2}}\right) \tag{1}$$

where ε is the initial electron energy, ω_h is the cyclotron frequency. In the limiting case $n \gg 1$ the value $W_{nn'}$ must go into well-known classical formulas for synchrotron radiation. Its transition, however proceeds in a rather peculiar manner. With $W_{nn'}$ the probability of a photon emission as a function of its energy can be obtained as

$$\frac{dP(\omega)}{d\omega} = \sum_{n'=0,}^{n-1}\int W_{nn'}(p_z,\Theta)\,\delta(\omega-\omega(n',\Theta))\,d\Omega \tag{2}$$

Shown in Fig. 2 is the spectrum for $n = 20$. The longitudinal impulse of an electron was assumed equal to zero. In the Figure distinct asymmetric peaks can be seen, which transform into separate lines in the region of low photon energies. The peaks broadening towards the region of lower energies can be explained by quantum recoil; the location of the peaks corresponds to the difference of the Landau energy levels.

The Figure also presents the results obtained from the formulas of qusiclassical approximation [7] (curve 1):

$$\frac{dP(\omega)}{d\omega} = -\frac{e^2m^2}{\sqrt{\pi}\,\varepsilon^2}\left\{\int_x^\infty \Phi(\xi)\,d\xi + \left(\frac{2}{x}+\frac{\omega\chi}{\varepsilon}\sqrt{x}\right)\Phi'(x)\right\} \tag{3}$$

where $x = (\omega/\chi\varepsilon')^{2/3}$, $\chi = e\varepsilon H/(mc)^3$, $\Phi(x)$ is the Eire function. For the calculations of an electromagnetic cascade in the neutron star crust, an energy intensity averaged over the intervals greater than the distance between individual lines is of importance. Curve 2 in Fig. 2 was obtained by averaging an exact spectrum over the intervals between the peaks. It is this curve that should be compared with a quasiclassical spectrum. An obvious difference between curves 1 and 2 is due to that curve 1 is constructed from formulas valid for ultrarelativistic case only whereas at $n = 20$ and $H = 0.1 \cdot h_o = 4.4 \cdot 10^{12}$ G the electron energy is $\varepsilon \sim 1$ MeV. The calculations show that at $n \geq 100$ this difference between the curves will be less than 20%. This allows using quasiclassical expression (3) when simulating an electromagnetic cascade. The expression for the intensity of collision radiation in the medium where repeated scatterings and the material polarization are taken into account has the form [7] (in the symbols of Ref [7])

$$d\omega = \frac{\Psi(s)}{3L}\frac{d\omega}{\omega}\left\{\frac{2(\varepsilon^2+(\varepsilon-\omega)^2)\,\Phi(s)}{\varepsilon^2(1+(\omega_p\varepsilon/\omega m)^2)} + \frac{\omega^2}{\varepsilon^2}G(s)\right\} \tag{4}$$

In a wide range of ω values, $\Phi < 1$ and $G < 1$, i.e., a substantial suppression of collision radiation occurs. The suppression may be accounted for by a decrease in the length of the radiation formation because of repeated scatterings. The effect of repeated scatterings will become noticeable when theangular deviation along the length of the radiation formation because of scattering becomes greater than $1/\gamma$ ($\gamma = \varepsilon/m$). Magnetic field may also change the direction of a particle movement and this strongly suggests a simultaneous effect of magnetic field and repeated scatterings on radiation generation. A quantitative solution of this problem is beyond the scope of this paper . Our aim is to give a qualitative analysis of the processes occurring.

In a classical limiting case ($\omega \ll \varepsilon$) the intensity is determined by an integral of the function

$$\exp[i(\omega t - \vec{k}(\vec{r}(t) - \vec{r}(-t)))] \tag{5}$$

where $\vec{r}(t)$ - is the radius-vector of a particle. Averaging over atom arrangement and using the time dependence of $\vec{r}(t)$ - in magnetic field in the form $\vec{r}(t) = \vec{v}(0) + \dot{\vec{v}}(0)\,t^2/2 + \ddot{\vec{v}}t^3/6$, we can obtain from equation (5)

$$\exp\left[-\frac{i\omega t}{2\gamma^2}\left(1+\Theta^2\gamma^2+\omega_\Theta\Theta\,\gamma^2 t+\gamma^2(\omega_\Theta/\omega)^2+\frac{1}{2}<\Theta_s^2>\gamma^2 t+\frac{1}{3}\omega_\Theta^2\gamma^2 t^2\right)\right] \tag{6}$$

where $\omega_\Theta = eH/(m\gamma)\sin\Theta$, $<\Theta_s^2> = (\varepsilon_s^2/\varepsilon^2)(c/L)$. The expression (6) is integrated over time t_{kog}, when the phase in the exponential index becomes of an order of unity. If t_{kog} is determined by magnetic field, i.e.,

$\omega_{\Theta}^{2} t_{cog}{}^{2} \gamma^{2} \gg <\Theta_{s}^{2}> \gamma^{2} t_{cog}$
then we obtain formula (3(for synchrotron radiation. If t_{kog} is determined by repeated scattering, then the expression for collision radiation results that includes the Landau-Pomeranchuck effect. Therefore, when one of the processes dominates, its intensity can be calculated from well-known formulas, and radiation can be presented as a sum of collision and synchrotron radiations. This approximation is valid in the whole range of the parameter variations except for the cases when both processes are of equal intensity.

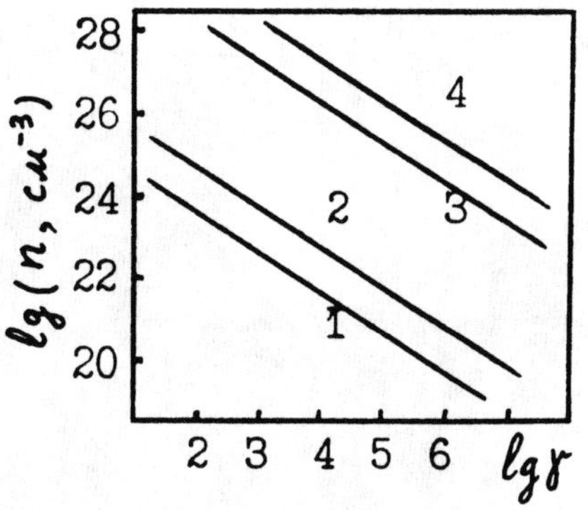

Figure 1

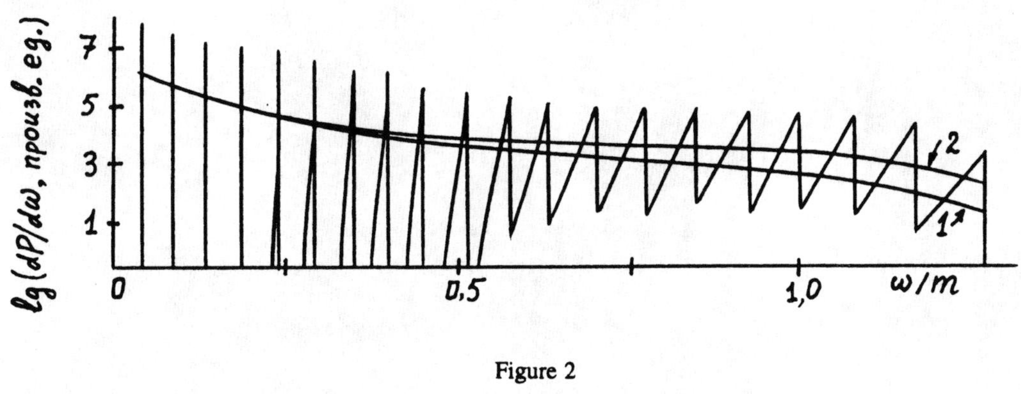

Figure 2

References

1. P.B. Jones, Astrophys. J., 1979, v. 228, p.536.
2. A.V. Gurevich and Ya.N. Istomin,ZhETF, 1985, v. 89, p. 3.
3. I.L. Rosental and V.V. Usov, Ap. Sp. Sci., 1985, v. 307, p. 4.
4. S.R. Kel'ner and Yu.D. Kotov, Pis'ma ZhETF, 1985, v.41, p. 200.
5. Yu.D. Kotov, S.R. Kelner, and S.V. Bogovalov, Pis'ma Astron. Zh., 1986, v. 12, p. 402.
6. A.A. Sokolov and I.M. Ternov, Relativistic Electron, 1974, Nauka, Moscow.
7. V.N. Bayer, V.M. Katkov, and V.S. Fadin, Radiation of Relativistic Electrons, Atomizdat, Moscow, 1973.

Physics of Neutron Stars
Edited by A. M. Kaminker

ISBN 1-56072-164-2

RAPID SPECTRAL VARIABILITY OF GAMMA-RAY BURSTERS

V.Sh. Dolidze, A.A. Kozlenkov, R.N. Kucherova, I.G. Mitrofanov, A.S. Pozanenko and A.M. Chernenko
Institute of Space Research, USSR AS, 117 810, Moscow
C. Barat, G. Vedrenne, M. Niel, and K. Hurley
Center for Space Radiation Research, 31029, Toulouse, France

Abstract - Major observational data on the spectra of space gamma-bursts are analyzed and compared by their main spectral characteristics: continuum, emission details in the energy range > 400 keV, probability of nuclear line emission and spectral "drop" at energies < 100 keV.

1. Gamma-bursts continuum

During the experiment "Konus" spectra were recorded with a time resolution of 0.5 s (for the first two spectra measurements on the stations "Venera" 13 and 14) and 4 s (for the rest of the measurements). From the analysis of the busts recorded by "Venera"11 and 12 the conclusion was drawn (see Ref. [1]) the spectral continuum according to the photon number can be approximated fairly well by the law

$$\frac{dN(E)}{dE} = CE^{-1}\exp(-E/kT) \qquad (1)$$

which corresponds to the thermal collision radiation of an optically thin plasma if Hound factor is not taken into consideration.

The experiment "Konus" enabled measuring integral fluxes of gamma rays in wide spectral channels 150 - 700 keV with a resolution of 0.25 s. Taking the approximation by law (1) valid for each short continuum interval, it was suggested in Ref. [2] that relative displacements of these channels accidentally formed on "Venera" 13 and 14 should be used to study rapid variability of the spectral parameter kT. Thus, the variability of kT was established for many powerful bursts by this method. Moreover, a correlation between variations in gamma-ray fluxes and the value of kT was found. Recording of gamma-bursts in the experiments HXRBS and GRS with the SMM showed that many events had hard spectra, their duration in some cases being up to 10 MeV [3]. For instance, photons with energies up to 30 MeV were registered for GB 820320. Another important result of the SMM is related to the spectrum evolution during separate peaks. It was shown in Ref. [4] that for most peaks the hardest spectrum corresponds to the peak increasing front followed by monotonous hardness decrease.

In the French-Soviet experiment SIGNE-2MS the spectra of bursts in a range of 40-60 KeV were measured up to 1-2 MeV, with a time resolution of 0.25 s on "Venera" 11 and 12 and 0.5 on "Venera" 13 and 14 [5]. In recostructing the spectra the main ambiguity is due to the vagueness of the photon flux in a spectral region above the upper limit of registration. To decrease the dependence of the result on the choice of the initial hypothesis to the maximum, the best fitting function was sought for among a wide class of functions for every time interval provided with statistical data

$$\frac{dN(E)}{dE} = CE^{-\gamma}\exp(-E/E_0) \qquad (2)$$

The functions include both the pure power law ($E_0 \to \infty$) and spectra of thermal collision radiation (1) and radiation due to Compton scattering.

Spectra of continuum with a time resolution of 0.25 - 0.5 s were reconstructed for most intensive bursts which were observed during the experiments in 1978-1983. Fig. 1 shows areas of reliable values for spectral parameters corresponding to the approximation by (2) for two intensive events GB 781104 and GB 820827C. The fact that these areas do not overlap means that a statistically meaningful changes in the shape of the continuous spectrum occurs in the time intervals corresponding to these areas.

A general character of the continuum evolution detected in separate peaks: the spectra become "softer" from the fore-front and the peak maximum towards its backward slope. The values of spectral parameters on the plane (γ, E_0) are shifted down and to the right within one-peak time. This evolution character is in agreement with the data obtained by SMM.

Whether the time resolution of 0.5s is "sufficient" for the spectrometry of gamma-bursts was analyzed on the examples of the intensive events GB 820827, 821104 and 821027. The analysis showed that approximation of the continuum by function (2) is unsatisfactory in 50% of case [6]. The causes may be spectral details (see below) and the variability with a characteristic time « 1 s. The experiment SIGNE-2MS allowed checking the latter possibility by measuring the hardness parameter a resolution time of 16 ms [6].

In many cases, observational data point to the spectrum variability with a characteristic time « 1 s [6]. Measurements of spectra in these moments require increasing time resolution up to milliseconds. On other hand, the main part of the burst profiles did not show any variability of the hardness parameter with a time « 1 s. It follows that the resolution time 0.25 - 0.5 s, which was used in the experiment SIGNE-2MS for the spectrometry of gamma-bursts, can be considered statisfactory.

2. "Annihilation" details 400 keV.

The spectral detail > 400 keV was first detected in the experiment "Konus" [1, 2] in the spectra of many intensive bursts. It was assigned to the annihilation of 511 keV e@+- pairs, whose line was shifted to the red region by the gravitational potential of the neutron star. However, there is a discrepancy concerning this spectral feature in different experiments (see [4]). Thus, in the experiment "Konus" the feature at 400 keV was clearly seen in the spectra of three intensive bursts GB 781104, 781119 and 8111232, whereas in other experiments it was not recorded. The experiment SIGNE-2MS allows an independent verification of the spectral feature at > 400 keV.

Analysis of the spectra of GB 781104 with a time resolution of 0.25 s revealed intervals in which spectra had a great excess of emission in an energy range of E > 400 keV [7]. Approximation of the spectra intervals by the law

$$\frac{dN}{dE} = CE^{-\gamma}\exp(-E/E_0) + B\exp\left(-((E-E_1)/\Delta E_1)^2\right) \qquad (3)$$

resulted in their much "better" fitting as compared to that by law (2). In this case χ^2 decreases greater than can be expected from an increased number of free parameters. Though for every interval the feature was established with a 95 - 98% reliability, the reconstructed spectrum from the sum of readings in these intervals supports the presence of the line with a 99% reliability [7].

All the three intervals whose spectra show the feature at 400 keV correspond to the peaks on the profile of GB 781104 burst. The data obtained in SIGNE-2MS suggest that annihilation emission during bursts is generated as short powerful flushes. In such a flash the ratio of the annihilation emission intensity to the continuum intensity is < 30%.

As the time of the spectra storing increases, the spectral feature "contrast", connected with a separate short flash, would decrease. Thus, the discrepancy concerning the feature at 400 kev in different experiments may be result from different possibilities to detect flashes during long spectra storing.

The SIGNE-2MS data pointed to the flash-like nature of the annihilation emission generation for other bursts as well, including GB 781119, 790305, 790418 and 811221. In the first intervals from GB 781119 and 790305 "strange" spectra were recorded which might be indicative of the annihilation emission predominance in these moments [9]. During the first 24 ms of the GB 790305 burst a "drop" was observed at energies < 300keV. Unfortunately, a very fast speed of counting gave erroneous results for the photon number, therefore, spectra for this interval were not restored. For GB 78119 the spectrum of readings within the first 126 ms had a "drop" in the region of 200 keV (Fig. 2) [10]. Its statistical reliability is 99.7%. Obviously, the interpretation of the spectrum allows two alternatives: "absorption lines" against a smooth continuum and a "two- component" spectrum. The latter alternative seems preferable because the nature of the "absorption" is not yet clear. In the case of a "two-component" spectrum the hard spectrum is naturally to be identified with the annihilation emission of a weakly relativistic electron-positron plasma.

3. "Nuclear" lines and hard emission of gamma-bursts.

In the experiment with a semiconductor Ge spectrometer on ISEE-3 a line at 740 keV at a significance level of 3.5σ was detected in the spectrum of an intensive gamma-burst [11].

It was suggested that it should be identified with a nuclear line of iron above 847 keV which was shifted towards the red region by the gravitational potential of the neutron star. According to the SMM data [12] nuclear lines are the main contributors to the gamma-ray flux in the energy range 1 - 8 MeV for many bursts.

In reconstructing the spectra gamma-bursts recorded in the SIGNE-2MS experiment, excessive emission relative to the continuum level was observed in some short intervals in the spectral region 500 - 600 keV, the excess reaching the level of $> 3\sigma$ in some cases. This excess was confirmed for many bursts by observations on two stations during synchronous intervals (GB 811225(A), 820104, 820313(A), 82053, 820331 and some others).

According to the SMM data [13] the bursts of GB 820320 had the hardest spectrum. "Scanning" of this burst in the SIGNE-2MS experiment with a resolution time 0.5 s showed that the flux of hard photons was not present all the time. In the first peak where the front increased, the spectra showed an excess of hard photons. When the peak and the backward slope were reached, the spectra became smooth and "softer". In the second peak an excess of hard emission was also observed with increasing front.

To verify the presence of this spectral feature the SIGNE- 2MS data and the SMM data for gamma-bursts of GB820530 and 821024 were compared [14]. The "flashes" in the hard spectral channels of the SMM detectors turned out to synchronously coincide with the excessive radiation in the 500 - 600 keV range of the SIGNE-2MS spectra.

The nature of the variable hard excessive emission in the spectra recorded in the SIGNE-2MC experiment is not clear. By analogy with the line at 740keV [11] it may be assumed to be of nuclear origin. However, by analogy with the interpretation of this line given in [15] the excess may be connected with a single photon annihilation of $e^{\pm}$ in a strong magnetic field.

4. Emission deficit at energies 100 keV

The feature in the spectrum of an "absorption line" type at energies < 100 keV was detected in the spectra of many gamma-bursts [1, 2]. The feature was proposed to be identified with the absorption in the main harmonics of an electron resonance. Comparison of the spectra recorded with a resolution of 4 s revealed a gradual disappearance of the spectral feature from the beginning of the burst to its end.

In the SIGNE-2Ms experiment the "deficit" of emission at energies < 100 keV with respect to the continuum level, extrapolated into the "soft" region, was observed in many bursts. The events GB 781119, 820320, 820331 and 82082(c) allow the evolution of this feature to be followed at a resolution time of 0.25 - 0.5 s. For many intervals the radiation deficit in the spectrum at energies 50 - 70 keV was established with a very high degree of reliability [16].

According to the data of the experiment SIGNE-2MS the radiation deficit in the range < 100keV changes nonmonotonously during the burst: intervals with a strong deficit alternate with intervals in which the radiation flux in the first channel corresponds to the continuum level [17]. The "absorption lines" detected in the experiment "Konus" are most likely a result of the superposition of spectra with a varying deficit. For instance, comparison consecutive 4-second spectra of GB 820827(c) recorded in the "Konus" experiment give the feeling of a gradual disappearance of the "absorption line" from the beginning of the burst to its end. However, resolution of one spectrum without the feature into 8 spectra for 0.5 s intervals, measured in the SIGNE-2NC experiment, shows that the deficit is clearly seen in some moments (Fig. 3) Suggesting itself is the conclusion that the monotonously varying deficit of radiation with energies 100 keV is a specific feature of many gamma-bursts. In some cases the deficit is at maximum on the forefront of the burst profile and decreases towards the back slope. The spectral feature in "instant" spectra of the SIGNE experiment is evidently due to the varying "drop" in the hard spectra of gamma-rays in the energy range < 100 keV rather than to an absorption line against the smooth continuum.

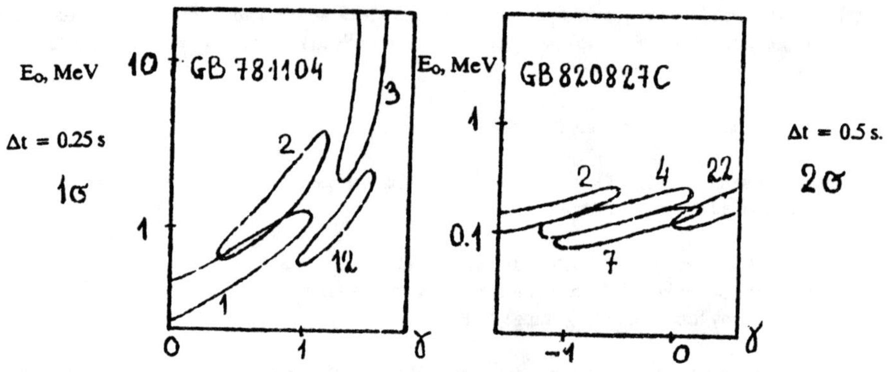

Fig. 1. Areas of reliable values of spectral parameters γ and E_0 for two gamma-bursts with time intervals of 0.25 and 0.5 s. The significance levels are 1σ and 2σ, respectively.

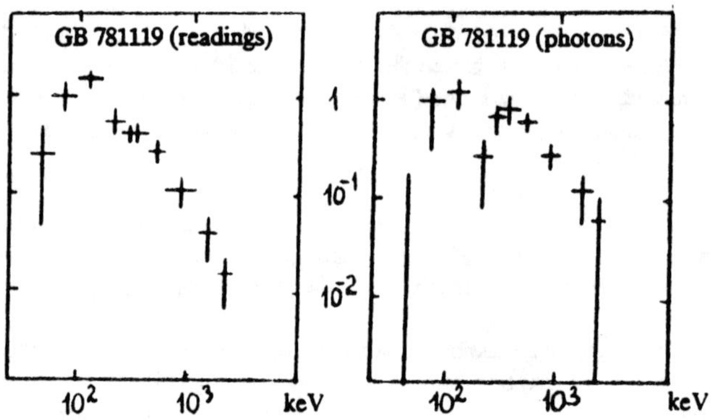

Fig. 2. Spectra of readings and photons for the first short interval (126 ms) of the GB 781119 burst.

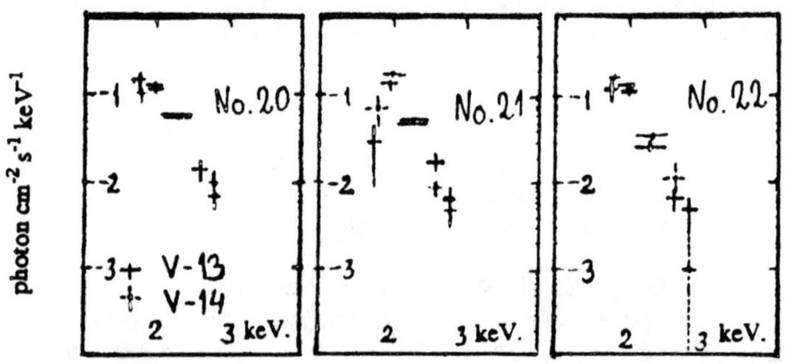

Fig. 3. Nonmonotonous behavior of the feature in a soft region of the spectra from the GB 820827C burst recorded in SIGNE-2MS experiment in successive 0.5 s intervals. These intervals are within a larger interval during which the spectrum of the "Konus" experiment does not contain this feature in the soft region.

References

1. E.P. Mazets et al., Nature, 1981, v. 290, p.378.
2. E.P. Mazets et al., in: Proc. of Conf. No. 101, 1983, N. Y., p. 36.
3. P. L. Nolan, G. H. Share et al., in: Proc. of AIP Conf. No. 115, 1984, N. Y., p. 399.
4. A.K. Harding, V. Perosian et al., Chapter 2 in: Proc. of AIP Conf. No. 141 "Gamma-Ray Bursts", N.Y., 1986.
5. C. Barat, G. Chambon et al., Space Sci. Instr., 1981, v. 5, p. 229.
6. E. Eveno, C. Barat et al ., Adv. Sp. Res., 1986, v. 6, n. 4, p. 27.
7. C. Barat, K. Hurley et al., Astrophys. J. (lett.), 1984, v. 286, p. L11.
8. P.L. Nolan, G.H. Share et al., Nature, 1984, v. 311, p. 360.
9. G. Vedrenne, Adv. Sp. Res., 1984, v. 3, n. 10 - 12, p. 97.
10. C. Barat, E. Eveno et al., Adv. Sp. Res., 1986, v. 6, n. 4.
11. B.J. Teegarden and T.L. Cline, Astrophys. J. (Lett.), 1980, v. 236, p. L67.
12. P.L. Nolan, G.H. Share et al., in: Proc. of AIP Conf., No. 115, N. Y., 1984,p. 339.
13. E. Rieger, in: Proc. of Workshop "Accreting Neutron Stars", 1982, p. 229.
14 J.P. Norris et al., Adv. Sp. Res. 1986, v. 6, n. 4.
15. J.I. Katz, Astrophys. J., 1982, v. 260, p.371. 16. A.S. Pozanenko, I.G. Mitrofanov et al., in: Proc. of IAU/COSPAR Conference "The Physics of Compact Objects: Theory Versus Observations", Bulgaria, 1987.
17. I.G. Mitrofanov, V. Sh. Dolidze et al., in: Proc. of 20th International Cosmic Ray Conference, Moscow, 1987.

Physics of Neutron Stars
Edited by A. M. Kaminker

ISBN 1-56072-164-2

MECHANISM OF CHARGED PARTICLE EJECTION IN RADIOPULSARS

A.G. Muslimov and A.I. Tzygan
Ioffe Physics-technical Institute, USSR AS, 194021, Leningrad

Processes resulting in intensive ejection of charged particles from the magnetic polar areas of neutron stars are discussed briefly.

After the discovery of radiopulsars a number of ideas were suggested concerning generation mechanisms of corpuscular emission by neutron stars. This report deals with some of these ideas prevailing in modern theoretical concepts of radiopulsars. The aim of the report is to touch upon the theoretical difficulties of the mechanisms and to discuss a possible physical situation with the magnetic pole of neutron stars.

An electric field is induced by a rotating neutron star (NS) near its surface. The component of the electric field $E_{||}$, parallel to magnetic field can accelerate charged particles along magnetic field up to relativistic energies. Depending on the physical conditions on the NS surface (temperature, magnetic field strength, atomic composition) both the regime of "free emission" of charged particles and the regime of emission with overcoming the final potential barrier (work function) are possible. In the case of "free emission" the rate of particle escape is affected by a space charge formed in the process (see, for example Ref. [1, 2]). A drop in the electric potential caused is sufficient to overcome the inertia of charged particles and to accelerate them. If the emission of ions (electrons) is impeded, a situation arises in which the space charge does not any longer affect the particle escape. Then, a "vacuum gap" is formed near the NS surface where a considerable drop of the electric potential occurs [3].

The structure of electric potentials and fields in the area of open lines of magnetic field (the area of charged particle acceleration) is strongly dependent on which of the above regimes of discharge takes place. In the case of "free emission", i.e. when the work function of charged particles is small (the surface is hot), the areas of closed lines of magnetic field are assumed to be filled by plasma. In the areas of open lines the rate of charged particle escape is close to the limiting, at which a complete screening of electric fields in this area occurs. The results of a corresponding electrodynamic analysis are presented in Refs [4 - 6]. At sufficiently high work function of charged particles (a cold surface) the structure of the electric field in the area of open lines of magnetic field differs from the purely vacuum [7] because of the screening effect of the NS magnetosphere area filled with plasma (the area with closed lines of magnetic field). This structure was calculated in Ref. [6].

Present relatively simple formulas for electric potentials and fields in the area of open lines of a NS dipole magnetic field. In the coordinate system rotating together with the electric field is potential $\vec{E} = -\nabla\Phi$. The potential can be found from the equation

$$\Delta\Phi + \frac{2}{c}\vec{\Omega}\cdot\vec{B} = -4\pi\rho \qquad (1)$$

where $\vec{\Omega}$- is the angular speed of the NS rotation, $\vec{B}$- is the NS magnetic field, ρ is the total density of an electric charge in the area of open lines of magnetic field.

The current density in the area of the electric potential drop can be described by

$$\vec{j} = c\rho\vec{B}/B \qquad (2)$$

and

$$\nabla\cdot\vec{j} \equiv c\vec{B}\cdot\nabla(\rho/B) = 0 \qquad (3)$$

Because charged particles in this area move with relativistic velocities along the lines of magnetic field. In the spherical coordinate system with the polar axis along the NS magnetic field, the bases of the last open lines of magnetic field are $v_0 = (\Omega a/c)^{1/2}$ wide, where a is the NS radius. The surface formed by the last open lines can be described by the equation $v = v_0 (r/a)^{1/2}$. This surface as well as the NS surface are equipotential.

Introduce dimensionless coordinates $\eta = r/a$ and $\xi = v/v_0\eta^{1/2}$.

Let us first consider the solution at $\eta - 1 \gg v_0$, i. e. when the hight above the NS surface is much greater than the horizontal size of the area of magnetic field open lines at the base. Since the angle v is small, a quasi- onedimensional approximation is valid. Solving equations (1) - (3) with a boundary condition $\Phi_{|\xi=1} = 0$, we obtain

$$\Phi = \frac{3}{8}a\left(\frac{\Omega a}{c}\right)^{5/2} B_0(\eta^{1/2} - 1)\,\xi\,(1 - \xi^2)\sin\chi\cos\varnothing \qquad (4)$$

$$\rho = -\frac{\Omega B_0}{2\pi c}\frac{1}{\eta^3}\left(\cos\chi + \frac{3}{2}v_0\eta^{1/2}\,\xi\sin\chi\cos\varnothing\right) \qquad (5)$$

where B_0 is the field value on the magnetic pole of the NS, χ is the angle between the vectors of the magnetic moment $\vec{\mu}$ and the angular velocity $\vec{\Omega}$, $\varnothing$ is the azimuthal angle. Here and below it is assumed that $\chi \gg v_0$. The component of the electric field along magnetic field is equal to

$$E_{||} \approx -\frac{3}{16}B_0\left(\frac{\Omega a}{c}\right)^{5/2}\eta^{-1/2}\,\xi\,(1 - \xi^2)\sin\chi\cos\varnothing \qquad (6)$$

When $\eta - 1 \le v_0$, and with the boundary condition $\Phi_{|\eta=1} = 0$, we have the following solution

$$\Phi = 3a\left(\frac{\Omega a}{c}\right)^{5/2} B_0 \sin\chi \cos\varnothing \sum_{t=1}^{\infty} \frac{1}{k_i^3 J_2(k_i)} \times \tag{7}$$

$$\times \left[\frac{v_0}{k_i} C_i\, sh \frac{k_i}{v_0}(\eta - 1) + \frac{k_i}{v_0}(\eta - 1)\right] J_1(k_i, \xi)$$

where k_i are zeroes of the function $J_i(k_i)$, $J_\nu(z)$ are cylindrical Bessel functions of the 1st order , c_i = const. "Crosslinking" solutions (4) and (7) at $v_0 \ll \eta - 1 \ll 1$, we find $C_i \equiv 0$. Considering that $\sum_{i=1}^{\infty} \frac{J_1(k_i \xi)}{k_i^3 J_2(k_i)} \equiv \frac{1}{16}\xi(1-\xi^2)$ we obtain instead of (7)

$$\Phi = \frac{3}{16} a \left(\frac{\Omega a}{c}\right)^{5/2} B_0 (\eta - 1)\xi(1-\xi^2)\sin\chi\cos\varnothing \tag{8}$$

On the basis of equation (8) we have

$$E_{||} \approx \frac{-3}{16}\left(\frac{\Omega a}{c}\right)^{5/2} B_0 \xi(1-\xi^2)\sin\chi\cos\varnothing \tag{9}$$

Now let us consider the solutions when the space charge in the area of open force lines is equal to zero ($\rho = 0$) At $\eta - 1 \gg v_0$

$$\Phi = \frac{1}{2} a \left(\frac{\Omega a}{c}\right)^{2} B_0 \left[(1-\xi^2)\cos\chi + \frac{3}{4} v_0 \eta^{1/2} \xi (1-\xi^2)\sin\chi\cos\varnothing\right] \tag{10}$$

$$E_{||} \approx \frac{-3}{16}\left(\frac{\Omega a}{c}\right)^{5/2} B_0 \eta^{-1/2} \xi(1-\xi^2)\sin\chi\cos\varnothing \tag{11}$$

Now let $\eta - 1 \leq v_0$. The resulting solution should be "crosslinked" with (10) at $v_0 \ll \eta - 1 \ll 1$. We obtain

$$\Phi = 2a\left(\frac{\Omega a}{c}\right)^{2} B_0 \sum_{i=1}^{\infty} \left\{ \cos\chi \left[1 - e^{-\frac{k_i}{v_0}(\eta-1)}\right] \frac{2 J_0(k_i \xi)}{k_i^3 J_1(k_i)} + \right. \tag{12}$$

$$\left. \frac{3}{2} v_0 \sin\chi\cos\varnothing \left[1 - e^{-\frac{æ_i}{v_0}(\eta-1)}\right] \frac{2 J_1(æ_i \xi)}{æ_i^3 J_2(æ_i)} \right\} (12)$$

where k_i, $æ_i$ - are zeroes of the functions $J_0(k_i)$ and $J_1(æ_i)$, respectively. At $\eta - 1 \ll v_0$, i.e. close to the NS surface, we have

$$E_{||} \approx -\left(\frac{\Omega a}{c}\right)^{3/2} B_0 \left[a_1(\xi)\cos\chi + \frac{3}{2} v_0 a_2(\xi)\sin\chi\cos\varnothing\right] \tag{13}$$

Where

$$a_1(\xi) = 4\sum_{i=1}^{\infty} \frac{J_0(k_i \xi)}{k_i^2 J_1(k_i)}, \quad a_2(\xi) = 4\sum_{i=1}^{\infty} \frac{J_1(æ_i \xi)}{æ_i^2 J_2(æ_i)}.$$

at $\chi \cong 0, \pi$ we can write down

$$\int_0^1 E_{||}\, d\xi \approx \mp \frac{\pi}{2} \left(\frac{\Omega a}{c} \right)^{3/2} B_0 \qquad (14)$$

All the above solutions are valid at r « c/Ω.

Now let us take a standard model of a radio pulsar work proposed by Ruderman and Sutherland [3]. The model assumes that the work function of ions (for instance, iron) from a NS surface is about 10 keV in a strong, ~ 10^{12} G, magnetic field (see also [8, 9]). At the NS surface temperature of about 0.3 ÷ 1.0 keV the thermal of emission of ions is mostly suppressed and a "vacuum" electric gap is formed near the surface. At the electric potential difference of $\Delta\Phi \geq 10^{12}$ a break-down may occur in this area. This break-down may be due either to a cascade birth of electron-positron pairs in the NS strong magnetic field [3] or to pair birth by thermal X-ray quanta from the star in the Coulomb field of relativistic ions [8]. Of importance is that an intensive flux of 10^{12} electrons towards the NS surface arises in each of these cases. As a result, a local warming-up of the surface to ≤ 1 keV occurs. Thus, this model allows a quasi-stationary regime of charged particle injection into the NS magnetoshpere.

However the results of theoretical calculations made in recent years [10 - 12] give rather low estimates of the ion work function from the NS surface, of an order of hundreds eV. These estimates make impossible or non-selfconsistent the models based on a great work function. Indeed, the model proposed by Ruderman and Sutherland shows that at the work function of ions from the NS surface of about 100 - 200 eV the surface warming-up by a reserve electron flux (arising from the electron-positron pair formation) would cause an exponential increase in the thermal emission of ions from the surface. The resulting space charge would screen the electric field in the area, the discharge would stop and a pause would set in. The probability of the discharge having a quasistationary character and the cyclic nature of charged particle generation in the model by Ruderman and Sutherland were discussed in Ref. [9].

Thus, the pulsar may be expected to "work" in the regime of "free emission" at the NS surface temperature (in the polar area) $T_s \leq 10$ eV.

For the space charge, formed by the ion thermal emission from the NS surface,not to screen the electric field near the surface (in the area from which positive ions escape) the surface temperature should be $T_s \leq 10$ eV if the work function is ~ 100 - 200 eV.

Theoretical calculations of a NS cooling down (see, for instance, Ref. [13]) show that the surface temperature of a NS would be $T_s \approx 10$ eV in $3 \cdot 10^6$ years after its birth a typical age of radiopulsars determined from the slowing down of their rotation is equal to $\tau = P/2\dot{P} \approx (3-4) \cdot 10^6$ years, where P and $\dot{P}$ are the NS rotation period and its time derivative, respectively).

In the polar area of a NS an electric field is induced when a "vacuum" electric gap is formed (see eq. (14)).

$$E \sim (\Omega a / c)^{3/2} B_0 \qquad (15)$$

Taking $B_o = 2 \approx 10^{12}$ G, we obtain

$$E \sim 2 \cdot 10^9 (P_m / P)^{3/2} \quad B / cm \qquad (16)$$

where $P_m = 0.8$ s, a median value of pulsar periods [14]. Under the above physical conditions on the NS surface, a break down of the "vacuum" gap can take place due to either the electron-positron avalanche formation [1, 3, 8] or as a result of electrohydrodynamic instability developing on the NS surface [15, 16]. Electrohydrodynamic instability of a conducting liquid surface in a strong external electric field in vacuum, was discovered in the first experiments with liquid cathodes. The ignition potential with a liquid metallic cathod was found to be much smaller (by about 20 times) that with a solid cathod. Works [17, 18] showed that a microprotrusion on the conducting liquid surface would elongate and become sharper (unlike on the solid surface) in an electric field exceeding some critical field. The electric field strength will increase on the top of the microprotrusion until the charge ignition starts. The phenomenon of electrohydrodynamic instability was first used in Ref. [15, 16] to interpret the process proceeding in NS polar areas. Most favorable conditions for the instability were found to be on the NS surface near the magnetic poles. In these areas the material (iron) must be molten (for instance, either all the star surface is liquid or in the respective areas some craters are formed filled with a liquid phase) and have a temperature $\leq$ 10 eV. The value for a critical electric field to induce electrohydrodynamic instability on the NS surface is equal to [15, 16]

$$E_c = 2 (2\pi)^{1/2} \left(g \rho \Sigma \right)^{1/4} \qquad (17)$$

where g is the gravity acceleration on the NS surface, ρ is the liquid density in the NS surface layer, Σ is the coefficient of the liquid surface tension. The values ρ and Σ depend on magnetic field and can be estimated from the results obtained in Refs [10, 11]. Taking $g = 10^{14}$ cm/s^2, we have

$$E_c \approx 2 \cdot 10^9 \rho_3^{5/12} W^{1/4} \quad B / cm \qquad (18)$$

where $\rho_3 = \rho/(10^3 \text{ g cm}^{-3})$, W is the ion work function in keV. For a characteristic wave number and the of the electrohydrodynamic instability development we have

$$k_c \approx 4 \cdot 10^4 \rho_3^{1/6} W^{-1/2} \quad cm^{-1} \qquad (19)$$

$$\tau \approx 10^{-9} B_{12}^{-1/3} \rho_3^{2/9} \, W^{1/3} \quad s \qquad (20)$$

where $B_{12} = B/(10^{12}$ G), B is magnetic field on the NS surface (gauss is most convenient in this context as a unit of magnetic induction).

The estimates from (16) and (18) support the idea of two possible mechanisms for the break-down of the "vacuum" gap.

By analogy with the experiments on high-current emitters of ions and electrons based on liquid metals (see Refs. [19, 20] and references therein, the NS surface may be expected to be covered by a great number of micropoints at a nonlinear stage of electrohydrodynamic instability. These micropoints will be unstable too, so that a whole spectrum of microdrops would break away from the micropoint tops (together with ion emission). The tendency of the swinging wave crests to elongate and sharpen can be observed even in the linear instability regime. Indeed, near the wave crests the strength of the electric field is greater than the intensity of the initial uniform electric field. As the strength of the electric field increases, the instability area will expand, with waves of smaller and smaller lengths becoming unstable. Therefore, the top of a wave crest will tend to sharpen and elongate. This process was fairly well followed in the experiments mentioned above.

Let microdrops eject from the NS surface as a result of electrohydrodynamic instability development. In the area of the electric potential drop, $< 5 \cdot 10^{-3}$ mkm charged microdrops can be accelerated up to a parabolic velocity ≈ 0.5 s. When flying through this area, microdrops have no time to evaporate, the evaporation time for microdrops being $\propto \exp(W/T)$. However, at a distance of one star radius from the surface microdrops must evaporate (here magnetic field is an order of magnitude smaller than on the NS surface and the ion work function approaches its normal value). If ejection of microdrops proceeds in a regime close to saturation, the microdrops may become a source of plasma in the NS magnetosphere at some distance from the surface where they evaporate. Their density is $\sim 10^3$ times greater than the limiting density of charged particles in the accelerating electric field [16].

Now one more effect should be considered which must be taken into account in those NS areas where positive ion outflow occurs. If the NS surface temperature is ≤ 0.1 keV, then only some electrons will "torn" off from the out most shells of iron ions escaping from the surface. Under the conditions, considered here, the ions can be accelerated up to relativistic energies (the Lorentz factor $\gamma_{ion} > 3$) in the area of the electric potential drop.

In the coordinate system accompanying relativistic ions, the maximum of the star thermal emission shifts to a harder spectral region and becomes near the star equal to $\sim 3 T_s \gamma_{ion} (1 - \vec{v}_{ion} \cdot \vec{n}/c)$, where $\vec{v}_{ion}$ is the ion velocity, $\vec{n}$ - is the unit vector for the direction of photon motion. The symbols without a prime are for the values in a laboratory reference system, the symbols with a prime are for the system accompanying ions.

The spectral density of energy radiation is transformed in the following way (see, for instance, [21]):

$$\omega'(\omega',\vec{n}') \equiv \hbar\omega' n'_{hp}(\omega',\vec{n}') = \gamma_{ion}^2\left(1-\frac{\vec{v}_{ion}}{c}\cdot\vec{n}\right)^2\omega(\omega,\vec{n}) \qquad (21)$$

with

$$\vec{n}' = \left[\vec{n}-\gamma_{ion}\frac{\vec{v}_{ion}}{c}+(\gamma_{ion}-1)\frac{\vec{v}_{ion}}{v^2}(\vec{v}_{ion}\cdot\vec{n})\right] \Big/ \gamma_{ion}\left(1-\frac{\vec{v}_{ion}}{c}\cdot\vec{n}\right) \qquad (22)$$

Thus, ions "torn" out of the surface and accelerated to relativistic energies can be additionally and effectively ionized by the star thermal emission with a temperature $T_s \leq 10$ eV owing to the above effect of aberration. It means that the space charge in the area of ion outflow can be multiply increased and is dependent of γ_{ion}. An increase in the space charge density due to post-ionization of iron ions by photons from the star will cause the electric field in the area to decrease, because in the regime of "free emission" electric field arises as a result of small decompensation of the densities of an electric charge ρ and $\rho_c = -\vec{\Omega}\cdot\vec{B}/2\pi c$. If this effect is taken into consideration in the model analyzed in Refs. [4, 22], then the condition required for the birth of an electron-positron plasma over the NS surface will not be satisfied at the parameters set here. Let us estimate the temperature to which the NS surface is warmed by a reverse flux of photoelectrons. For this, let us take the energy, acquired (per a unit of time) by electrons through the electric field effect, equal to the solid body luminosity of the surface, i.e. write down the condition

$$\sigma T_s^4 = \int_1^{\eta_{max}} (\rho-\rho_c)\frac{\partial\Phi}{\partial\eta}\,d\eta \qquad (23)$$

Where σ is the Stefan-Boltzmann constant. For a pulsar with P = 1 s we will obtain $\eta_{max} \approx 1.3$ .if the increase of the space charge due to the above discussed effect is taken into account. In this case $T_s \sim 14$ eV, i.e. the NS surface will warm up less efficiently than it would according to the model analyzed in Ref. [4] where T_s was estimated equal to ~ 70 eV at the same pulsar parameters. This result means that the stationary regime of charged particle escape from the pulsar polar area will become impossible as soon as the surface cools down to ~ 10 eV. In such a situation the electric potential drop near the NS surface nay be expected to increase at $T_s \leq 10$ eV after plasma density decreases say, because of its escape along the open lines of magnetic field. This will create conditions favorable for the development of electrohydrodynamic instability which will initiate the break-down. The latter can proceed via the mechanism of the electron-positron avalanche formation. The electro-hydrodynamic instability in the area of the electric potential drop will produce relativistic ions. The flux of photoelectrons directed to the NS surface will sharply increase. The NS surface in this area may warm up to $T_s > 10$ eV, at which the potential drop decreases and the regime of "free emission" will again "switch on" for the pulsar. Thus, ejection of charged particles from the NS surface will be essentially of non-stationary character.

References

1. P.A. Sturrock, Astrophys. J., 1971, v. 164, No. 3, p. 529.
2. F.C. Michel, Astrochys. J., 1974, v. 192, No. 3, p. 713.
3. M. Ruderman, P.G. Sutherland. Astrophys. J., 1975, v. 196, no. 1, p. 51.
4.J. Arons, E.T. and Scharlemann, Astrophys. J., 1979, v. 231, No. 3, p. 854.
5. A.I. Tzygan, "Proceedings of the XIII All-Union Conference on the Galactic and Beyond the Galaxy Radioastronomy" (September 21 - 23, 1979, Moscow) ISR, 1979 p. 14.
6. A.I. Tzygan, Astron. J., 1980, v. 57, No. 1, p. 73.
7. A. Deutsch, Ann.d' Astrophys., 1955, v. 18, p. 1.
8. A.F. Cheng and M.A. Rudermann, Astrophys. J., 1977, v. 214, no. 3, p. 598.
9. A.F. Cheng and M.A. Rudermann, Astrophys. J., 1980, v. 235, no. 3, p. 576.
10. P.B. Jones, Phys. Rev. Lett., 1985, v. 55, p. 1338.
11. P.B. Jones, Mon. Not. . Astr. Soc., 1986, v. 218, p. 477.
12. D. euhauser, K. Langanke, and S.E. Koonin, Phys. Rev. a, 1986, v. 33, p. 2084.
13. K. Nomoto and . Tsuruta, Astrophys. J., 1987, v. 312, no. 3, p. 711.
14. A.G. Lyne, .N. Manchester, and J.H. Taylor, Mon. Not. R. Astr. Soc., 1985, v. 213, p. 613.
15. A.G. Muslimov and A.I. Tsygan, Preprint FTI im. A.F. Ioffe, AN SSSR, 1987, no. 1095.
16. A.G. Muslimov and A.I. Tsygan, Astrofizika, 1988, v. 28, no. 43, p. 483.
17. L. Tonks, Phys. Rev., 1935, v. 48, p. 562.
18. Ya.I. Frenkel', ZhETF, 1936, v. 6, p. 347.
19. G.A. Mesyats, In: Proc. of the X Intern. Symp. on Discharge and Electrical Insulation in Vacuum, New York, USA, IEEE, 1982, p. 37.
20. G.A. Mesyats and D.I. Proskurovskii, Impul'snyi Elektricheskii Razryad v Vakuume (Pulsed Electrical Discharge in Vacuum), 1984, Nauka, Novosibirsk.
21. L.D. Landau and E.M. Lifshits, Teoriya Polya (Field Theory), 1973, Nauka, Moscow.
22. J. Arons, Astrophys. J., 1983, v. 266, no. 1, p. 215.

Physics of Neutron Stars
Edited by A. M. Kaminker

ISBN 1-56072-164-2

TWO-STREAM INSTABILITY IN INHOMOGENEOUS ELECTRON-POSITRON PLASMA OF PULSAR MAGNETOSPHERES

V.N. Ursov[1] *and V.V. Usov*[2]
[1] The Earth Magnetism, Magnetosphere and Investigation of Radio-Wave Propagation, USSR Academy of Sciences, 142092, Troitsk, Moscow District
[2] Institute of Space Research, USSR Academy of Sciences, 117810, Moscow

One-dimensional spread of initially isolated clouds of relativistic electron-positron plasma during its efflux from a pulsar has been analyzed; mutual overlapping of adjacent clouds and development of two-stream instability are discussed.

1. Introduction

Presently, the most popular hypothesis [1] in the theory of pulsars is that generation of pulsar radio-frequency radiation is due to two-stream instability in their magnetospheres. Some causes were discussed [1 - 3], which can lead to the development of two-stream instability in the magnetosphere of pulsars. However, further detailed analysis showed that in all the cases the two-stream instability either becomes stable (for example, because of a wide energy distribution of plasma particles due to high temperatures [4]), or its increment is so small that oscillations with the density, which would account for the pulsar radio-frequency radiation [5], cannot be generated within the time period when plasma leaves the pulsar magnetosphere.

Non-stationary formation of electron-positron plasma in the vicinity of a pulsar [1, 2, 7] was recently shown [6] to cause development of a two-stream instability in the pulsar plasma. The instability may be accounted for by the following. The non-stationary character of electron-positron pairs birth makes the plasma, discharged from the neutron star, inhomogeneous and localized in separate clouds. The clouds move along the lines of magnetic force. As they move off far from the pulsar, the electron-positron clouds disperse, because the particle Lorentz-factors range from $\gamma_{min} \sim 10$ to $\gamma_{max} \sim 10^3 - 10^4$. The high-energy particles rush faster ahead. At a distance of $\sim 10^8$ cm from the pulsar they overtake low-energy particles of another cloud moving in front. In this sector of the pulsar magnetosphere the distribution of particles in energy correlates with two plasma streams penetrating each other. The estimations [6] show that instability increment in a system of two plasma streams is quite sufficient for its development in a pulsar magnetosphere. In this work we will analyze the spread of plasma clouds during their efflux, their mutual overlapping, and development of two-stream instability (for more details, see [6]).

2. Kinematics of plasma cloud movements

Non-steady state of plasma efflux will be simulated as follows. Assume that particles (electrons and protons) with Lorentz-factors from γ_{min} to γ_{max} and $\gamma_{max} \gg \gamma_{min}$ n » 1 are ejected in a plane (z = 0) along axis z for the time T. Then for the time τ no ejection occurs, after which the cycle repeats again and again. It can thus be stated that ultrarelativistic electron-positron plasma flows out as separate clouds isolated from each other.

The distribution functions of electrons and protons born in the pulsar magnetosphere are essentially identical. Therefore, plasma clouds remain, on average, uncharged during their efflux. The pulsar plasma can be regarded as collisionless. In this case the spatial distribution of flowing out particles (z > 0) is determined from simple kinematic relationships. To study the plasma clouds spreading and their mutual overlapping, it will suffice to analyze the movement of two clouds ejected in succession. Let the instant of the completion of the particle ejection from the second cloud be the initial instant of time (t = 0).

At a moment t_0 determined from the relationship

$$t_0 = \frac{v_{min}(T+\tau) - V_{max} T}{v_{max} - v_{min}} \approx 2\tau\gamma_{min}^2 \gg \tau \qquad (1)$$

the fastest particles from the second cloud overtake slower particles from the first cloud to cause mutual overlapping of the clouds. We will analyze here only the case when the fastest particles of the second cloud had no time to overtake all slowest particles from the first cloud. This can occur in a time interval t:

$$t_0 < t < t_1 = \frac{v_{min}(2T+\tau) - V_{max} T}{v_{max} - v_{min}} \qquad (2)$$

This limitation is justified because, as will be discussed below, an increment of the instability analyzed is great, and the instability starts to develop at the initial stage of cloud overlapping ($t_o < t < t_1$).

At the moment t the clouds overlap in the region described by

$$v_{min}(T+\tau+t) < z < v_{max}(T+t) \qquad (3)$$

and its wide is equal to

$$\Delta = \Delta_1 + \Delta_2 = v_{max}(T+t) - v_{min}(T+\tau+t) \qquad (4)$$

where $\Delta_1 = z - v_{min}(T+\tau+t)$ and $\Delta_2 = v_{max}(T+t) - z$ are distances from the plane with a given z, satisfying the condition (3), to the back and the front boundaries of the cloud overlapping region.

The particle distribution function $F(\gamma,z,t)$ in the overlapping region if of the form:

$$F(\gamma,z,t) = \begin{cases} 0 & \gamma < \gamma_1 \\ F_0(\gamma) & \gamma_1 < \gamma < \gamma_2 \\ 0 & \gamma_2 < \gamma < \gamma_3 \\ F_0(\gamma) & \gamma_3 < \gamma < \gamma_4 \\ 0 & \gamma_4 < \gamma \end{cases} \qquad (5)$$

where $F_o(\gamma)$ is the particle distribution function at the moment of efflux.

$$\gamma_i = \left[1 - \left(\frac{V_i}{c}\right)^2\right]^{-1/2}, \quad i = 1,2,3,4,$$
$$V_1 = v_{min}, \quad V_2 = \frac{z}{T+\tau+t}, \quad V_3 = \frac{z}{T+t}, \quad V_4 = v_{max} (6) \qquad (6)$$

3. Two-stream instability

In each point of the cloud overlapping region there is a group of fast and a group of slow particles, that is, there are practically two plasma streams. It is known that such a situation can lead to the development of two-stream instability. To describe the instability we will use the simplest model of one-dimensional steady-state plasma and investigate a linear stage of the instability development. It was shown in Ref. [8] that this model is valid for a pulsar plasma.

In this approximation the dispersion equation for plasma potential oscillations is:

$$1 - \frac{4\pi e \vec{N}}{m} \int \frac{F\, d\gamma}{\gamma^3 (\omega - \vec{k}\vec{v})^2} = 0 \qquad (7)$$

where ω and $\vec{k}$ are the oscillation frequency and wave vector, respectively.

From (5) it follows that $F(\gamma,z,t)$ function describes two groups of essentially low-energy particles with the velocities V_2 and V_4, at the initial stage of the cloud overlapping when $\Delta \ll \tau c$. Taking this into account, equation (7) can be rewritten as:

$$1 - \omega_p^2 \left[\frac{\alpha}{(\omega - kV_2)^2} + \frac{\beta}{(\omega - kV_4)^2}\right] = 0 \qquad (8)$$

where $\omega_p = \left(\frac{4\pi e^2 N}{m}\right)^{1/2}$ is the plasma frequency,

$$\alpha = \int_{\gamma_1}^{\gamma_2} F_0(\gamma)\gamma^{-3}d\gamma, \quad \beta = \int_{\gamma_3}^{\gamma_4} F_0(\gamma)\gamma^{-3}d\gamma \qquad (9)$$

Assuming $|\delta| \ll |k V_4 - k V_2|$, we will seek for a solution of equation (8) in the form:

$$\omega = k V_4 + \delta \qquad (10)$$

Substituting (10) into (8) gives:

$$1 - \omega_p^2 \left[\frac{\alpha}{k^2(V_4 - V_2)^2} - \frac{2\alpha\delta}{k^3(V_4 - V_3)^3} + \frac{\beta}{\delta^2} \right] = 0 \qquad (11)$$

The variable δ, which is the solution of dispersion equation (11), has a non-zero imaginary part at some k values; the part is responsible for the development of the two-stream instability.

The maximum value of the oscillation growth rate is equal to

$$Im\,\omega_{max} = \omega_p\,\alpha^{1/2}\frac{\sqrt{3}}{2}\left(\frac{\beta}{2\alpha}\right)^{1/3} \qquad (12)$$

The function $F_0(\gamma)$ can be removed from an integral in (9) and the integration can be made, because γ values for the plasma particles fall within a narrow range when the cloud overlapping is small ($\Delta \ll \tau c$). Substituting the obtained α and β values into (12) gives:

$$Im\,\omega_{max} = \frac{\sqrt{3}}{2^{4/3}}\omega_p \left(F_0(\gamma_{min})\right)^{1/6} \left(F_0(\gamma_{max})\right)^{1/3} \left(\frac{\Delta - \Delta_2}{t_0 c}\right)^{1/6} \left(\frac{\Delta_2}{t_0 c}\right)^{1/3} \qquad (13)$$

The rate of oscillation growth (13) depends on the distance Δ_2 to the front boundary of the cloud overlapping region and reaches its maximum value

$$(Im\,\omega_{max})_{max} = \frac{\omega_p}{2}\left(F_0(\gamma_{min})\right)^{1/6}\left(F_0(\gamma_{max})\right)^{1/3}\left(\frac{\Delta}{t_0 c}\right)^{1/2} \qquad (14)$$

at $\Delta_2 = (2/3)\Delta$.

4. Application to pulsars

The distribution function for particles in an electron-positron plasma, flowing out from a pulsar vicinity, can be roughly approximated by the following law [9]:

$$F_0(\gamma) = \begin{cases} A\gamma^{-3/2} & \gamma_{min} < \gamma < \gamma_* \\ A\gamma_*^{-3/2} & \gamma_* < \gamma < \gamma_{max} \end{cases} \qquad (15)$$

where A is the constant determined from the normalizing condition.

Substituting (15) into formula (14) and taking $t_0 c \approx 2\tau c \gamma_{min}^2$, a relationship for a characteristic time τ_i of the instability development can be derived provided high-energy particles of the second cloud penetrate into the region of low-energy particles of the first cloud as deep as Δ:

$$\tau_i \equiv (Im\,\omega_{max})_{max}^{-1} = 2^{3/2} A^{-1/2} \omega_p^{-1} \gamma_{min}^{5/4} \gamma_*^{1/2} (\tau c / \Delta)^{1/2} \qquad (16)$$

ity in any system has enough time to develop strongly, if the life time of the system exceeds the time τ_i of the instability development by more than an order of magnitude. Using a similar criterion $10\,\tau_i = \Delta_i/c$, we can determine Δ_i at which instability develops:

$$\Delta_i \approx \left[30\,c A^{-1/2}\omega_p\,\gamma_{min}^{5/4}\gamma_*^{1/2}(\tau c)^{1/2}\right] \qquad (17)$$

Let us estimate the magnitudes characterizing the development of a two-stream instability in pulsar magnetospheres. If the parameters of the flowing out plasma [1, 6, 9] are $\gamma_{min} = 5, \gamma_* = 10^2, \gamma_{max} = 10^3, \tau c = T_c = R = 10^6$ cm; then the concentration of electron-positron pairs in the vicinity of a neutron star surface is about $\sim 10^{15}$ cm^{-3}. Equation (17) gives a distance from a pulsar to the region of instability development as $\sim t_0 c \approx 2\tau c \gamma_{min}^2 \approx 0.5 \cdot 10^8$ cm and $\Delta_i \approx 0.8 \cdot 10^5$ cm. It follows that a two-stream instability can develop even at a small overlapping of the clouds ($\Delta_i \ll T_c$)

References

1. M. Rudermann and P. Sutherland, Astrophys. J., 1975, v. 196, no. 1, p. 51.
2. A. Cheng and M. Rudermann, Astrophys. J., 1977, v. 212, no. 3, p. 800.
3. A. Cheng and M. Rudermann, Astrophys. J., 1980, v. 235, no. 2, p. 576.
4. . Buschauer and G. Benford, Monthly Notices Roy. Astron. Soc., 1977, v. 179, no. 1, p. 99.
5. G. Benford and R. Buschauer, Monthly Notices Roy. Astron. Soc., 1977, v. 179, no. 2, p. 189.
6. V. Usov, Astrophys. J., 1987, v. 320, no. 2, p. 333.
7. P. Sturrock, Astrophys. J., 1971, v. 164, no. 3, p. 529.
8. V.N. Ursov and V. Usov, Astrophys. Sp. Sci., 1988, v. 140, p. 325.
9. J. Arons, Workshop Plasma Astrophys., Varenna, Italy, 1981, p. 273.

Physics of Neutron Stars
Edited by A. M. Kaminker

ISBN 1-56072-164-2

GENERATION OF WAVES IN PULSAR MAGNETOSPHERES

A.Z. Kazbegi, G.Z. Machebeli and G.I. Melikidze
Abastuman Astrophysical Observatory,
GSSR Academy of Sciences, 380060, Tbilisi

Abstract -- A possibility of wave generation in a magnetosphere plasma has been studied for the cases of: (1) transverse pulse $p_\perp = 0$ and (2) drift motion of particles taken into account. In the first case instability does not arise. In the second case the amplification of waves was found possible.

The key problem in interpretation of pulsar radio-frequency radiation is generation and propagation of waves in the plasma of their magnetospheres. Dense electron-positron plasma is presumed in the majority of theoretical models of pulsars to exist in the region of open-circuit lines of magnetic force of pulsars. The plasma has a density $n_p = 10^{17}$ cm^{-3}, the Lorentz-factor of particles $\gamma_p \approx 10$ and is usually penetrated by a primary relativistic beam of similar charged particles with $n_b \approx 7 \cdot 10^{10}$ cm^{-1} and $\gamma_b \approx 10^7$ [1]. Under these conditions the distribution function becomes one-dimensional, because the components of particle $p_\perp$ pulse, transverse relative to $\vec{B}_p$, are nullified by emitting radiation in strong magnetic fields of pulsars. The magnetic field is assumed to be bipolar and decreases with a distance from a star center according to the law $B_p = B_o (r_o/r)^3$, where $r_o \approx 10^6$ cm is the radius of a neutron star, $B_o \approx 10^{12}$ G. This law also describes density change $n = n_o(r_o/r)^3$. The aim of this work is to study possible plasma instabilities in pulsar magnetospheres.

Let us consider wave generation, ignoring transverse motion of particles at $B_p \to \infty$. Introduce cylindrical coordinates x, r, and ϕ to take into account the curvature of the lines of magnetic force. Here, x is the axis perpendicular to the plane comprising the lines of magnetic force, r and ϕ are a radial and an azimuthal coordinates, respectively. In the approximation of geometrical optics the field of excited disturbances may be as follows:

$$\vec{E} = \vec{E}_0 \exp\left\{-i\omega t + i(k_r r + k_x x + s\varphi)\right\}$$

Here, ω is the frequency, k_x and k_r are the components of wave vector $\vec{k}$-- along the x and the r axes, respectively; S is the harmonics number. Then, the dispersion relationship, taking the magnetic field curvature into account, transforms into two equations in the zero approximation by parameter $\Delta R/R_B \ll 1$ (where ΔR is the excitation size, R_B is the curvature radius):

$$k^2 - \omega^2/c^2 = 0 \qquad (1)$$

$$(k_\varphi^2 c^2 - \omega^2)(k_1^2 c^2 - \omega^2 + A) + k_1^2 k_\varphi^2 c^4 = 0 \qquad (2)$$

where $A = \sum_a \omega_{Pa}^2 \int_{-\infty}^{+\infty} \frac{\omega v_\varphi}{\omega - k_\varphi v_\varphi} \frac{\partial f_{0b}}{\partial p_\varphi} dp_\varphi$, $k_\varphi = \frac{s}{r} = \frac{s}{R_B}$, $k_1^2 = k_x^2 + k_r^2$

P_ϕ is the normalized to mc ϕ-component of the pulse, $v_\varphi = \frac{p_\varphi}{\gamma}$, $\omega_{Pa}^2 = 4\pi e^2 n_a / m_a$, e and m are the electron charge and mass. The summation is taken to particle types (to the plasma electrons and positrons and to the beam electrons as well).

Equation (1) describes purely transverse t-waves, which are not excited in this approximation. Taking the frequency as $\omega = \omega_o + i\Gamma$, an increment of the "potential-nonpotential" (lt) waves from (2) will be

$$\frac{\Gamma}{\omega_0} \approx \frac{\pi}{8} \frac{1}{\gamma_b^2} \left(\frac{\omega_b^2}{\omega_0^2}\right) \left(\gamma^3 \frac{\partial f_{0b}}{\partial p_\varphi}\right)_{P_\varphi = P_\varphi^{res}} \sim \frac{\pi}{8} \frac{1}{P_T^2} \qquad (3)$$

Here, ω_b is the beam Langmuir frequency, P_T is the normalized to mc half-width of the beam,

$$f_{0b} = \frac{1}{\pi^{1/2}} \frac{1}{P_T} \exp\left\{-\frac{(P - P_b)^2}{P_T^2}\right\} \qquad (4)$$

The disturbances grow at $\Gamma > 0$, when $\frac{\partial f_0}{\partial P_\varphi} > 0$, i.e. a beam must exist. Otherwise, only damping would occur, which is in agreement with the results of Refs. [2, 3]. In the hydrodynamics limit, $\Gamma = 0$ in the context of the geometrical optics approximation and the beam excluding [4]. Therefore, the statement of the authors of work [5] about magnetodrift amplification of the waves in the plasma considered does not seem justified enough.

Relationship (3) holds true when a resonance width exceeds the increment, giving

$$\frac{\Gamma}{\omega_0} \ll \frac{1}{P_b^2}\left(\frac{P_T}{P_b}\right) \qquad (5)$$

Condition (5) is incompatible with expression (3) for real parameters of pulsar magnetospheres, and a kinetic instability cannot develop.

Let us consider a hydrodynamic instability in the presence of the beam. Write ω as $\omega = k_\varphi v_b + \Delta$, where

$$v_b = c\left(1 - \frac{1}{2\gamma_b^2}\right), \quad |\Delta| \gg |\omega - k_\varphi v_\varphi|$$

At $\gamma_p^2 \Delta \ll k_\varphi c$

$$\frac{\Gamma_{lt}}{\omega_0} = \frac{\mathrm{Im}\,\Delta}{k_\varphi c} \approx \frac{\omega_b}{\omega_p} \frac{1}{\gamma_b^{3/2}} \frac{1}{\gamma_p^{1/2}} \qquad (6)$$

Increment (6) is too small for relation $|\Delta| \gg |\omega - k_\varphi v_\varphi|$ to be valid; therefore, $\Gamma = 0$. Thus, in the approximation $p_\perp = 0$ and $B_p \to \infty$ no wave amplification takes place whether or not the beam exists.

Let us analyze the wave generation when particles are in drift motion due to a magnetic field nonuniformity. The motion rate is perpendicular to the plane of a distorted line of magnetic force, and is given by

$$v_1 / c = v_\varphi P_\varphi / (\omega_B R_B) \qquad (7)$$

For the plasma particles the rate is seen to be practically zero, and for the beam particles it is $v_1/c \approx 10^{-2}$ at a distance of $r \approx 10^9$ cm. In this case the imaginary part of the dispersion relationship cannot be divided into expressions describing t- and lt-waves. Therefore, in the generation region both types of the waves are excited simultaneously, the waves giving equal contributions to the imaginary part. However, t- and lt-waves should leave the region separately, because in the real part of the dispersion relationship a t-wave would have a frequency

$$\omega_0 \approx kc\left(1 - \frac{1}{2}\delta\right) \qquad (8)$$

where $\delta = (\omega_p / \omega_B)^2 / \gamma_p^3$, $\omega_B = e B_p / mc$ and an lt-wave frequency is

$$\omega_0 \approx k_\varphi c\left(1 - \frac{1}{16} k_1^2 c^2 / \gamma_p \omega_p^2\right) \qquad (9)$$

Formula (9) is true when $\omega_0^2 < 2\omega_p^2 \gamma_p$. From the imaginary part of the dispersion relationship, the instability increment for these waves will be

$$\frac{\Gamma}{\omega_0} \approx \pm \left(\frac{\omega_b}{\omega_0}\right)^2 \left(\frac{k_r}{k}\right) \gamma_b \left(\frac{\partial f_0}{\partial P_\varphi}\right)_{P_\varphi = P_\varphi^{res}} \qquad (10)$$

The condition for the kinetic instability to be excited is

$$\frac{\Gamma}{\omega_0} \leq \left(\frac{v_1}{c}\right)^2 \left(\frac{P_T}{P_b}\right) \qquad (11)$$

The resonance condition for t-waves is then:

$$\left(\delta - \frac{k_r^2}{k^2}\right) - \left(\frac{v_1}{c} - \frac{k_x}{k}\right)^2 = 0 \qquad (12)$$

and for lt-waves:

$$\frac{v_1^2}{c^2} = \left(\frac{k_1}{k}\right)^2 \frac{\omega_0^2}{2\omega_p^2 \gamma_p} \qquad (13)$$

It is easy seen that the drift motion in the nonuniform magnetoactive electron-positron plasma can cause both lt- and t-wave generation under the conditions discussed here. It should be noted that the crucial role of the drift motion in the lt-wave generation was first pointed out by the authors of work [6].

References

1. J. Arons, Proc. of Varenna Summer School and Workshop on Plasma Astrophys., Paris, ESA, 1981, p. 273.
2. R.D. Landford, Mon. Not. R.A.S., 1975, v. 170, p. 551.
3. D.B. Melrose, Ap. J., 1978, v. 225, p. 557.
4. E. Asseo, R. Pellat, and H. Sol, Ap. J., 1983, v. 266, p. 201.
5. V.S. Beskin, A.V. Gurevich, and Ya.N. Istomin, UFN, 1986, v. 150, p. 257.
6. V.V. Zheleznyakov and V.E. Shaposhnikov, Austral. J. Phys., 1979, v. 32, p. 59.

Physics of Neutron Stars
Edited by A. M. Kaminker

ISBN 1-56072-164-2

RADIO- AND OPTICAL RADIATION OF PULSARS CAUSED BY THE MODULATION INSTABILITY OF HIGH-FREQUENCY TRANSVERSE WAVES

A.D. Pataraya and T.A. Pataraya
Abastuman Astrophysical Observatory,
GSSR Academy of Sciences, 380060, Tbilisi

Abstract - Radio- and optical radiation from a pulsar has been analyzed. The radiation is caused by modulation instability of high-frequency transverse waves with circular polarization propagating in an electron-positron plasma along the magnetic field. Under these conditions a charged bunch is formed, whose motion along a curved line of the magnetic field brings about radio-frequency radiation of pulsars. Optical emission is due to the interaction between fluxes of primary electrons (or positrons) and the charged bunches.

Consider a high-frequency transverse electromagnetic wave with a circular polarization propagating in an electron-positron plasma along magnetic field $\vec{B}_0$--aligned with the axis z. The wave with a frequency higher than the Langmuir plasma frequency should exist in the region of open-circuited lines of a pulsar magnetic force owing to the radiation of a primary electron (or positron) beam [1, 2]. A linear dispersion relationship for the waves is as follows [3]:

$$\omega^2 - k^2c^2 = \omega_p^2 \int_{-\infty}^{+\infty} \gamma(\omega - k v_z)^2 \left[\gamma^2(\omega - k v_z)^2 - \Omega^2\right]^{-1} F(p_z)\, dp_z \qquad (1)$$

where ω and $\vec{k}$ -- are the frequency and the wave vector of the linear wave, respectively; $\Omega = e B_0 / m c$ is the cyclotron frequency of electrons (or positrons); $\omega_p = (4\pi n_0 e^2 / m)^{1/2}$ is the Langmuir frequency; R_o is the nondisturbed electron density; m and e are the electron mass and charge; c is the velocity of light; $f = F(p_z)\,\delta(p_\perp)(2\pi p_\perp)^{-1}$ is the nondisturbed function of electron (positron) distribution; P_x, P_y, P_z are the x-, y-, and z-component of the electron pulse, respectively; $p_\perp = (p_x^2 + p_y^2)^{1/2}$, $\delta(p_\perp)$ is the Dirac function; $\gamma = (1 + p_z^2/m^2c^2)^{1/2}$, $v_z = \frac{p_z}{m\gamma}$. For a cold plasma dispersion equation (1) has two solutions:

$$\omega^2 = 0.5\left\{\Omega^2 + \omega_p^2 + k^2c^2 + \left[(\Omega^2 + \omega_p^2 + k^2c^2)^2 - 4k^2c^2\Omega^2\right]^{1/2}\right\} \qquad (2)$$

$$\omega^2 = 0.5\left\{\Omega^2 + \omega_p^2 + k^2c^2 - \left[(\Omega^2 + \omega_p^2 + k^2c^2)^2 - 4k^2c^2\Omega^2\right]^{1/2}\right\} \qquad (3)$$

Formula (2) is the dispersion relationship for transverse waves, whose frequency is higher than the cyclotron electron frequency and phase velocity is higher than the velocity of light. It follows from (3) that the wave frequency is lower than the cyclotron electron frequency, and its phase velocity is less than the velocity of light.

Using the method of the theory of disturbances, developed in Refs. [4, 5], the following generalized nonlinear Schredinger equation can be derived for E_1 together with an expression for the charge density ρ:

$$i\frac{\partial}{\partial t}E_1 + 0.5 v_g' \frac{\partial^2}{\partial \xi^2}E_1 + \alpha|E_1|^2E_1 + \alpha_1 E_1 \int_{-\infty}^{\xi} |E_1|^2 \sin\left[\omega_p v_g^{-1}(\xi - \xi')\right] d\xi' = 0 \qquad (4)$$

$$\rho = A_0 |E_1|^2 \qquad (5)$$

where $E_1 = (E_x - i E_y)\exp\left[i(\omega t - k z)\right]$, E_x and E_y are the x and the y components of an electric field strength; $v_g = d\omega / dk$, $v'_g = d v_g / dk = d^2\omega / dk^2$, $\xi = z - v_g t$, $i = (-1)^{1/2}$ α and α_1 are α, ω, k, ω_p, and F dependent, α_1 is proportional to A_o,

$$A_0 = \frac{e\,\omega_p^2}{2m\,\Omega\,\omega^2} \int_{-\infty}^{+\infty} \frac{\gamma^{-1} F(p_z)\,dp_z}{(1-\alpha^2)(v_g - v_z)^2} \times \qquad (6)$$

$$\times \left\{ 2\alpha^2 \left[\left(k + \frac{\omega}{c^2} v_g \right) v_z - \omega - k\,v_g \right] - \left[\omega - kv_g + (\omega + kv_g)\,a^2 \right] \gamma^{-2} \right\}$$

Here, $a = \gamma\,(\omega - k\,v_z)\,/\,a$, and at $B_o = 0$ $A_o = 0$, that is $\rho = 0$.

Equations (4) and (5) for a cold electron plasma at $B_o = 0$ were derived in work [6]. For au electron-proton plasma with stationary protons and $B_o \neq 0$ the equations are valid as well, but coefficients α, α_1, and A_o take the form:

$$\alpha = A_1 \frac{e^2(\omega + \Omega)^4}{2m^2\omega^2 a_0^4 v_g^2} \left\{ \omega^3 \frac{v_g^T}{c^2} - \left(k\,v_g + \omega - \Omega \right) \left(\omega^2 - \Omega\,\omega + k\,v_g\,\Omega \right) \right\} \qquad (7)$$

$$A_1 = \omega\,\omega_{pe}^2\,(\omega - \Omega)^2 \left[2\,\omega\,(\omega - \Omega)^2 + \Omega\,\omega_{pe}^2 \right]^{1/2}$$

$$a_0 = \omega^2 - \Omega^2$$

$$\alpha_1 = A_1 A_0\,(\omega + \Omega)^2 (k\,v_g + \omega - \Omega)\,c^2 \omega_{pe}^{-1} v_g^{-1} a_0^{-2} \qquad (8)$$

$$A_0 = 0.5\,e^2 \omega_{pe}^2\,(\omega + \Omega)^2 \left[\omega^2 + (k\,v_g - \Omega)\,\Omega \right] a_0^{-2} m^{-2} c^{-2} v_g^{-2} \qquad (9)$$

$\omega_{pe}^2 = 4\pi\,n_0\,e^2/m$. The dispersion relationship for transverse waves propagating along the magnetic field of a cold electron plasma is

$$\omega\,(\omega^2 - \omega\,\Omega - \omega_{pe}^2) = k^2 c^2\,(\omega - \Omega) \qquad (10)$$

A solution of this equation behaves as follows: for $k \to 0$

a) $\omega = \Omega\,k^2c^2/\omega_{pe}^2$, b) $\omega = 0.5\left[\Omega + (\Omega^2 + 4\,\omega_{pe}^2)^{1/2}\right]$,

c) $\omega = 0.5\left[\Omega - (\Omega^2 + 4\,\omega_{pe}^2)^{1/2}\right]$;

for $k \to \infty$ a) $\omega \le \Omega$, b) $\omega \approx k\,c$, c) $\omega \approx k\,c$.

Let us analyze the modulation instability of transverse waves using equation (4). Using the standard method [7], a modulation instability increment in a long-wave range ($\omega_p \Delta \gg v_g$, $\chi \ll k$) is:

$$\nu^2 = (v_g')^2 \chi^2 \Delta^{-2} \qquad (11)$$

where $\Delta = \left[(\alpha + \alpha_1\,v_g\,\omega_p^{-1})\,|E_0^2|\,/\,v_g' \right]^{-1/2}$, χ is the disturbance wave vector, E_o is the constant amplitude.

For $\omega_p \Delta \gg v_g$ equation (4) has a soliton solution

$$|E_1| = E_{max}/ch \left[(z - v_g t)\,/\,\Delta_0 \right] \qquad (12)$$

where $\Delta_0 = \left[(\alpha_1 + \alpha_2\,v_g\,\omega_p^{-1})\,E_{max}^2\,/\,v_g' \right]^{-1/2}$.

Increment ν has a great value at $\omega \approx \Omega$. For a cold plasma

$$\nu^2 = 0.25\,?^2\,|E_0|^2 \chi^2 m^{-2} \omega_p^{-2} D \qquad (13)$$

where $D = (4\,a_1 - 3)\,(4\,a_1^2 - 4\,a_1 - 1)\,(a_1 - 1)^{-1} a_1^{-1}$, $a_1 = (\omega^2 - \Omega^2)\,/\,\omega_p^2$.

Assuming $\Omega \gg \omega_p$, $\omega_p = 10^{12}\sqrt{n_*}\, c^{-1}$, $\chi = 1\,\mathrm{cm}^{-1}$, $D_o = 0.04$, $E_o = 200$ CGS, we can obtain:

$$\nu = 10^{7}\sqrt{n_*}\, c^{-1} \qquad (14)$$

Here, n_* is the density of the particle number in units 10^{14} cm^{-3}.

An increment of modulation instability for an electron plasma is always positive, i.e., $\nu^2 > 0$.

When the modulation instability does not arise ($\nu^2 < 0$), pulsing charged bunches can be obtained, which are arranged along magnetic field. If V_g' is of a small magnitude the term with the coefficient V_g' can be neglected in (4). Then, assuming $E_1 = E(\xi)\exp(i\,\omega_1\,\tau)$, the charge density is given by

$$\rho = A_0 E_{max}^2 \left\{ 1 + \varepsilon \sin\left[\left(1 + \alpha_1 v_g \alpha^{-1} \omega_p^{-1}\right)^{1/2} \left(\omega_p v_z^{-1} z - \omega_p t\right) \right] \right\} \qquad (15)$$

It follows from this formula that the pulse frequency is equal to the Langmuir plasma frequency.

Optical radiation from a pulsar may be accounted for as follows. At $\Delta\omega_p \sim V_g$ a characteristic frequency of amplitude oscillations for nonlinear waves is of the order of the Langmuir plasma frequency. Under these conditions a Langmuir wave is formed, when a primary beam passes through a series of charged bunches. When moving through an inhomogeneous medium, the wave emits in the optical range. The radiation intensity, the Langmuir solution including, is given by

$$dJ = A\,\omega_0^2 \eta^2 \beta^{-2} ch^{-2}\left[0.5\,\eta\,\beta_1^{-1}\left(1 + \eta^2 - \omega_0\omega_p^{-1}\right)\right] do \qquad (16)$$

Here, $\beta = 0.75\,\gamma_0\,\omega_p\sqrt{p_1/r_p}$, $\beta_1 = 1.5\,v_g\gamma_0\,r_p^{-1}\omega_p^{-1}$, $p_1 = 0.5\,v_g'$, $v_g \le c$,

$$v_g' = 6\,T\,m^{-1}\omega_p^{-1}, \gamma_0 = \left(1 - \frac{v_0^2}{c^2}\right)^{-\frac{1}{2}},$$

V_o is the plasma velocity relative to the pulsar, r_p is the pulsar radius, μ is the dimensionless amplitude of the Langmuir solution, A is the constant, ω_o is the radiation frequency. It follows from (16) that the radiation intensity in the spectrum has a maximum at $\omega_0 = \omega_p(1 + \mu^2)$. The weaker the effect of the plasma nonuniformity in terms of the β and β_1 coefficients, the narrower is the maximum.

It should be noted in conclusion that the modulation instability of equation (4) was investigated for $\Delta\omega_p \gg V_g$. In this case charged bunches are formed in an electron-positron plasma in the region of open-circuited lines of magnetic force, which move along magnetic field with a velocity V_g. For a cold plasma the statement holds
when 1) $\Omega^2 + 0.5(\sqrt{2} + 1)\omega_p^2 < \omega^2 < (2 + \sqrt{5})\Omega^2$, 2) $\Omega^2 - 0.5(\sqrt{2} - 1)\omega_p^2 < \omega^2 < \Omega^2$.
The first condition is for the waves with dispersion (2), and the second, with dispersion (3). For the latter the linear Landau damping is small at $\Omega - \omega > k v_T$. The nonlinear Landau damping appears at a group velocity of the waves if plasma temperature is taken into account, the damping being small at $F \neq 0$, $|v_z| > v_g > 0$. Formula (15) is true in the frequency range, where the modulation instability does not arise. This holds for the waves with dispersion (2) at

$$\Omega^2 + \omega_p^2 < \omega^2 < \Omega^2 + 0.5(\sqrt{2} + 1)\omega_p^2$$

The approach presented in this paper is valid when $\lambda > n^{-1/3}$, where λ is the wave length of transverse waves. The condition holds well for (15) if $\omega \approx (\Omega^2 + \omega_p^2)^{1/2}$ and is true for $\nu^2 < 0$. We showed above that $\nu^2 < 0$ and ν is great at $\omega \approx \Omega$, which leads to the relation $c/\Omega > n^{-1/3}$. If the cyclotron frequency and the density of the particle number near the pulsar surface are denoted as Ω_* and n_*, respectively, the above expression takes the form:

$$\frac{c}{\Omega_*} > \left(\frac{2}{2p}\right)^2 n_*^{-1/3}$$

The inequality holds for millisecond pulsars with $B_* = 10^9$ G and $n_* = 10^{18}$ g/cm^{-3}. Other pulsars obey the relation far from their surfaces, at $r \gg r_p$.

References

1. R. Manchester and J. Teilor, Pulsars, Russian translation, Mir, Moscow, 1980, p. 205.
2. P.G. Smith, Pulsars, Russian translation, Mir, Moscow, 1979, p. 196.
3. J.G. Lominadze, G.Z. Machabeli, G.I. Melikidze, and A.D. Pataraya, Fizika Plazmy, 1986, v. 12, n. 10, p. 1233.
4. J.G. Lominadze and A.D. Pataraya, Physica Scripta, 1982, v. 2/1, p. 215.
5. G.I. Melikidze and A.D. Pataraya, Astrofizika, 1984, v. 20, n. 1, p. 157.
6. L.M. Gorbunov and V.I. Kirsanov, ZhETF, 1987, v. 93, n. 2(8), p. 509.
7. B.B. Kadomtsev, Collective Phenomena in Plasma, Nauka, Moscow, 1976, p. 130.

Physics of Neutron Stars
Edited by A. M. Kaminker

ISBN 1-56072-164-2

COMPARISON OF SOME PULSAR MODELS WITH OBSERVATIONAL DATA

I.F. Malov
Lebedev Physical Institute, USSR Academy of Sciences, 117924, Moscow

Abstract - It has been shown that an angle increase between the rotation axis and magnetic field predicted by the eddy current loss model was not observed. It is suggested that the Melrose model should be used to account for the central component in complex profiles.

A radiopulsar activity is believed to be mainly due to the rotation energy loss by a neutron star with magnetic field of about 10^{12} G. This fundamental statement is the basis of all modern models of pulsars (PSR). All other arguments concerning the magnetosphere structure, the nature of emission mechanisms and localization of emitting regions, as well as PSR evolution cannot be considered well-grounded. Therefore, it is important to determine statistical relationships on the basis of data observed and to compare them with conclusions and predictions of theoretical models.

1. The magnetic field structure

One of the fundamental problems in determining the PSR magnetosphere structure is that of the nature of the magnetic field in the radio-frequency radiating region.

Calculations in the terms of the hollow cone model, evaluations of an observed profile shape, of the frequency dependence of the position angle for linear polarization, and of the superdispersion delay effect allow some conclusions. The assumption that the magnetic field of some pulsars is of a dipole nature does not contradict the whole set of data observed. Some pulsars may be considered to have the dipole magnetic field only in an emission frequency range from 100 to 1000 MHz. At frequencies lower than 100 MHz multipole components contribute essentially, and in the region of frequencies higher than 1000 MHz the field is distorted due to PSR rotation. And, lastly, in many of PSR the magnetic field is likely to be nondipole throughout the magnetosphere.

2. Theory of eddy current loss in comparison with observational data

Experimentally determined values of the position angle of linear polarization enable verification of the theory of Beskin et al. [1, 2]. According to this theory the maine delay mechanism of a neutron star rotation is a ponderomotive interaction between the surface currents and the magnetic field. Using relationship (3) from Ref. [2],

$$J_r \Omega \dot{\Omega} = -\frac{f^*(\beta)}{8} \frac{B^2 \Omega^4 R^6}{c^3} i \cos\beta \qquad (1)$$

where J_r is the inertia moment of a neutron star, $f^*(\beta)$ is the weak function of angle β between the rotation and the magnetic axes (for $\beta = 0 - 90^\circ$ $f^* = 1.55 - 1.94$), i is the dimensionless current flowing off the star surface, we can obtain in terms of this model:

$$\dot{P} \propto \frac{i \cos\beta}{P} \qquad (2)$$

that is maximal rotation delay should take place near the rotators having the common axis. Hence, pulsars with small β may be expected to have great values of $\dot{P}$. These pulsars can be detected with help of the data on $\left(d\Psi/d\varphi\right)_{max}$ derivatives because, according to Ref. [3], the maximum rate of polarization plane revolution is

$$\left(d\Psi/d\varphi\right)_{max} = \frac{\sin\beta}{\sin(\xi-\beta)} \qquad (3)$$

where ξ is the angle between the rotation axis and the direction toward the observer. On the average, the smaller β, the smaller is the derivative $(d\Psi/d\varphi)_{max}$ according to (3). An apex angle of the cone of open-circuited lines of magnetic force is

$$\Theta \approx \sqrt{\frac{r}{r_{LC}}} \qquad (4)$$

where r_{LC} is the radius of the light cylinder. For the sake of estimation let us take $r / r_{LC} \leq 0.1$, i.e., $\Theta \leq 18^{\circ}$. The mean distance of the sight beam from the cone center is

$$(\xi - \beta) \sim \frac{\Theta}{2} \leq 9^{\circ}\text{ , that is, } \sin(\xi - \beta) \sim 0.16 \text{ and}$$

$$(d\Psi/d\varphi)_{max} \sim 6 \sin\beta \qquad (5)$$

We will consider angles to be small if $\beta < 30^{\circ}$, which leads to $d\Psi/d\varphi \leq 3$. When the distances between the sight beam and the rotation axis are very small, that is, at small angles $(\xi - \beta)$, $d\Psi/d\varphi$ values may increase even for small β. In this case the profiles would be two-component with a pronounced decrease of intensities in the center. PSR with these profiles are also included in the analysis. The Table lists data on PSR with small $(d\Psi/d\varphi)_{max}$. It can be seen that averaged over 13 PSR, the $\bar{\dot{P}}$ value is equal to $2.11 \cdot 10^{-15}$, which is well below the mean value averaged over all the PSR: $\bar{\dot{P}}_{-15} = 9.36$. Using the data on $(d\Psi/d\varphi)_{max}$ for all PSR, we will see (Fig. 1) that $\bar{\dot{P}}$ increases with $d\Psi/d\varphi$ (with increasing β). This disagrees with the predictions of the theory of eddy current loss. Moreover, in the theory the parameter

$$Q = 2P^{1.1}\dot{P}^{-0.4} \qquad (6)$$

characterizes "young" (Q < 1) and "old" (Q > 1) PSR, i.e., PSR with small β (small $(d\Psi/d\varphi)_{max}$ would have Q < 1. However, the data in the Table point to the opposite: only 4 from 13 PSR have Q < 1, and the other 9 have Q > 1. The averaged value for the 12 PSR is $\bar{Q} = 1.95$.

The analysis shows that PSR with small $\dot{P}$, which may be regarded as an indication of the PSR old age, have small $(d\Psi/d\varphi)_{max}$ (in other words small angle β) as well. Thus, the observational data confirm that angle β decreases with increasing PSR age, as is predicted in the traditional models based on magneto-dipole loss.

3. Classification of profiles

A phenomenological scheme of PSR classification according to the shape and characteristics of observed profiles was proposed in the works by Ranckin. They state that radiation features of the majority of PSR may be explained in terms of the hollow cone model. These works, however, do not discuss the mechanism of the central component generation. The central radiation component and especially S_t-profiles may be described in terms of the model of relativistic diagram formation (the Smith model). Typical of these profiles are circular and linear polarizations (which are characteristic of cyclotron radiation), small and random change of the position angle, an abrupt spectrum, and the lack of zeros and mode switchings. All these features are observed in short-period PSR. However, some of the central components in the complicated profiles differ from the single ones (S_t): low-frequency intensity fluctuations are more often observed and the components themselves of PSR radiation are detected in a wider range of P and $\dot{P}$, that is, both in young and old PSR. This radiation is more difficult to explain in terms of the relativistic model, and may be better described by the Melrose mechanism [4]. Indeed, the radiation may be due to a periodic (with the ω_0 frequency) acceleration of charges with medium Lorentz-factors ($\gamma \leq 100$) along the magnetic field and could play a certain role at frequencies

$$\frac{c\gamma^3}{\rho} \ll \omega \leq 2\omega_0\gamma^2 \qquad (7)$$

If the left-hand inequality does not hold, bending radiation would dominate. The right-hand inequality determines the condition the absorption coefficient to be negative or the radiation to be amplified because of a longitudinal acceleration of charges. The radiation spectrum, as it follows from (7), must be abrupt. At $\gamma = 10$, which is optimal for the Melrose mechanism to operate, and $\omega_0 \sim 10^{8}$ s^{-1} a limiting radiation frequency is about 3 GHz. This value is characteristic of central components of the complicated profiles. In the cone center, where ρ is great, the intensity of bending radiation is considerably weaker than at the edges, and the role of the longitudinal acceleration mechanism may be important here. For $\gamma = 10 - 100$ an angular size of the mechanism operation ($\sim \gamma^{-1}$) is from 6° to $0^{\circ}.6$ near the cone center. Moreover, the mechanism is associated with an alternating acceleration of the charges and maser amplification of radiation; so that perturbations of any parameter (n_e, γ, ω_0) will bring about distinct intensity fluctuations. All the features discussed are actually characteristic of the central radiation part in complicated profiles, which advocates the use of the Melrose mechanism for their explanation. Verification of the idea requires further accumulation of observational data concerning the central radiation as well as a more detailed analysis of the profiles with a complicated structure.

Tables

N	PSR	P (c)	$\dot{P} \cdot 10^5$	$\|d\Psi/d\varphi\|_{max}$(100-2700), MHz	Q
1	0031-07	0.94	0.41	0.6-1.4	2.67
2	0149-16	0.83	1.30	0	1.47
3	0809+74	1.29	0.17	2.6	5.38
4	0919+06	0.43	13.72	2.8	0.28
5	0943+10	1.10	3.53	2.5-3.0	1.34
6	0950+08	0.25	0.23	1.2-2.9	0.78
7	1541+09	0.75	0.43	0	2.04
8	1907+10	0.28	2.64	2.8	0.33
9	1929+10	0.23	1.16	1.0-1.9	0.37
10	1944+17	0.44	0.02	0.7	3.88
11	1112+50	1.66	2.49	4.1-8.1	2.42
12	1919+21	1.34	1.35	3.3-10.6	2.45
13	1952+29	0.43	0	9.3-10.2	∞

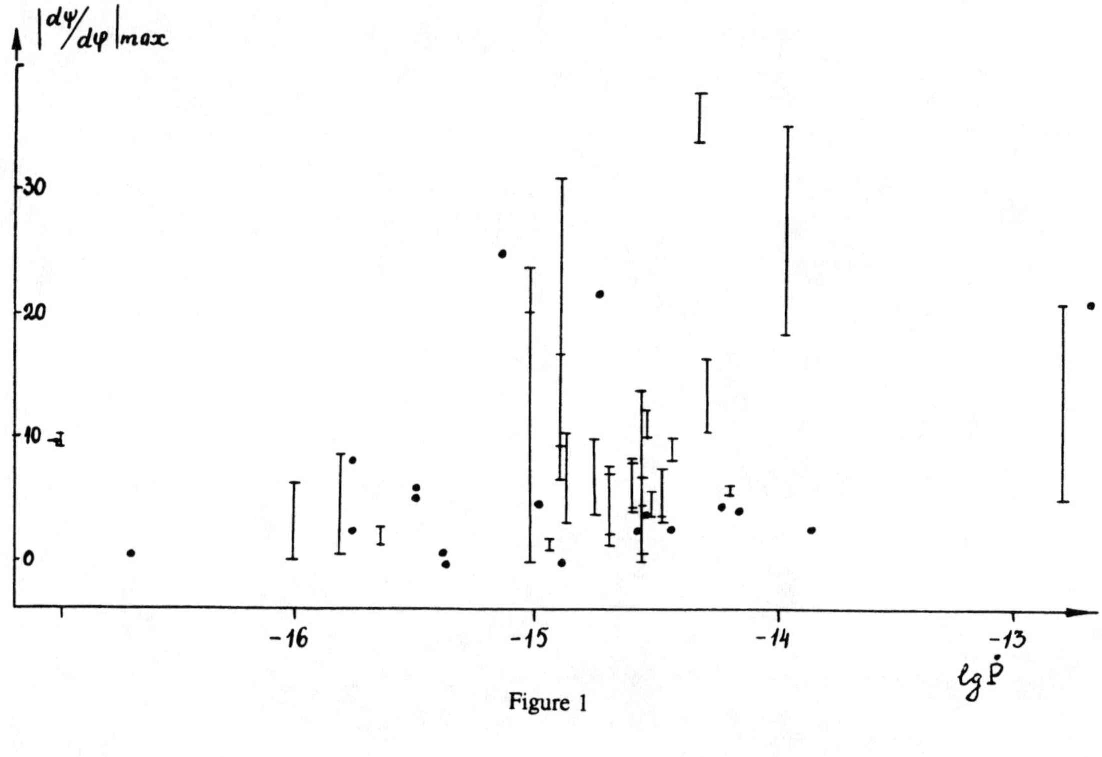

Figure 1

References

1. V.S. Beskin, A.V. Gurevich, and Ya.N. Istomin, Zh. Experim. i Teor. Fiziki, 1983, v. 85, n. 2, p. 401.
2. V.S. Beskin, A.V. Gurevich, and Ya.N. Istomin, Astrophys. Sp. Sci., 1984, v. 102, n. 2, p. 301.
3. R. Manchester and J. Teilor, Pulsars, 1980, Russian translation, Mir, Moscow, 1980, p. 244.
4. D.B. Melrose, Astrophys. J., 1978, v. 225, n. 2, p. 557.

Physics of Neutron Stars
Edited by A. M. Kaminker

ISBN 1-56072-164-2

A POSSIBLE EXPLANATION OF THE CORRELATION BETWEEN TRANSVERSE VELOCITIES AND MAGNETIC FIELDS OF PULSARS

B.M. Vladimirskii
The Crimean Astrophysical Observatory, USSR Academy of Sciences,
334413, The Crimea, P.O. Nauchnyi

Abstract - A correlation between transverse velocities and magnetic fields of pulsars is found only for pulsars of the second generation. They appear after the first explosion of a supernova in binaries and renew their radio-frequency radiation after the completion of an accretion stage (they are type 0 pulsars according to the earlier given classification). The correlation arises as a consequence of the anticorrelation between an orbital period and a pulsar rotation period and of the pulsar equilibrium period dependence on magnetic field.

Introduction

The dependence found between spatial speeds of pulsars V and the strength of the magnetic field on their surface B [1] has not been yet unambiguously interpreted. Evidently, it cannot be completely due to the effects of selection [2]. The proposal that the dependence results from acceleration of pulsars by magnetic field faces some problems (see, for example, Ref. [3]), though arguments are likely to exist [4] supporting hypotheses of this sort. On the other hand, analysis of pulsar distribution in their spatial speeds [3, 5] seem to indicate that they should form mainly in binaries, which would make these hypotheses unnecessary. In this work data and considerations are presented, which point how the correlation can arise if the majority of pulsars do form in binaries.

2. Data and major proposals

The initial data for the analysis were measurements of transverse velocities of pulsars, obtained by investigation of interstellar scintillations. According to the approaches developed in work [7], the pulsar population observed can be divided in two types: First, pulsars passing their accretion stage in tight binaries and restoring their radio-frequency radiation (type 0). Second, neutron stars formed as a result of evolution of single stars or in binaries after the 2nd explosion of Supernova (type Π). The classification follows the idea about two pulsar generations, proposed in Refs. [8, 9]. Pulsars are assigned to the first or the second type on the basis of a parameter $\eta = L_R/E$, which characterizes radioemitting properties of their magnetospheres. Here, L_R is the radio-frequency luminosity, E is the rotation energy of a neutron star.

According to the "standard" scenario, after the 1st explosion of Supernova a binary system remains bound. The pulsar formed spins down and then again accelerates with accretion. Its period after accretion is close to the equilibrium and proportional to magnetic field: $P \propto B^{6/7}$. After this "initial" period, it slows down again. In about $5 \cdot 10^6$ years, the 2nd component explodes in the system as aSupernova. The system disintegrates to form a couple of pulsars: an accreting pulsar that restors its radio- frequency radiation (the second generation, type 0), and a pulsar of the first generation (type Π). The distribution of type 0 pulsars, according to the spatial velocities, must reflect the distribution of binaries in their orbital periods before the 2nd explosion of Supernova. The distribution of type Π pulsars is more complicated: in part, it should reflect the distribution according to the values of a pulse received by a neutron star in Supernova explosion (for single stars). One of the variants of this scenario includes a case when the system orbital period decreases during the evolution after the formation of the 1st pulsar. Under these conditions an anticorrelation may appear between the system orbital period and a pulsar rotation period [10], so that the older systems would have a smaller orbital period and a greater period of the pulsar rotation. For such objects (tupe 0) an anticorrelation between the orbital period and the pulsar magnetic field or a correlation V(B) may be expected because even the initial rotation period for pulsars of the 2nd generation is proportional to magnetic field. For type Π pulsars (the first generation) similar correlation should not arise in terms of the above proposals. Objects with maximal velocities should belong to type 0 pulsars.

3. Basic results of the analysis

The above proposals were checked for the catalog of transverse velocities of pulsars [6] using the list of classification objects [7]. The dependence V(B) was found only for type 0 pulsars. For these objects with known velocities [6] the correlation between period P and magnetic field B is 0.65 ± 0.06, so that the dependence V(P) is also valid for them (Fig. 1). The dependence is not observed for the 1st generation pulsars of type ΠI (r = 0.07 ± 0.13). Examination of the dependences of various parameters as a function of velocities of type 0 pulsars allows some conclusions:

1. The line approximating the dependence P(V) in Fig. 1, which is constructed by the least-squares method through points and circles corresponding to the cases of uncertain classification, crosses the horizontal axis for V = 10 km/s at the period value P_o = 0.56 s. The value corresponds to the mean equilibrium period for the given set of objects. It seems to be overestimated because the selection effects lead to the diagram depletion in Fig. 1 of objects in the left- hand bottom angle.

2. The value of $(P_f - P_o)/\langle\dot{p}\rangle$, where P_f is the "maximum" value of the mean period and $\langle\dot{p}\rangle$ is the weighed mean of a period derivative for a given pulsar group, should be of an order of the time elapsed between Supernova explosions. In our case it equals $\approx 2 \cdot 10^7$.

3. The value of the mean distance of type 0 pulsars from the galactic plane z shows distinctly tends to increase with the velocity. If the tendency does exist, it is in agreement with the idea that pulsars in the upper part of the diagram in Fig. 1 are, on the average, older than those in its bottom.

4. The angles β between the rotation axis and the axis of the magnetic dipole for pulsars of type Π tend to increase with velocity (the values of β were taken from [11]). The nature of this dependence is still unclear.

Figure 2 shows the velocity distribution for both types of pulsars. If the selection effect is taken into account, the distribution is very similar to the velocity spectrum of pulsars formed in massive tight binaries [5]. An unexpected feature is equidistant maximums (showed by arrows in Fig. 2). If this effect does exist (its significance is minor, about 10%), it might be taken to point to the existence of unusually stable orbits of binaries. Clusters in period distributions are known for spectral binaries [12] and polars [13]. The mean values of pulsar periods for the three maxima are 0.60, 1.04, and 1.44 s, respectively. It can be seen that they differ by a "discrete" value 0.4 s. It is possible that this effect would permit interpretation of modulation in distributions of pulsars according to their periods [14].

Conclusion

The main conclusion of the analysis made is that the correlation dependence V(B) is likely to exist only for that part of pulsar population that belongs to the 2nd generation. The dependence follows from the anticorrelation between an orbital and a pulsar periods for a certain class of binaries.

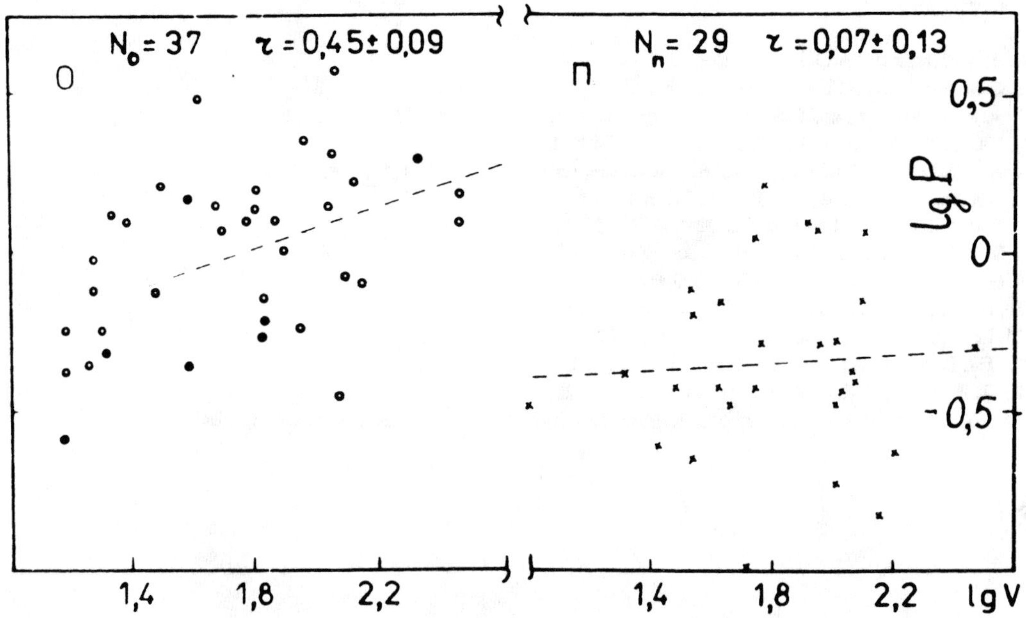

Figure 1

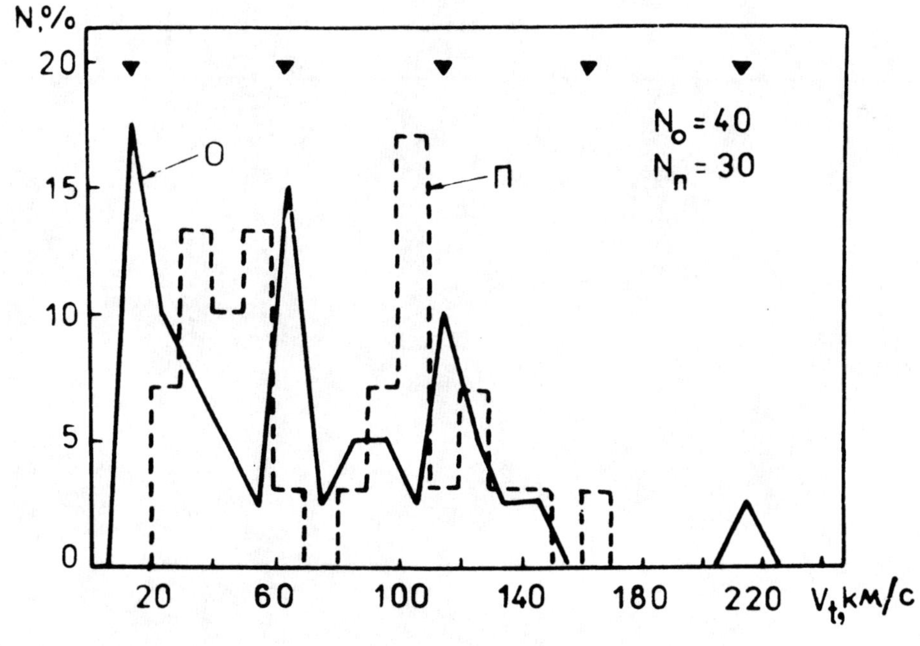

Figure 2

References

1. B. Anderson and A.G. Lyne, Nature, 1983, v. 303, p. 597.
2. G.M. Stollman and E.P.J. van der Heuvel, Astrophys. J., 1986, v. 162, n. 1/2, p. 87.
3. V. Radhakrishnan and C.S. Shukre, Astrophys. Space Sci., 1986, v. 118, n. 1/2, p. 329.
4. Yu.P. Pskovskii and O.F. Dorofeev, Astron. Tsirk., 1987, n. 1504.
5. A.V. Tutkov, N.N. Chugai, and L.R. Yungel'son, Pis'ma AZh., 1984, v. 10, n. 8, p. 586.
6. J.M. Cordes, Astrophys. J., 1986, v. 311, n. 1, part 2, p. 183.
7. B.M. Vladimirskii, Izv. Astrofiz. Obs., 1988, v. 78.
8. V.F. Shvartsman, Izv. VUZ, adiofizika, 1970, v. 13, n. 12, p. 1852.
9. G.S. Bisnovatyi-Kogan and B.V. Komberg, Astron. Zh., 1974, v. 51, p. 373.
10. R.H.D. Corbet, in: The Evolution of X-ray Binaries, Reidel, 1986.
11. I.F. Malov, Astrofizika, 1986, v. 24, n. 3, p. 507.
12. E.I. Popova et al., Ap. Space Sci., 1982, v. 88, n. 1, p. 55.
13. B. Warner and M. Livio, Astrophys. J., 1987, v. 322, n. 2, p. L95.
14. V.A. Kotov and B.M. Vladimirskii, Izvestiya Krymskoi Astronomicheskoi Observatorii, 1987, v. 76, p. 93.

Physics of Neutron Stars
Edited by A. M. Kaminker

ISBN 1-56072-164-2

RESOLUTION OF RADIATING REGION OF THE MAGNETOSPHERE OF PSR 1919+21

O.A. Kuz'min
Institute of Space Research, USSR Academy of Sciences, Moscow

Abstract - Analysis of scintillation spectra of radio- frequency radiation of pulsar PSR 1919+21 at a frequency of 102 MHz revealed their quasiperiodic structure. The structure is likely to be an interference picture due to the interference of several beams in a refraction multibeam propagation of pulsar radio-frequency radiation in interstellar medium with large-scale nonuniformities. A displacement of emitting regions, corresponding to various components of the pulse was measured from a periodicity phase change of the scintillation spectra along the pulse. The displacement was found to be $\sim 6.2 \cdot 10^8$ cm, assuming that a refraction shield is placed half way between the pulsar and the Earth.

Introduction

Early investigations of scintillations of pulsar radio- frequency radiations revealed regular structures both in frequency and time in scintillation spectra. Presently, 12 pulsars are known, whose scintillation spectra comprise a rather regular structure. Shortly after their detection, Ewing [1] suggested, on the basis of relatively simple shapes of quasiperiodic scintillation spectra, that the regular spectra should be interference pictures caused by interference of several ($2 \div 3$) beams - incident and refracted. Under these conditions refracted beams can explained by the refraction of pulsar radio-frequency radiation by large-scale (characteristic size is $\sim 10^{13}$ cm) nonuniformities of electron concentration of the interstellar medium. Thus, a focusing "shield" is presumed to exist, which consists of large-scale nonuniformities. Each pair of beams (incident and refracted) reaching the observer will form a characteristic regular structure (in time and frequency) of a dynamic spectrum (a spectrum in frequency - time coordinates). Distances between the interference petals on the time and the frequency coordinates of the interferometer will be $P_t \sim \lambda / (\Delta \Theta \cdot v)$ and $P_\nu \sim c / \Delta d$, respectively. Here, $\Delta \Theta = |\Theta_1 - \Theta_2|$ is the difference in the angles of an incident and a refracted beams (Fig. 1), v is the rate of the interference picture motion relative to the observer, and Δd is the difference between the way lengths of an incident and a refracted beams. It was proposed in work [2] and experimentally confirmed in Ref. [3] that the "interstellar interferometer" may exist with a base of several 10@11 m and an angular resolution $< 10^{-6''}$ for meter wavelengths. The resolution is high enough to resolve pulsar magnetospheres, because the angular size of a light cylinder of the pulsar relatively close to the Earth (D = 330 pc and period $P \approx 1$) is $\sim 10^{-6''}$.

Displacement of the radiation source from position 1 to position 2 (Fig. 1) for a distance of δX_s will change the difference in the way lengths Δd and will be detected by the interstellar interferometer as a phase change of interference petals [3]:

$$\delta \varphi \sim \nu \cdot \delta x_s \cdot \left[\frac{(D/D_s) - 1}{D \cdot P_\nu} \right]^{0.5} \qquad (1)$$

where ν is the frequency of radio-frequency radiation, D is the distance to the pulsar, D_s is the distance from the pulsar to the shield.

As was first shown in work [4] the pulsar rotation through an angle Ψ, would cause an instantaneous radiating region to displace in the direction perpendicular to the observer by a distance of $\delta x_s \approx R_{em} \cdot \Psi \cdot \sin\beta/3$ as suggested by the "hollow-cone" model. Here R_{em} is the height of a radiation level, β is the angle between the rotation and the magnetic axes of the pulsar. The displacement is due to that the direction of the pulsar magnetic field does not coincide with the pulsar radius (the magnetic field is considered to be dipole). Figure 2 from work [4] illustrates the displacement effect of an instantaneous radiating region.

Thus, various parts of the pulsar pulse should come to the observer from spatially separated regions of the magnetosphere and, therefore, must have different phases of interference petals of the interstellar interferometer. In work [3] a periodicity phase change in scintillation spectra recorded in various components of the pulse was detected for the first time for pulsar PSR 1237+25.

The aim of this work was to detect the dependence of a periodicity phase of a scintillation spectrum on pulse components, i.e., the resolution of the pulsar magnetosphere, for PSR 1919+21 at frequencies of about 100 MHz.

Observations and processing

Records of four observational sessions for PSR 1919+21 on February 1980 using the antenna array of BSA FIAN in Pushchino are prcessed. A high-frequency signal in a band of 100 kHz at a base frequency of about 102 MHz in the "window" of the pulsar radiation was numbered every 5 Ms and tape recorded as 80-Ms blocks. A time resolution of 10Ms was reached by preliminary processing according to the technique of predetector compensation of dispersion smearing [7].

The aim of the first stage processing was to determine scintillation parameters of radio-frequency radiation from pulsar PSR 1919+21. For each day of observations dynamic spectra of scintillations construeted (the spectra were computed along all the pulse length); a regular structure of the scintillation spectrum was detected in all the four observational sessions.

The values for frequency periods P_ν in different days varied from ~ 10 to ~ 50 kHz. Periods P_ν and P_t were determined quantitativelyfrom autocorrelation functions of spectrum and intensity, respectively, averaged over a session. Figure 3a shows the scintillation dynamic spectrum of the session on February 8, 1980 together with the averaged scintillation spectrum in 75-kHz band (Fig. 3b) and the mean pulse intensities (Fig. 3c). Regular spectrum bands with a period $P_\nu \approx 16.7$ KHz are easily seen in the dynamic spectrum (Fig. 3a). The values of time and frequency periods P_t and P_ν of the scintillation spectra are listed in the Table for each session of observations.

At the second stage of procesing the dependence was determined of a frequency periodicity phase of the scintillation spectra on the position in the pulse. For this purpose the mean pulse profile (Fig. 4a) was divided into 10 longitudinal intervals 5.12 Ms long, and in each of these intervals power spectra were calculated. To determine the longitude dependence of the frequency periodicity phase, cross-correlation functions (CCF) were computed between the spectra of all longitudinal intervals and the spectrum of a reference interval. To compute CCF only spectra, whose energy was 3 times the mean energy of the spectrum, calculated outside the pulsar radiation pulse were used. The CCF, computed in this manner for various longitudes, were then averaged over all pulses of a given observational session. The spectrum periodicity phase was determined from maximum positions of mean CCF in two ways. The first was to fit a parabola by the least-square method, whose maximum was analytically found, and the second was to determine the center of gravity of the central maximum of CCF. Both methods gave similar results. In this way phase curves were constructed for all the four observational sessions. A meaningful change of the periodicity phase of scintillation spectra along the pulsar pulse was observed in sessions on February 8, 1980 (Fig. 4a) and February 24, 1980 (Fig. 4b); amplitudes of the phase change were o.4 ± 0.15 and 0.6 ± 0.25 rad, respectively. Within the limits of phase determination error the relationship between the phase change and the pulse longitude, obtained in two observational days (on February 8 and 24, 1980), were of similar character. An additional confirmation of the result obtained is that total amplitudes of the phase change obey the relation $\delta\varphi \propto P_\nu^{-0.5}$, predicted theoretically [3]. The lack of a meaningful phase change along the pulse in sessions on February 22 and 25, 1980 with periods P_ν 26 and 50 kHz, respectively, is likely to be a consequence of the relation $\delta\varphi \propto P_\nu^{-0.5}$.

Discussion

Let us first summarize the major experimental results obtained in this work.

1. Scintillation spectra of radio-frequency radiation of pulsar PSR 1919+21 at a frequency of 102 MHz have a quasiperiodic character with periods in frequency from 9.8 to 48.7 kHz in all the four observational sessions.

2. In two observational sessions a phase periodicity change of scintillation spectra along the pulsar pulse was observed; the phase change amplitudes are 0.4 ± 0.15 and 0.6 ± 0.25 rad.

Phase change amplitudes obtained enable determination of the transverse displacement δX_s of an instantaneous radiating region along the pulsar period from formula (1). Estimates of δX_s will depend on the ratio D/D_s. Values of the transverse displacement δX_s for sessions on February 08 and 24, 1980 are $(5.2 \pm 2) \cdot 10^8$ and $(6.4 \pm 2.7) \cdot 10^8$ cm, respectively. The values were obtained on the assumption that $D/D_s = 2$, that is the refraction shield is half way between the pulsar and the Earth. The values obtained are less than one tenth of the radius of the light cylinder R_{LC}. For pulsar PSR 1919+21 with a period P = 1.3373 s the radius is about $6.7 \cdot 10^9$ cm. As was stated above, for a dipole magnetic field the dependence of the periodicity phase of scintillation spectra on the pulsar pulse longitude at small values of Ψ would be linear: $\delta\varphi \sim \delta x_s = R_{em} \cdot \Psi \cdot \sin\beta/3$. Phase curves obtained on February 8 and 24, 1980 do not contradict the linear dependence (within the phase determination errors). Hence, knowing δX_s and a pulse width $\psi \sim 8^\circ$ the height of a radiation level above the neutron star surface can be determined. The values of the radiation levels R@em for sessions on February 8 and 24, 1980 are $(1.1 \pm 0.4) \cdot 10^{10}$ and $(1.3 \pm 0.54) \cdot 10^{10}$ cm, respectively. Estimates were also made on the assumption that $D/D_s = 2$ and the angle between the rotational and the magnetic axes is 90°. As can be seen the values of the radiation levels obtained are greater than the value of the light cylinder radius R_{LC}. The conclusion can hardly be accepted, because the results of the majority of observations on pulsar radio-frequency radiation (for example, observations of frequency dependence of a mean profile width) indicate that radio-frequency radiation is generated inside the pulsar magnetosphere, that is, $R_{em} < R_{LC}$. Two explanations can be proposed to resolve the contradiction. First, if the magnetic field structure is not dipole (for example, quadrupole), the dependence of the transverse displacement δX_s on the radiation level R_{em} would be nonlinear, which would probably decrease the estimate for R_{em}. Second, as was already mentioned, estimates of the values of thetransverse displacement δX_s and radiation levels R_{em} depend on the ratio D/D_s, characterizing the shield position between the pulsar and observer. Bringing the shield closer to the pulsar, that is, increasing D/D_s, the values δX_s and R_{em} can be lowered. Independent estimates of the radiation level R_{em} indicate that $R_{em} = 0.1\ R_{LC}$. This means that the shield must be at a distance of $\sim$ 1 pc from the pulsar. Then, the ratio D/D_s for sessions on February 8 and 24, 1980 will be $4 \cdot 10^{-3}$ and $2.5 \cdot 10^{-3}$, respectively. The refraction shield, so close to the pulsar, may be due to the inherent activity of the pulsar.

I wish to express my gratitude to M.V. Popov and V.A. Soglasnov for a gift of observation record and to N.S. Kardashev for the helpful discussions.

Table

Date of session	P_ν kHz	P_t, min
08.02.80	16.7	1.4
22.02.80	48.7	1.7
24.02.80	9.8	4
25.02.80	26.5	3

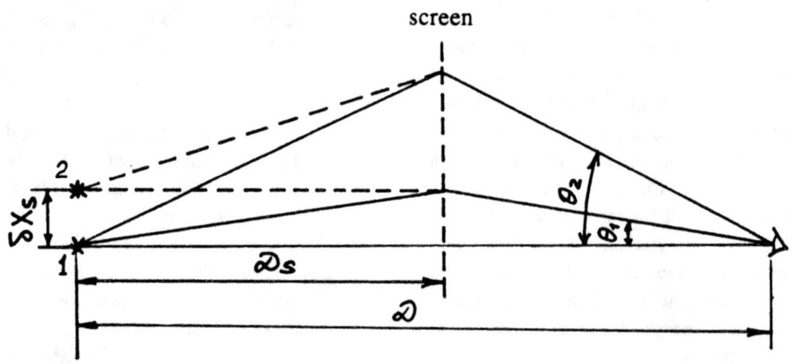

Figure 1

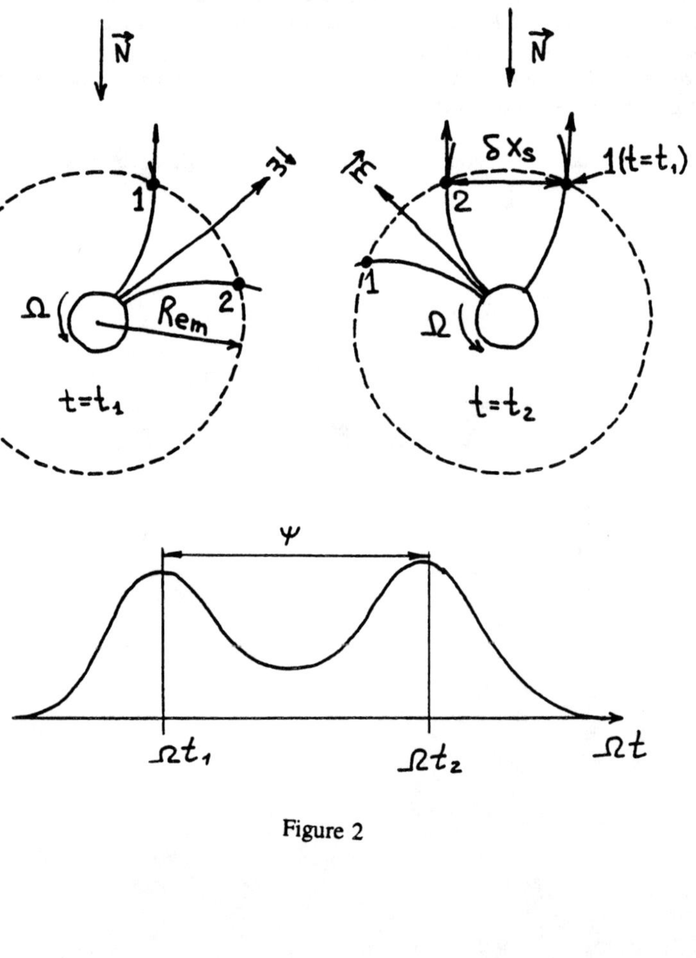

Figure 2

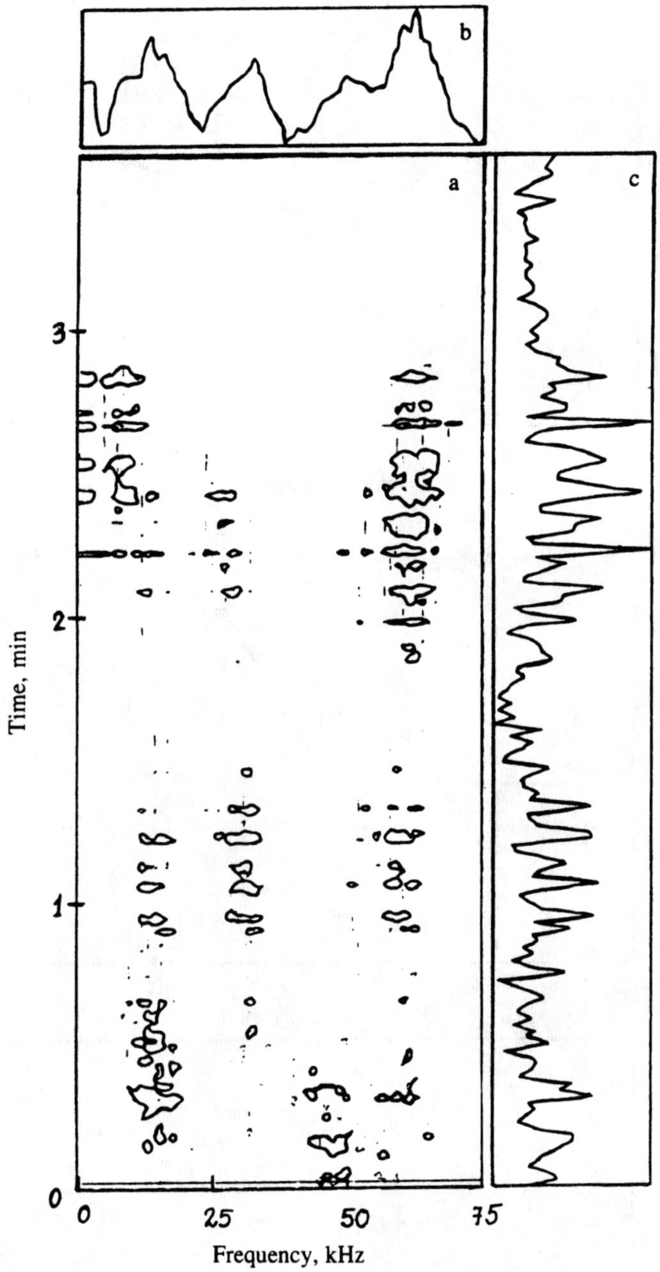

Figure 3

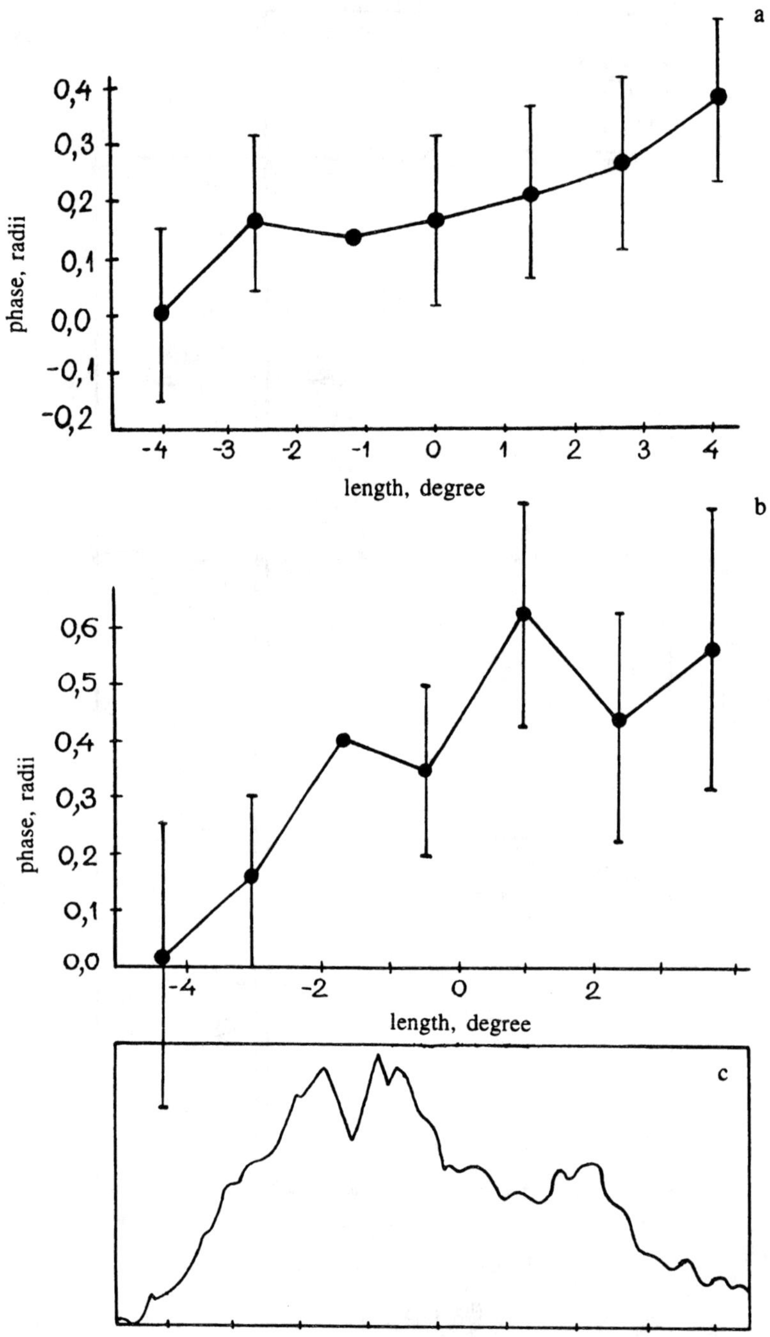
a
0,4
0,3
0,2
0,1
0,0
-0,1
-0,2
phase, radii
-4
-3
-2
-1
0
1
2
3
4
length, degree
b
0,6
0,5
0,4
0,3
0,2
0,1
0,0
phase, radii
-4
-2
0
2
length, degree
c

Figure 4

References

1.M.S. Ewing, R. Batchelor, R. Friefeld, R. Price, and D. Staelin, Astrophys. J., 1970, v. 162, p. L169.
2. J. Cordes, A. Pidwerbetsky, and R. Lovelace, Astrophys. J., 1986, v. 310, p. 737.
3. A. Wolszozan and J. Cordes, Astrophys. J., 1988.
4. J. Cordes, J. Weisberg, and V. Boriakoff, Astrophys. J., 1983, v. 268, p. 370.
5. M. Ruderman and P. Sutherland, Astrophys. J., 1975, v. 196, p. 51.
6. J. Cordes and A. Wolszozan, Astrophys. J., 1986, v. 307, p. L27.
7. T. ankins, Astrophys. J., 1972, v. 177, p. 1.

Physics of Neutron Stars
Edited by A. M. Kaminker

ISBN 1-56072-164-2

A GLITCH IN THE PULSAR PSR 0355 + 54

T.V. Shabanova
Lebedev Physical Institute, USSR Academy of Sciences, 117924, Moscow

Abstract - Results are presented of time observations of pulsar PSR 0355+54 at a frequency of 102.5 MHz. Two glitches of the pulsar period were detected. Before the first glitch the derivative $\dot{P}$ was observed to decrease by $(0.03 \pm 0.01)\%$. The decrease may be regarded as a manifestation of pulsar activity associated with the next glitch. Investigation of the behavior of the rotation speed after the first glitch leads to the conclusion that after the glitch the pulsar slows down faster than before it, and the deceleration rate $\dot{P}$ increases progressively but not immediately.

Information from the time observations of pulsars is important for understanding the physical nature of pulsars. Magnetosphere regions generating the radiation observed rotate synchronously with the neutron star; therefore, the times of the pulse arrival give information on the star rotation character. Together with a regular increase of periods due to the star losing its rotation energy by radiation and to the charged particles escape along open- circuited lines of magnetic force, random fluctuations of the periods may be observed. They may be of two types: discrete period changes as great as $\Delta P/P = 10^{-6} \div 10^{-10}$ and small noisy period variations manifesting themselves as random fluctuations of pulse arrival times.

Observations of discrete variations of the pulsar rotation speed (period leaps or glitches) are of great interest, because they give information about the internal structure of stars. The events are rather scarce and were observed from only 10 pulsars, from four of them they were observed repeatedly from 2 to 7 times. The character of changes in the pulsar rotation rate during leaps is well explained in terms of the two-component model of neutron stars, where the components are loosely connected with each other. Analysis of period leaps detected from pulsars PSR 0531+21 and 0833- 45 shows that during the leaps the pulsar rotation speed abruptly increases and then exponentially drops to some value. An increase in the rotation rate is accompanied by an increase in the period derivative $\dot{P}$, i.e., after leaps pulsars start to decelerate faster than before them. In some time, the deceleration rate of $\dot{P}$ returns to its initial value.

This work presentes the results of time observations of pulsar PSR 0355+54. The pulsar is of interest because during a year two period glitches were detected at a frequency of 408 MHz, the second glitch having a maximum value ($\Delta P/P = 4.4 \cdot 10^{-6}$ [1]) known now. We also observed the two period glitches at a frequency of 102.5 MHz, and we were fortunate enough to obtain new results that were not reported before.

Pulsar PSR 0355+54 was detected by antenna array BSA FIAN at a frequency of 102.5 MHz for 4.5 years from April 1982 to November 1986. To receive signals 32 channels (each of 20 kHz) of a multichannel radiometer with a time constant 3 ms were used. The observations were performed in a computerized regime of individual pulse accumulation in each channel, synchronized with the observable pulsar period. An averaged pulse profile was obtained by the addition of the pulses in all the channels, with allowance made for the extent of dispersion. Pulse arrival times were determined from the maximal value of a cross-correlation function. The function was computed from an averaged pulse obtained in one observation session and a standard profile averaged over 31 000 individual pulses. A mean error in the determination was 900 Ms. The measured arrival times were reduced to the barycenter of the Solar System using ephemeris JPLDE 200/ 200 and the known position of the pulsar [2]. The stability of the pulsar rotation was characterized by residual deviations, which were variations of the detected pulse arrival times as compared to the calculated. The calculated time was computed on the assumption that time dependence of the period is $P(t) = P_0 + \dot{P}(t - t_0)$, where P and $\dot{P}$ are the period and its derivative, respectively. To investigate the character of the pulsar rotationchanges with time, the residual deviations weredifferentiated successively. The differentiation was performed by extension of 80-day deviation segments to obtain straight lines.

Figure 1a shows the deviations of the observed pulse arrival times from calculated, obtained from April 27, 1982 to February 13, 1986. The circles show deviations averaged over 2 - 4 runs of observations. The observation errors are not depicted because of the small scale. The line connects the points subsequent in time. To compute the time the values of P and $\dot{P}$ were taken from Ref. [2]. The first detected pulsar glitch was occured on January 1985 and caused a rapid negative growth of deviations, as can be easily seen in Fig. 1a. The glitch took place between two adjoining measurements on January 11 and 17, 1985 and, therefore, the glitch data was determined rather precisely by a simple linear extrapolation of the data before and after the glitch. The glitch was found to be on January 12, 1985 JD 2446 077.6 $\pm$ 1.1). The period decreased by (0.88 $\pm$ 0.02) ns, and a relative change of the period was $\Delta P/P = -(5.62 \pm 0.08) \cdot 10^{-9}$.

The value of the first period leap ($5.6 \cdot 10^{-9}$) is by three orders of magnitude smaller than of the second. The leap is small, but we were able to observe its shape, which implies that it does belong to simple abrupt changes of the period. Figure 1b shows how the pulsar rotation rate chahged. The result was obtained by successive differentiation of the residual deviations in Fig. 1a. The errors are equal to root-mean-square errors in the determination of an inclination of the corresponding segments of the straight lines. Modified Julian date MJD = JD - 2 400 000.5 is plotted on the X axis. Figure 1b shows that during the glitch a drastic increase in the pulsar rotation rate was observed, though the rate reached its maximum value approximately in 60 - 70 days after the beginning of the event. Then, the rotation rate started to decrease not exponentially but almost linearly, and the decrease was interrupted by the second glitch.

Investigation of the residual deviations preceding the first period glitchgave an interesting result. Figure 1a shows that right before the first glitch a slow negative growth of the deviations was detected. The growth was shown by the measurements to be associated with a decrease in the period derivative $\dot{P}$, that is, with an increase in the pulsar rotation rate because of braking spin-down. The result follows from Fig. 1b, where the last points preceding the glitch are above the first ones, which points to some increase in the rotation rate. The derivative $\dot{P}$ decrease was determined by calculating the $\dot{P}$ value in two 500-day intervals. the resulting values were used to divide the data obtained before the first glitch. A decrease of the derivative $\Delta \dot{P}/\dot{P}$ in the second interval is (0.03 $\pm$ 0.01)% relative to the original value $\dot{P} = 4.3878 \cdot 10^{-15}$. The decrease can be considered as the manifestation of the star activity preceding a glitch.

The observed type of the period leap - a progressive increase of a pulsar rotation rate followed by a smooth slowing down - does not agree with an exponential decay, predicted by the theory of starquakes in terms of the two- component model [3].

To account for this leap a mechanism of spinning-up a star rotation may be invoked, which is associated with thermal instabilities inside neutron stars [4]. A sudden brief increase in the internal temperature can result in a stronger interaction between the crust and the superfluid core rotating faster than the solid crust. Then, the crust would start rotating faster because of a transfer of angular momentum from the superfluid to the solid crust. Decreases in $\dot{P}$ may result from small disturbances of the star temperature before the leap. The braking torque applied to the solid crust by the magnetic field around the star can, in principle, change P as well. The analysis showed that during all the observations of pulsar PSR 0355+54 pulse shapes and intensities remaind essentially unchanged.

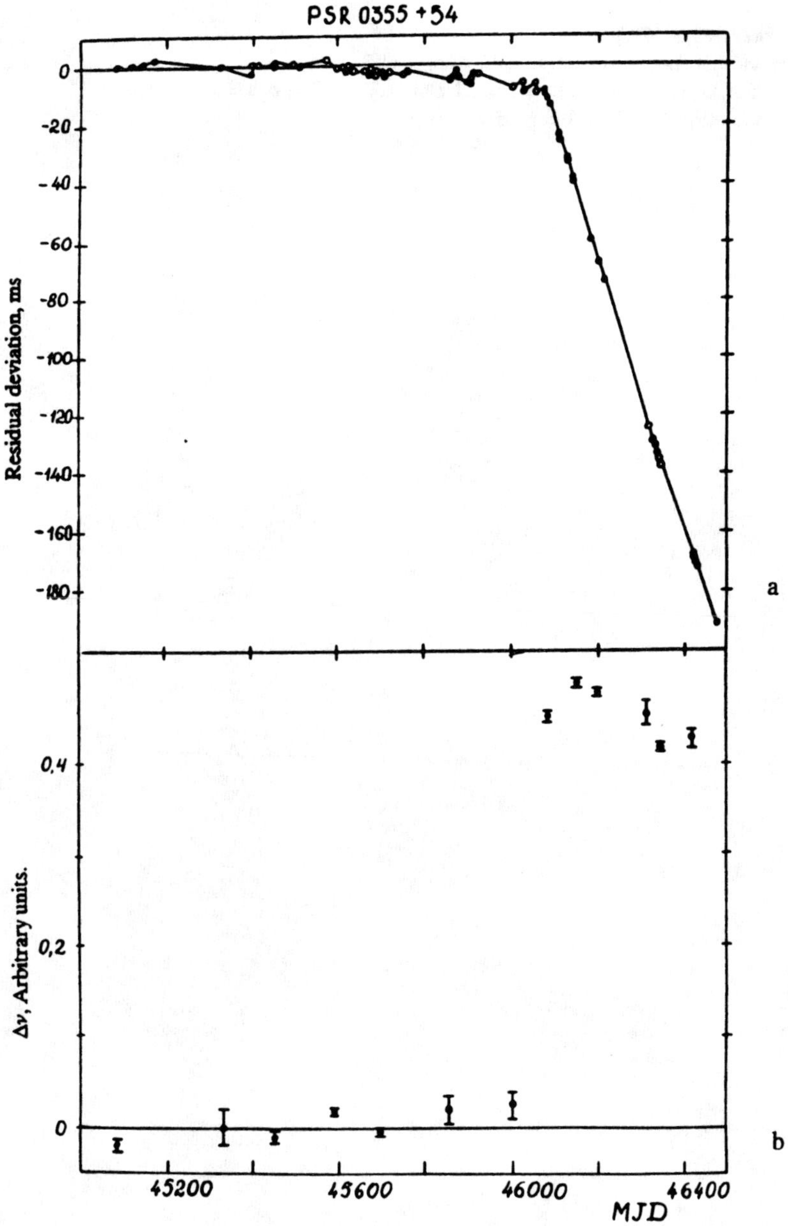
PSR 0355 +54
Residual deviation, ms
0
-20
-40
-60
-80
-100
-120
-140
-160
-180
a
Δν, Arbitrary units.
0,4
0,2
0
b
45200
45600
46000
46400
MJD

References

1. A.G. Lyne, Nature, 1987, v. 326, p. 569.
2. .N. Manchester and J.H. Taylor, Astron. J., 1981, v. 86, p. 1953.
3. G. Baym, C. Pethick, D. Pines, and M. Ruderman, Nature, 1969, v. 224, p. 872.
4. G. Greenstein, Astrophys. J., 1976, v. 208, p. 836.

Physics of Neutron Stars
Edited by A. M. Kaminker

ISBN 1-56072-164-2

A DECREASE OF THE ROTATION DECELERATION OF PSR 0355 + 54 MAY BE CAUSED BY CHANGE OF THE PHASE STATE OF THE NEUTRON STAR CRUST

A.D. Kuz'min
Lebedev Physical Institute, USSR Academy of Sciences, 117924, Moscow

Abstract - An interpretation is proposed of a preglitch decrease of the rotation deceleration rate $\dot{P}$ for pulsar PSR 0355+54 observed by Shabanova [1]. The decrease may be caused by a change in the phase state of the neutron star crust: by transition from an elastic deformation region to a region of plastic deformations. Relationships are proposed to estimate a characteristic interval between the period changes and the change values as a function of the pulsar parameters.

One of possible interpretations of glitchlike decreases in a pulsar rotation period, proposed by Baym back in 1969 [2], is a glitchlike decrease of the inertia momentum I. The decrease is caused by cracking of a neutron star crust (the so-called starquakes) when a critical value of mechanical stresses in the crust is exceeds and the star shape flattening decreases.

In terms of this interpretation the decrease in the rotation deceleration rate $\dot{P}$, preceding a glitch with a period P, detected by Shabanova [1], may indicate that a rapid but not disrupting (as in a glitch) deformation of the neutron star crust precedes crust cracking. This deformation could decrease the star flattening and its inertia momentum.

The assumption agrees with the theory of solid state deformation. According to the theory a transition from elastic deformations to a plastic deformations precedes a material failure when a critical stress σ_{cr} is exceeded. Plastic deformations are characterized by a rapid growth of deformation ε at minor increases of load σ or even at a constant load (Fig. 1).

In the context of this interpretation a decrease in the period deceleration rate $\dot{P}$ correlates with the transition from elastic to plastic deformations.

Deformation stresses in the crust arise under a secular deceleration of a neutron star rotation, because the rigid crust cannot take the shape of an equilibrium spheroid and the centrifugal force decreasing with deceleration does not counterpoise the gravitational force.

Let us assume that at the moment of cracking the crust has the shape of an equilibrium rotation figure. Then, the centrifugal F_C and the gravitational F_G forces are equal, and the stresses in the crust are zero:

$$F = F_G - F_C = 0$$

The centrifugal force, acting on mass m, rotating with a frequency Ω at a distance R from the center, is

$$\tilde{F}_C = m\,\Omega^2 R \qquad (1)$$

Let us isolate a unit volume for further consideration. The centrifugal force acting on it is:

$$F_C = \rho\Omega^2 R = \frac{4\pi^2 \rho R}{P^2} \qquad (2)$$

Under a secular deceleration of rotation the centrifugal force will decrease, violating the balance of $F_C = F_G$ and giving rise to a deformation force

$$\sigma(t) = \Delta F_C(t) = 8\pi^2 \rho R \frac{\dot{P}}{P^3} t \qquad (3)$$

At the linear dependence of σ(t) the time dependence of deformation ε follows the dependence $\varepsilon(\sigma)$ scaled:

$$\varepsilon(t) = \varepsilon(\delta) \qquad (4)$$

Hence, three qualitatively different states in the rotation deceleration should be expected (Fig. 2).

1. The first is from the formation of an equilibrium shape (at $\sigma = 0$) after a preceding glitch to a state of increasing stresses up to $\sigma = \sigma_{cr}$. This is an area of small elastic deformations. The rotational moment remains practically unchanged. A deceleration rate is $\dot{P}_1$.

2. When $\sigma = \sigma_{cr}$ deformations become plastic, and the star flatting begins to decrease, thus decreasing the rotational moment and accelerating the star rotation. This positive acceleration $\Delta\dot{P}$ reduces the rate of secular deceleration rotation:

$$\dot{P}_2 = \dot{P}_1 - \Delta\dot{P} < \dot{P}_1$$

3. When approaching the ultimate strength $\sigma = \sigma_u$, the crust undergoes cracking and the star shape rapidly transforms into equilibrium. The corresponding glitchlike decrease in the inertia moment gives rise to a glitchlike decrease in the rotation period.

The radius of the isolated unit volume decreases by glitches (approaching the rotation axis) by

$$\varepsilon = \Delta R = \frac{\sigma_{lim} P^2}{4\pi^2 \rho} \qquad (5)$$

which leads to a glitchlike decrease in the period by

$$\Delta P = P\frac{\Delta I}{I} = P\frac{2\Delta R}{R} = \frac{\sigma_{lim} P^3}{2\pi^2 R\rho} \qquad (6)$$

Then the whole cycle must repeat.

A characteristic cycle time T is determined by the stress growth (3) to the critical value σ_{cr}:

$$T = \frac{\sigma_{cr}}{8\pi^2 \rho R}\frac{P^3}{\dot{P}} = a\frac{P^3}{\dot{P}} \qquad (7)$$

that is, a characteristic time between period glitches is proportional to $P^3/\dot{P}$. The ratio is minimal for pulsars PSR 0531+21 and 0833-45, where period glitches are often observed indeed. Pulsars PSR 1953+29, 1951+32, 1930+22, and 0355+54 have a rather small ratio $P^3/\dot{P}$, and only for PSR 0355+54 period glitches were observed.

An expected value for glitches, according to (6), is proportional to P^3. The observational glitches of the pulsar PSR 0531+21 period are in fact far less than those of PSR 0833-54. Therefore, it is difficult to isolate them from casual period changes, and only strongest glitches can be detected [3].

Most reliably measured are period glitches for pulsar PSR 0833-45. The interval between the glitches is, on average, 2.5 year. From this the strength (the critical stress) of the material of a neutron star crust can be estimated:

$$\sigma_{cr} = 8\pi^2 \rho R (\dot{P}/P^3) T$$

Taking T = 2.5 year, $\rho = 10^{14}$ g/cm^{-3}, and R = 10^6 cm^3 [3], we will obtain:

$$\sigma_{cr} = 10^4 \ kG/mm^2 \approx 10^{22} \ dyne/cm^2$$

It follows from the model that period glitches ΔP and a preglitch decrease of the deceleration rotation rate $\dot{P}$ must be a common property of all pulsars. The lack of observational data on a preglitch change in $\dot{P}$ for pulsars PSR 0531+21 and 0833-45, which have clearly observable glitches, may be due to that the time of a plastic phase for the pulsars is less than an interval between observations.

Special observations are necessary to clarify the problem.

References

1. T.V. Shabanova, This issue, p. 168.
2. G. Baym, Nature, 1969, v. 224, p. 872.
3. R. Manchester and J. Tailor, Pulsars, Russian translation, Mir, Moscow, 1980.

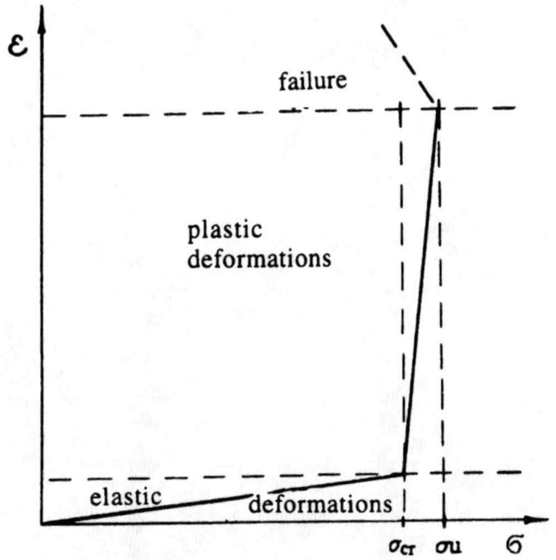

Fig. 1.

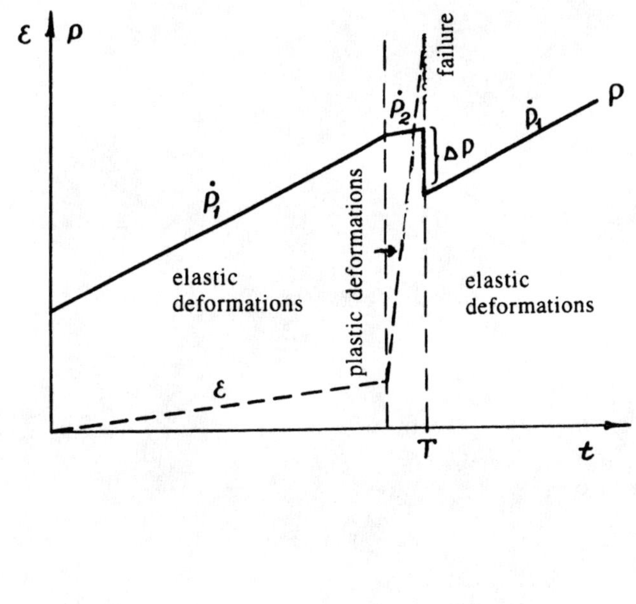

Fig. 2.

Physics of Neutron Stars
Edited by A. M. Kaminker

ISBN 1-56072-164-2

OPTICAL LIGHT CURVE OF THE CRAB PULSAR WITH RESOLUTION 3.3 μs

V.F. Shvartsman, G.M. Beskin and V.L. Plokhotnichenko
Specialized Astrophysical Observatory, USSR Academy of Sciences,
357147, Zelenchukskaya, Stavropol Territory

Abstract - An optical light curve of PSR 0531+21 with a time resolution of 3.3 Ms has been obtained with a 6-meter telescope using the program-controlled apparatus complex in the experiment MANIYa. The peak of the main pulse is flat, its width is 230 Ms at the height 0.95 to the maximum. Limitations were found for the fine time structure of the main pulse of the phased light curve; its relative amplitude for 3.3 Ms does not exceed 10% (the confidence coefficient is greater than 99%). Simulation of y_2-functions enables determination of the limitations for the relative power of sporadic bursts for $5 \cdot 10^{-7}$ - $6 \cdot 10^{-5}$ s, which are 67 to 18%, respectively.

Introduction

A search for a fine time structure in an optical light curve of the Crab Nebula pulsar is important for elucidation of mechanisms, by which high-frequency pulsar radiation is generated. Indeed, a fine spatial structure in a pulsar megnetosphere, which is due to nonuniformity of magnetic fields and density of emitting particles [1, 2], must form a fine time structure of a pulse in the phased light curve. On the other hand, variations of the brilliance temperature in sources of coherent radiation would cause rapid stochastic variability at smallest times [3, 4]. Investigations of the fine time structure of pulsar radiation in radio frequencies confirmed high efficiency of this approach for understanding the nature of pulsar radiation and, naturally, encouraged us to attempt similar investigations in the optical range.

1. Observations

A photometer and specialized engineering and mathematical complex of the experiment MANIYa were used to perform observations in the primary focus of the 6-meter telescope of the Specialized Astrophysical Observatory (SAO) of the USSR Academy of Sciences [5 - 7]. The time of any photon registration was recorded using a converter "time-code" (or, the so-called, "quantochron") and sent to the computer SM 4 to be stored on magnetic disks. Then they were re-recorded on a magnetic tape. The dead time of the recording devices was 300 ns.

During the night from September 6 to 7 we accumulated 5 data files of photocounts from the pulsar and the nebula region within the diaphragm (the image size of starlike sources was 3"), see Table 1.

Table 1.

	Observational information					
N	UT start	Duration, s	Diaphragm	Filter	Flux counts/s	Volume counts
1	00 39 00	268	4".3	R	2304	$6 \cdot 10^5$
2	00 44 23	350	4".3	B	1770	$5 \cdot 10^5$
3	00 51 58	259	4".3	R	2286	$5 \cdot 10^5$
4	00 57 58	310	4".3	B	1885	$5 \cdot 10^5$
5	01 27 20	788	6".9	R	2757	$20 \cdot 10^5$

Note: The fluxes were attenuated to decrease the apparatus distortions at the main pulse top.

A flux of reference marks, viz. standard pulses from the Observatory time service, was registered simultaneously with photocounts from the pulsar. The reference marks were used to correct nonuniformities of the internal timepiece of the "quantochron". The correction resulted in the improvement up to 2 Ms the accuracy with which the time of the event appearance in the registration channel circuit of the quantochron was measured. This is 3 times better than the accuracy in our experiments in 1982 [8].

$$\varphi(t) = \varphi_0 + \Omega(t_0)(t - t_0) + \frac{1}{2}\dot{\Omega}(t_0)(t - t_0)^2 + \frac{1}{6}\ddot{\Omega}(t_0)(t - t_0)^3 + \dots \quad (1)$$

where $\Omega(t_0) = 1/P(t_0)$ is the observed pulse frequency at t_0 (P is the period), $\dot{\Omega}(t_0)$ and $\ddot{\Omega}(t_0)$and are its first and second derivatives. $\Omega(t_0)$, $\dot{\Omega}(t_0)$ and $\ddot{\Omega}(t_0)$were determined from the time files of photocounts registration by minimization of a pulse phase displacement during their storing.

Figure 1 shows the comparison of our measured periods with the results of preliminary computations made by T.V. Shabanova (FIAN). The difference in the curves behavior may be explained by the time nonuniformity of the timepiece of the time service itself. The deviation swing of the period values from the approximating line is ΔP ü 0.14 ns for the time of file 5 storing, which provides the pulse phase conservation as accurate as 0.7 Ms. This accuracy combined with the accuracy, with which the arrival time of pulses of a standard frequency is recorded, give the limiting time resolution. From it, a light curve summarized over the periods can be constructed with an accuracy as good as 3 Ms.

2. Results of observations

The main pulse of the light curve with a 3.3-Ms window is presented in Fig. 2. The roughness of the peak top does not exceed the level permissible by Poisson statistics, which allows the conclusion that a regular fine structure of the radiation modulation is less than 10% of the main pulse amplitude (the confidence coefficient is 0.99).

In the context of the experiment MANIYa we developed a technique for searching a stochastic variability against the background of regular variations. For this purpose we used y_2-function formalism described in works [5, 9]. Calculated y_2-functions of the Poisson flux [10] are subtracted from the y_2-functions, obtained from processing the files of photon registration time. Thus, a possible contribution of stochastic bursts to the total pulsar radiation can be isolated. The power of the stochastic variability in the form of bursts with a relative burst duration 1/10 in the R-band was found to be smaller than the data in Table 2. Here, the relative burst duration is the ratio of the mean burst duration to the mean interval between them.

Table 2.

Limitations on the power of the radiation variable.

Time, Ms	0.45	0.90	1.8	3.6	7.2	14	29
Power limit, %	67	47	38	30	27	23	18

3. Discussion

The lack of a fine time structure in the main pulse of a phased light curve points to a relatively uniform character of the field and the particle distributions in the region of optical radiation. The data are in contradiction with models [4] based on coherent plasma mechanisms and predicting the existence of an extremely narrow feature in the main pulse maximum.

At the same time the lack of observable variability at smallest resolution times cannot exclude, in our opinion, coherent mechanisms of optical radiation. The variability times characteristic of these mechanisms may be smaller than the dead time of our instruments, and their detection would require further improving of the time resolution of the complex MANIYa.

We wish to express our gratitude to T.V. Shabanova for computations of the pulsar periods by the time of our observations; to S.I. Neizvestnyi, A.V. Zhuravkov, and S.N. Mitrofanov for the assistance in pre-starting procedure and observations; and to L.A. Pustil'nik for the assistance in the preparation of this work.

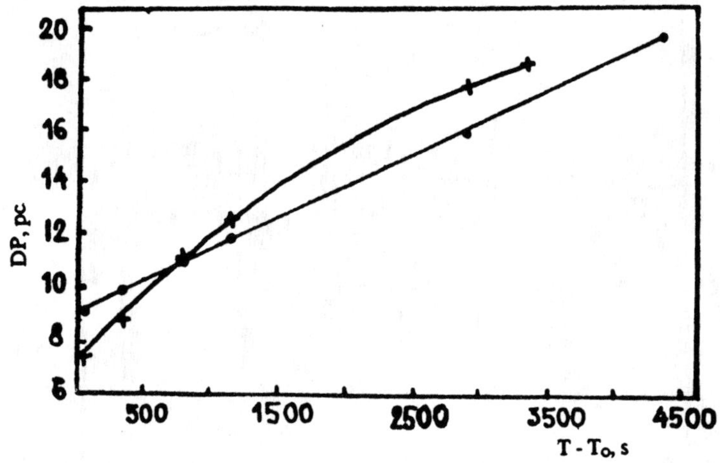

Fig. 1. Comparison of time curves of the observed (+) and the calculated (.)periods. $P(t) = 0.33325000$ s + DP $(t - t_0)$, $t = 00^h39^m00^s$.

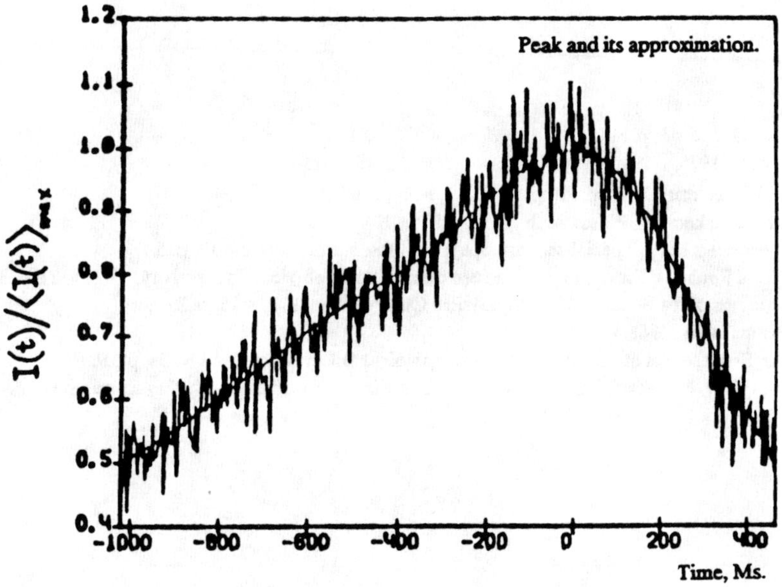

Fig. 2a. The light curve peak of pulsar PSR 0531+21 with a resolution of 3.3 Ms.

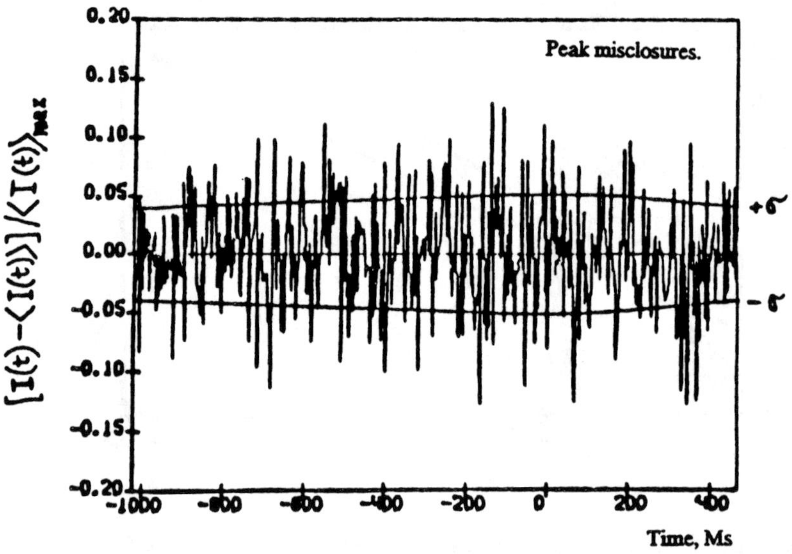

Fig. 2b. The normalized errors of closure between the light curve and its approximation by splines. Also shown is the dependence of Poisson root-mean-square deviations on the light curve phase.
Supplement: Time, Ms. Peak misclosures.

References

1. F. Pacini, LAU Symp., 1971, n. 46, p. 394.
2. G. Benford, Astrophys. J., 1975, v. 201, n. 2, p. 419.
3. V.L. Ginzburg and V.V. Zheleznyakov, Sp. Phys., 1970, v. 2, n. 5, p. 167.
4. G.Z. Machabeli and R.G. Sakhokiya, Pis'ma AZh., 1982, v. 8, p. 78.
5. V.F. Shvartsman, Proceedings of the Specialized Astrophysical Observatory, 1977, v. 19, p. 5.
6. S.I. Neizvestnyi and A.A. Pimonov, Proceedings of the Specialized Astrophysical Observatory, 1978, v. 23, p. 39.
7. A.A. Pimonov, Proceedings of the Specialized Astrophysical Observatory, 1979, v. 25, p. 39.
8. G.M. Beskin et al., Pis'ma AZh., 1983, v. 9, n. 5, p. 280.
9. V.L. Plokhotnichenko, Proceedings of the Specialized Astrophysical Observatory, 1983, v. 38, p. 29.
10. V.G. Lipovetskaya, V.L. Plokhotnichenko, and V.P. Chuprina, Proceedings of the Specialized Astrophysical Observatory, 1988, v. 57.

Physics of Neutron Stars
Edited by A. M. Kaminker

ISBN 1-56072-164-2

A SEARCH FOR PULSED OPTICAL RADIATION FROM THE MILLISECOND PULSAR PSR 1937 + 21

V.F. Shvartsman, G.M. Beskin, S.I. Neizvestnyi and V.L. Plokhotnichenko
Special Astrophysical Observatory, USSR Academy of Sciences, 357147, Zelenchukskaya, Stavropol Territory

Abstract - Pulsed optical radiation from the millisecond pulsar PSR 1937+21 was searched for in 1983 during several sessions with the 6-meter telescope of Specialized Astrophysical Observatory (SAO) of the USSR Academy of Sciences. The engineering and mathematical complex of the experiment MANIYa was used, it allowed time resolution $3 \cdot 10^{-7}$ s. Approximately $5 \cdot 10^6$ photocounts, obtained in two sessions without the filter and in the R-band, were processed by a special program of "fan searching for periods." A pulsed radiation component was not found, with the confidence coefficient greater than 99%. Limitations on the pulsar brightness in the B- and R-bands were $26^m.2$ and $24^m.6$, respectively. The results obtained appear to include that the distance to the pulsar is 5 kpc.

1. Introduction

The theory of ejecting pulsars gives, probably, the only relationship between pulsar parameters and its optical luminosity. It is the Pacini equation derived from simple considerations back in 1971 [1]:

$$L_{opt} \propto B^4 P^{-10} \qquad (1)$$

where B is the magnetic field strength near the neutron star surface, P is the pulsar period.

The prediction of such a strong dependence of pulsar optical luminosity on its rotation speed immediately singled out pulsars with short periods and stimulated searches for young rapid pulsars in supernova explosions [2]. It should be noted that the ratio of the luminosities of the pulsars in Crab (P = 33 ms) and Sails (P = 89 ms) as well as the limitations on the brightness of a binary pulsar PSR 1913+16 (P = 69 ms) are in agreement with expression (1).

As soon as in 1982 the pulsar PSR 1937+21 with P = 1.56 ms was discovered, searches started for its optical radiation. However, the hope to detect optical pulses was almost lost, when it was found that its $\dot{P} = 10^{-19}$ s/s and, hence, PSR 1937+21 has a weak magnetic field ($B \approx 5 \cdot 10^8$ G). It follows from (1) that
$L_{1937} / L_{vela} = 75$

However, the millisecond pulsar is much farther: the distance to it is estimated to be 2.5 or 5 kpc depending on the estimation method. Correspondingly, its brightness must coincide with that of PSR 0833-45 (24^m) or be by 6^m weaker (when absorption is taken into account) [3]. Shortly after the discovery of PSR 1937+21, Manchester et al {3] made several sessions of its observations to find pulsed optical radiation. But they succeeded only in determination of the upper limit for the object R-brightness (25^m). PSR 1937+21 was also studied with a 6-meter telescope in the SAO of the USSR Academy of Sciences in the framework of the experiment MANIYa. The results are presented in this work.

2. Observations

Search for PSR 1937+21 optical radiation was performed using the electrophotometer of the primary focus of the 6-meter telescope at SAO of the USSR Academy of Sciences and the engineering and mathematical complex for the experiment MANIYa [4 - 7]. During the observations times of individual photons registration were determined; coded by the converter "time-code" or, the so-called, "quantochron"; sent to the operative memory of the computer SM 4; and recorded on a magnetic tape. The resolution time of the instrument was $3 \cdot 10^{-7}$ s. A system of successive second-duration marks coming from the Time Service generator was introduced into the whole mass of photons. This allowed all registered photons to be phased with the pulsar period with the maximum precision during their subsequent processing. Thus, the reference mark was obtained to correct the work of the system master generator. The method was tested during the observations of the Crap pulsar and showed that phased light curves can be constructed with a resolution time 3 Ms.

The observations were made without a filter (photocathode S20, FEU 106) and in the R-band using diaphragms 6".9 and 4".3. Under these conditions the diaphragm was centered by offsetting either to the star of Djorgovski or to the VLA coordinates of PSR 1937+21 [9]. The difference between these positions is 2".5. During one continuous session of observations approximately $2.5 \cdot 10^6$ photocounts were obtained. All in all 8 sessions were made in March, June, and August. Unfortunately, only in 2 of them the marks were registered. The information obtained in other sessions could not be properly treated.

3. Treatment of observational data and analysis of the results

The Table lists information on the observational sessions, during which data appropriate for processing were obtained.

Table

Information on observations

Session	Date	UT registered	Diaphragm	Filter	volume	position
1	7.08.83	$21^h37^m - 21^h50^m$	6".9	no	2.04 mill.	a star
2	8.08.83	$00^h46^m - 01^h22^m$	6".9	R	2.42 mill.	VLA

The search for optical radiation was made using a special program "PULSAR" [10] on the computer ES 1035. The program is based on the so-called "method of fan period search." Eventually, it consists in constructing of light curves phased with different periods, when their evolution with observation time is taken into account. Under these conditions the number of windows, into which the light curve is divided, can vary. In our case their number was 50, because the relative duration of the main pulse in radio frequencies is close to 1/40 (the half-width is 40 Ms).

The criterion for pulse detection is a meaningful difference in the photocounts number in one or several windows from the permissible under their Poisson distribution condition.

When processing data accumulated in each session, about 350 test periods were included. The number was chosen taking into account a topocentrical period computed before the beginning of the session. The precomputations were made by T.V. Shabanova on the basis of work [11].

The processing showed that no meaningful deviations from the Poisson distribution of photocounts were detected in any of the phased light curves. The confidence coefficient of the result of the result is higher than 99%. Thus, we did not detect pulsed optical radiation from radiopulsar PSR 1937+21.

Our results enable determination of the upper limit of pulsar optical brilliance. It is easy to see that

$$\Delta m = -2.5 \lg (\alpha \sigma / N) \qquad (2)$$

where Δm is the difference in the stellar values between the background in the diaphragm used and the limitation on the pulsar brilliance;
α is the number of standard deviations taken to determine the guaranteed upper brilliance limit;
N is the total volume of photocounts registered;
σ is the Poisson standard for the photocount number in a window of the phased light curve.

Here we assume that a characteristic duration of the pulsar pulse coincides with the window width. If k is the window number in the phased light curve, then

$\sigma = \sqrt{N/k}$

hence

$\Delta m = -2.5\, lg\, (\alpha / \sqrt{kN})$

Assuming that the pulsar optical spectrum is similar to the Crab pulsar spectrum we can use stellar values in the B-band:

$\Delta m = -2.5\, lg\, (\alpha / \sqrt{kN}) - 0.^{m}47$

Taking $22^{m}.55$ per a squared second (session 1) to estimate the stellar value for the sky background in the B-band, for the diaphragm 6".9, we have

$m = -2.5\, lg\, (\alpha / \sqrt{kN}) + 17.^{m}89$

For $\alpha = 5$, k = 50, N = $2.04 \cdot 10^{6}$ the upper limit of the pulsar brilliance is

$m_B = 26.^{m}2$

Similarly, for the R-band (session 2):

$m_R = 24.^{m}2$

(brilliance per a squared second is $20^{m}.39$).

4. Conclusion

In the observations made no optical radiation from the millisecond pulsar PSR 1937+21 was detected.

The limitation on the pulsar brilliance in the B-band exceeds the predicted value by 2^{m} at a distance to the object of 2.5 kpc, hence, the distance is likely to be 5 kpc [12].

It follows that if the Pacini relationship is valid, the long distance to PSR 1937+21 and interstellar absorption do not allow detection of optical radiation from the object even in observations from satellites.

We wish to express our gratitude to T.V. Shabanova for the period precomputation and to A.P. Polyntsev for The assistance in making the program of "fan period search."

References

1. F. Pacini, Ap. J., 1971, v. 163, p. L17.
2. G.M. Beskin, S.I. Neizvestnyi, A.A. Pimonov, V.L. Plokhotnichenko, and V.F. Shvartsman, Pis'ma AZh., 1979, v. 5, n. 10, p. 508.
3. R.N. Manchester, B.A. Peterson, and P.T. Wallace, Nature, 1984, v. 310, p. 569.
4. V.F. Shvartsman, Proceedings of the Specialized Astrophysical Observatory, 1977, n. 19, p. 5.
5. S.I. Neizvestnyi and A.A. Pimonov, Proceedings of the Specialized Astrophysical Observatory, 1978, n. 23, p. 56.
6. A.A. Pimonov, Proceedings of the Specialized Astrophysical Observatory, 1979, n. 25, p. 56.
7. V.L. Plokhotnichenko, Proceedings of the Specialized Astrophysical Observatory, 1983, n. 38, p. 29.
8. S. Djorgovski, Nature, 1982, v. 300, p. 618.
9. D.C. Backer, R. Srinias, Kulkarni, C. Heiles, M.M. Davis, and W.M. Goss, Nature, 1982, v. 300, p. 615.
10. V.L. Plokhotnichenko, Report of the Specialized Astrophysical Observatory of the USSR Academy of Sciences, 1986, n. 132.
11. T.V. Shabanova, Proceedings of Lebedev Physics Institute, 1977, v. 93, p. 119.
12. C. Heiles, S.R. Kulkarni, M.A. Stevens, and D.C. Backer, Ap. J., 1983, v. 273, p. 175.

Physics of Neutron Stars
Edited by A. M. Kaminker

ISBN 1-56072-164-2

GENERATION OF THE MAGNETIC FIELD OF THE CRAB NEBULA BY THE PULSAR PSR 0531 +21

E.G. Tsikarishvili, G.D. Chagelishvili, O.V. Chedia, J.G. Lominadze and Z.A. Sokhadze
Abastuman Astrophysical Observatory,
GSSR Academy of Sciences, 380060, Tbilisi

Abstract - Generation of the large-scaled toroidal-shaped magnetic field in the Crab Nebula by the dynamo-mechanism in the immediate vicinity of the light cylinder of pulsar PSR 0531 + 21 has been studied.

The activity of the Crab Nebula is known to be wholly due to pulsar PSR 0531 + 21 in it. The pulsar energy is emitted into the nebula as low-frequency magnetically bipolar electromagnetic waves, relativistic particle fluxes, and static toroidal magnetic field. A toroidal Shape of the magnetic field in the central region of the Crab Nebula is supported by numerous observations.

The synchrotron nature of the Crab continuous radiation is presently beyond any doubt. To explain it, the constant magnetic field of about 10^{-3} G must exist in the Nebula. A generally accepted model for the generation of the field is nowadays the model proposed by Piddington [1], which was later improved by Kardashev [2], Rees and Gunn [3], and Kennel and Coroniti [4]. The model is based on the so-called "twisting" of toroidal components of the magnetic field of pulsar PSR 0531 + 21 in the nebula center.

The buildup of the nebula toroidal magnetic field is supposed in the works cited to occur within the time of the Nebula existence (about 1000 years) owing to the rotation of the central object. However, we believe that the toroidal magnetic field is rapidly generated in the immediate vicinity of the light cylinder of pulsar PSR 0531 + 21 from the hollow-shaped magnetic field by the differential rotation of the relativistic electron-proton plasma flowing from the pulsar. Then the toroidal magnetic field is "transferred" by the plasma flux into the Nebula. Hence, we assume that the magnetic field observed in the nebula is generated continuously, giving a quasi-stationary picture. The quasi-stationary character of the magnetic field generation requires a "withdrawal" of the magnetic flux from the nebula, which is, in our case, released by reclosure of lines of magnetic force at the nebula fibres.

The shape of the periodic signal from pulsar PSR 0531 + 21 points to that the pulsar is an inclined rotator with the inclination angle (the angle between the magnetic and the rotational axes) close to the right ($\Theta \approx \pi/2$). Following works [5, 6] we assume that cascades of electron-positron pairs are formed in the pulsar magnetosphere, to give a sufficiently dense electron-positron plasma. Particles of the plasma move in a narrow cone of the polar cap in the radial direction with an ultrarelativistic speed in the pulsar rotation system. The plasma rotates together with the pulsar because the energy density of the magnetic field dipole (W_b) dominates the energy density of the directional plasma motion (W_n) up to the light cylinder ($r \leq r_c$). Hence, particles in the pulsar rotation system have only radial velocity. In the observer's system the particles, in addition to radial velocity, rotate together with the magnetosphere at the angular speed of the neutron star. In this system an azimuth component of the particle velocities is close to the velocity of light ($v_\varphi \leq c$) near the light cylinder. Particle velocities are known to be restricted to $v^2 = v_\varphi^2 + v_r^2 \leq c^2$. Hence, it follows that $v_r^2 \ll c^2 V$ in the vicinity of the light cylinder, i.e., plasma ejected beyond the light cylinder moves at a tangent to the cylinder. Moving in the magnetosphere the plasma gains on energy of azimuthal rotation at the expense of the magnetic field energy.

Beyond the light cylinder plasma obviously cannot corotate with the neutron star, because the angular velocity of the plasma rotation (ω) should decrease with increasing r ($r \geq r_c$). This decrease beyond the light cylinder may be described by the power law

$$\omega = \Omega \left(\frac{r_c}{r}\right)^n \quad \text{at} \quad r \geq r_c, \quad 1 \leq n \leq 2 \qquad (1)$$

In the vicinity of the light cylinder the particles are subject to the influence of the magnetic-dipole wave, whose energy density at $\Theta \approx \pi/2$ is much greater than that of the nonoscillating magnetic field. Because the energy transfer by electromagnetic waves occurs intensively, the inequality $W_n > W_B$ should hold in the immediate vicinity of the light cylinder. Therefore, the wind motion can be considered independent of the mean magnetic field, resulting in the conservation of the angular momentum ($v_\varphi \cdot r = \text{const}$) and n = 2. Thus, kinematics of the relativistic wind of electron-positron plasma beyond the light cylinder may be described as follows:

$$v_\varphi^2 \approx v_{\varphi c}^2 \left(\frac{r_c}{r}\right)^n \qquad (2)$$

$$v_{zc} \ll v_{rc} \ll v_{\varphi c}, \quad v_{\varphi c}^2 + v_{rc}^2 + v_{zc}^2 \approx c^2, \quad \Omega r_c = c \qquad (3)$$

$$v_r^2 \approx \Omega^2 r_c^2 \left[1 - \left(\frac{r_c}{r}\right)^n\right] + v_{zc}^2 \qquad (4)$$

$$v_z \approx v_{zc} \qquad (5)$$

It can be seen that beyond the light cylinder plasma moves along a rather ordered path. For n = 2 the path for a given small volume of plasma is a straight line drawn practically tangent to the light cylinder ($V_{zc} \ll V_{rc} \ll V_{\phi c}$). Recall that beyond the light cylinder plasma is ejected from both poles of the dipole, and the rotator inclination angle is close to $\pi/2$. Hence, the geometry of the ultrarelativistic plasma wind may be described as two cone-shaped disks. The discs acquire this shape because $v_z \neq$ 0V@z =/0.

Using formulas (1) - (5) and neglecting dissipation effects, an averaged (waveless) magnetic field beyond the light cylinder may be described by

$$\frac{\partial \vec{B}}{\partial t} = rot\left[\vec{v}, \vec{B}\right] \qquad (6)$$

Choosing axis z along the neutron star rotation axis, let us rewrite the azimuth component of (6) in cylindrical coordinates (r, ϕ, z) ignoring ϕ dependence of the parameters φ ($\partial\vec{v}/\partial\varphi$, $\partial\vec{B}/\partial\varphi = 0$):

(7)

In the equation the term $\left(r\frac{\partial\omega}{\partial r}\right)B_r$, due to the differential plasma rotation with respect to r, represents the generation of $B\phi$ from B_r. The term is of importance in the immediate vicinity of the light cylinder ($r_c \ll r \ll 2r_c$), because $V\phi$ and, hence, ω rapidly vanish with increasing r. Taking an averaged magnetic field B_c (B_{rc}, $B_{\varphi c}$, B_{zc}) for the light cylinder, let us solve equation (7) for $r \geq r_c$. We will consider a steady-state case ($\partial B_\varphi / \partial t = 0$) when generation of the magnetic field $B\phi$ is compensated for by its transfer. The stationary character of the averaged magnetic field in the observer's system implies that it does not depend on the azimuth coordinate on the light cylinder in the pulsar rotation system ($\partial B'_{\varphi c}/\partial\varphi = 0$, where the prime denotes that the variable is written in the pulsar rotation system).

Setting r dependence of B_r as $B_r = B_{rc}\left(\frac{r_c}{r}\right)^2$ and remembering that z dependence of $V\phi$ is weak, the solution of (7) for $r \gg r_c$ is

$$B_\varphi \approx \left(-\frac{n}{n+1}B_{rc}\right)\left(\frac{r_c}{r}\right) \qquad (8)$$

It follows from the formula that $B_r \ll B_\phi$ for long distances; therefore, in the central region of the Crab Nebula the averaged magnetic field is toroidal. Formula (8) for the reported n values shows that the toroidal magnetic field at the inner Nebula boundary ($r \approx 10^{17}$ cm from the pulsar) must be $B_\varphi \approx 10^{-3}$ G, which is in agreement with the observed value of the magnetic field in the Crab Nebula.

References

1. J.H. Piddington, Australian J. Phys., 1957, v. 10, p. 530.
2. N.S. Kardashev, Astron. Zh., 1964, v. 41, p. 807.
3. M.J. Rees and J.E. Gunn, M. N. R. A. S., 1974, v. 167, p. 1.
4. C.F. Kennel and F.V. Coroniti, Astrophys. J., 1984, v. 283, p. 694.
5. P.A. Sturrock, Astrophys. J., 1971, v. 164, p. 529.
6. A. Cheng and M.S. Ruderman, Astrophys. J., 1977, v. 216, p. 365.

Physics of Neutron Stars
Edited by A. M. Kaminker

ISBN 1-56072-164-2

HIGHLIGHTS OF THE KVANT MISSION

R. Sunyeav, E. Ghurazov, V. Efremov, M. Gilfanov, S. Grebenev, A. Kaniovsky, D. Stepanov, S. Yunin, A. Kuznetsov, A. Melioransky, N. Jamburenko, S. Kiselev, L. Lapshov, N. Pappe, M. Boyarsky, V. Loznikov, A. Prudkoglyand, O. Terekhov (USSR), C. Reppin, W. Pietsch, J. Englhauser, J. Truemper, W. Voges, E. Kendziorra, M. Bezler, R. Staubert (FRG), A.C. Brinkman, J. Heise, W.A. Mels, R. Jager (The Netherlands), G.K. Skinner, O. Al-Emam, T.G. Patterson, A.P. Willmore (UK), A. Smith and A. Parmar (ESTEC)

The International X-ray observatory ROENTGEN, aboard the KVANT module docked to the MIR space station, performed already more than 1000 sessions of observations of HER X-1, CYG X-1, CYG X-3, NGC 4151, CEN A, 3C 273, ASM 2000+25, and some other sources. The main target was SN1987A and the Large Magellanic Cloud field. Some results of observations (hard X-ray spectra, X-ray imaging and timing) are presented.

Introduction.

The International X-ray observatory ROENTGEN, aboard the KVANT module was put in orbit by a rocket-carrier Proton in April 1987. Several weeks later its X-ray telescopes started observations. Since that time more than 1000 sessions have been made. Among its targets were such sources as SN1987A, ASM 2000+25, HER X-1, CYG X-1, CYG X-3, Crab Nebula, 3C 273, and some other. In this review we are going to describe the Observatory X-ray telescopes and briefly outline some results of the first year observations.

Devices of the Observatory.

Characteristics of the devices forming the Observatory ROENTGEN aboard the KVANT module are given in Table 1. Simultaneous observations of all the devices allow coverage of a widest range of energies with a high time resolution and imaging of a sky region under observation in a soft (2 - 32 keV) range.

Cyclogram of observations.

The period of MIR space station revolution around the Earth is 90 minutes. Because of a high orbit inclination ($\sim 57^{\circ}$) the observation time on the dark part of the orbit, beyond the radiation bands is ~ 20 min [1]. Observations of bright X-ray sources by a TTM telescope showed that the stabilization of the complex holds to an accuracy up to ~ 2′ during a 20-minute session (Fig. 1). With Pulsar X-1 and GSPC instruments, high quality spectra can be obtained only by comparing a flux from the source (source + background) with a flux from the background. Therefore, the majority of the KVANT module observations were made by swinging of the whole complex (several subsequent trainings on the source and the background area, which is 4 - 12 degrees off the source and has no other bright sources). In this regime, orientation was changed with gyrodines without switching on the engines of the space station. The results of the TTM telescope show (Fig. 2) how precisely the huge 55-ton MIR-KVANT complex performed this manoeuvre on August 17, 1987. The complex reorientation usually takes 90 - 100 s. Thus, observations of the background areas took from 1/2 to 2/3 of all the session time.

The main targets for the observatory ROENTGEN during the first year of observations and the number of sessions are given in Table 2.

Below we are going to present some most interesting results obtained with the KVANT module.

The Large Magellanic Cloud field.
Since the first sessions, the supernova 1987A and the Large Magellanic Cloud (LMC) field were among the main targets of the observatory. To localize the source of hard radiation first detected on August 10, 1987, by the HEXE instrument, a telescope with a coding TTM aperture was supposed to be used. The LMC map (its $2.5^\circ \cdot 8.5^\circ$ section) in the range of 2 - 32 keV is shown in Fig. 3 [2, 3}. The map was constructed on the basis of 40 000 s observations of the TTM telescope during 50 observation sessions of this section of the LMC on August 10 - 23, 1987. No significant emission from the SN1987A area was recorded. Using the observational results of the 50-ms pulsar 0540-693 to graduate the efficiency of the instrument, an upper limit for the supernova emission was determined [2, 3] to be 0.5 mCrab in the range of 2 - 32 keV (for the period from august 10 to 23, 1987). This value together with the upper limits for SN1987A emission in the range 2 - 32 keV registered in June - August 1987, and soft X-ray emission from SN1987A detected by the GINGA satellite in September 1987 are presented in Fig. 4. According to the TTM telescope data no bright sources exist near the supernova (except the LMC X-1 and 0540-693). This allowed localization of the hard emission source. For the purpose, use was also made of other devices with simple collimators, setting the observatory axis with respect to different points so that the effective areas of SN1987A and other sources were different. Some observations made at a guidance by 1° apart from the supernova helped localize the hard emission source from the HEXE data. Its calculated position is in agreement with the supernova coordinates to an accuracy of about $10'$. The positions of the LMC X-1 and 0540-693 deviate from the calculated by a value greater than a $3\,\sigma$ error of a source position determination.

Detection of a hard component in the X-ray spectrum of ASM 2000+25 source.
On April 26, 1988, the monitor to all-sky survey on the GINGA satellite detected an extremely bright source flared up in the Fox constellation. It was given the name ASM 2000+25 [4]. At the moment of detection the source was about ten times brighter than the Crab Nebula and one of the brightest in the sky. Soon its optical companion was found: a star of the B class [5].

To study the evolution of the source properties as it was fading down, the ROENTGEN observatory performed a series of observations. From May 15 to June 8, 1988, the ASM 2000+25 was the target of 25 sessions. Fig. 5 shows the source spectrum covering an energy range of 4 - 300 keV. It was obtained with the help of Pulsar X-1 devices and HEXE operating in the hard energy range, and with GSPC instrument. At energies below 10 keV the spectrum is mainly that of a very soft component with $kT \approx 2$ keV. The GSPC instrument did not detect any traces of the Fe line: the 3σ upper level per an equivalent width of this line is 50 eV. At energies higher than 10 keV, the spectrum shows a hard component [6, 7]. The hard component spectrum can be adequately described as a collision thermal radiation spectrum with a kT temperature of $\approx 95 + 2$ keV. Figure 6 has a drop in the high-energy region of the spectrum, which supports the thermal nature of the hard component. The hard component in the spectrum implies a similarity of this source to another ultrasoft transient (A 0620-00). Optical observations of A 0620-00 led to the conclusion [8] that this source is a candidate for a black hole. Two other candidates, CYG X-1 and LMC X-3, also give similar ultrasoft spectra with hard components. These findings suggest that a compact object in ASM 2000+25 might be one more candidate for a black hole. Worth mentioning is a striking similarity between the observed spectrum of this source and the spectrum of the accretion disc near the black hole calculated [9] for a luminosity close to the critical Eddington's and at the disc substance viscosity $\alpha \approx 1$.

Observations of the Crab Nebula.
To graduate the instruments a series of sessions was made to observe the Crab Nebula. The results obtained allowed checking the response of the instruments aboard KVANT. Preliminary graduation results for the HEXE instrument are given in Fig. 7 [10].

At the same time the observations of the Crab were used to check the precision of the clock aboard KVANT. Figure 8 presents the measurements of an X-ray pulse period by the HEXE in September - October 1987. The solid line in the Figure shows extrapolation of radioobservations.

Sources beyond the Galaxy. During KVANT observations in 1987 both NGC 4151 and Cen A were in the state of low luminosity. When in JUly 1988 Cen A was in the state of high luminosity, its hard X-ray spectrum was reliably recorded by HEXE and Pulsar X-1 instruments. Observations of other sources beyond the Galaxy were also made (see Table 2). Figure 9 shows, as an example, a hard X-ray spectrum of 3C 273 quasar recorded by HEXE during 3600 s observations in December 1987.

Observations of Hercules X-1 pulsar. A bright X-ray pulsar Her X-1 was several times detected by module KVANT instruments. Analysis of HEXE and GSPC data allowed determination of an X-ray pulse period of the source (Table 3, [11, 12]).

These measurements clearly indicate that Her X-1 has again passed onto a stage of an accelerated rotation (Fig. 10). The Figure reveals a substantial acceleration of the neutron star rotation as compared with the EXOSAT observations in 1984. This comparison helped to determine an averaged acceleration rate P/P = 4 · 10^{-6} $year^{-1}$, which is in agreement with previous EXOSAT observations [13].

Gamma-bursts.
Several gamma-bursts were recorded by the IRA instrument. The instrument enables obtaining energy spectra of gamma-bursts in 63 energy channels of a quasi-logarithmic scale and time resolution of 1.44 s. In addition, the detector allows investigation of the burst evolution with time at a resolution of 20 ms. One event (October 6, 1988) was simultaneously registered by the burst IRA detector aboard KVANT and two Soviet-French detectors LILAS aboard the interplanetary station FOBOS. The time history of this burst in the spectral regions 46 - 223 and 223 - 1384 keV at a resolution of 1.44 s is presented in Fig. 11. The energy release in this event was 3.3 · 10^{-6} erg/cm^2 in the range 46 - 1384 keV.

Table 1. Roentgen observatory instruments.

Instrument	Device type	Energy range keV	Geometrical area of detectors, cm^2	Collaboration
HEXE	4 · phoswich detectors with 1.6 · 1.6 FWHM collimators	15 - 200	800	MPE (FRG) AIT (FRG) IKI (USSR)
TTM	Coded Mask Imaging Spectrometer 7.8 · 7.8 FWHM	2 - 32	600	SPU (Netherlands) Birmingham University (GB) IKI (USSR)
Pulsar X-1	4 · phoswich detectors with 3.0 · 3.0 FWHM collimators	30 - 1000	4 · 314	IKI (USSR)
GSPC	Gas scintillation proportoinal counter with 3.0 · 3.0 FWHM collimators	4 - 100	300	ESTEC IKI (USSR)
IRA	Phoswich gamma-burst detector, 2π	40 - 1400	314	IKI (USSR)

Table 2. The list of main targets of bRoentgen observatory after the first year of operation.

Target	Number of sessions
SN1987A	370
Cyg X-1	110
Cyg X-3	36
ASM 2000+25	25
Her X-1	72
Cen A	10
Crab Nebula	98
GX 1+4	7
GX 301-2	10
3C273	49
NGC4151	9
SMC X-1	8
AO538-66	12
H2252-035	13
Cep X-4	13

Table 3. The Hercules X-1 period determination by Roentgen observatory instruments in August, 1987 - October, 1988. Julian dates corresponds to the beginning of observations. Errors are 3σ values.

Date	JED	Ψ_{35}	T_{obs} sec	Period (sec) HEXE	Period (sec) GSPS
13-15.08.87	2447021	0.3	6600	1.237 772 6 ±0. 000 000 8	1.237 772 8 ±0.000 000 5
21.05.88	2447303	0.39	500	1.237 77 ±0.000 04	1.237 77 ±0.000 06
25.07.88	2447367	0.25	1140	1.237 767 0 ±0.000 003 5	1.237 767 ±0.000 010
7-09.10.88	2447441	0.35	1440	1.237 764 3 ±0.000 001 2	———

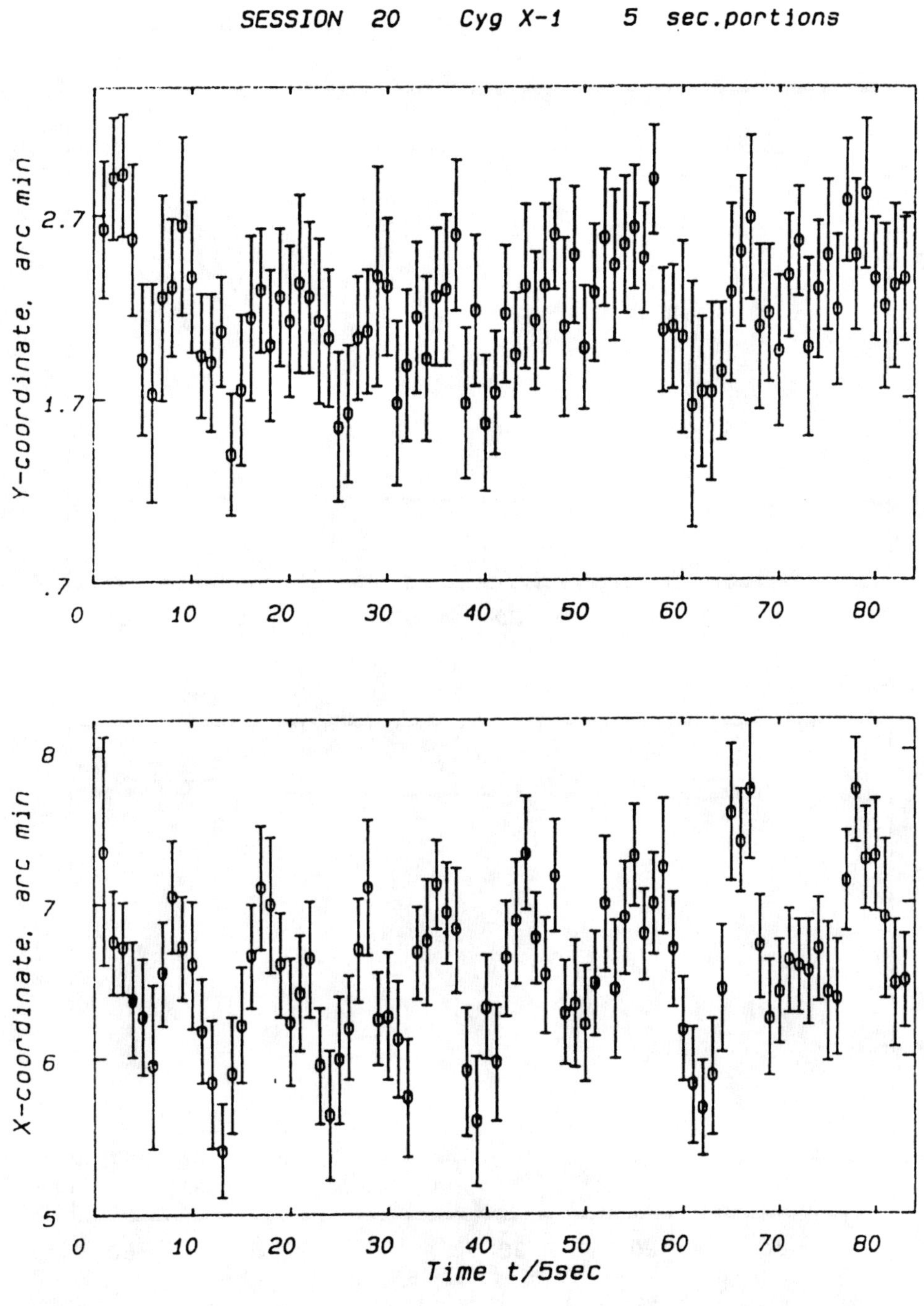

Fig. 1. Changes in the position of the Cygnet X-1 source observed with the TTM telescope within a 7-min observation (June 18, 1987). The exposure for each point was 5 s. The source position is given in the telescope coordinates, the point x = 0, y = 0 corresponds to the center of the visual field. An error in the position determination (1 σ) of the source is given [1].

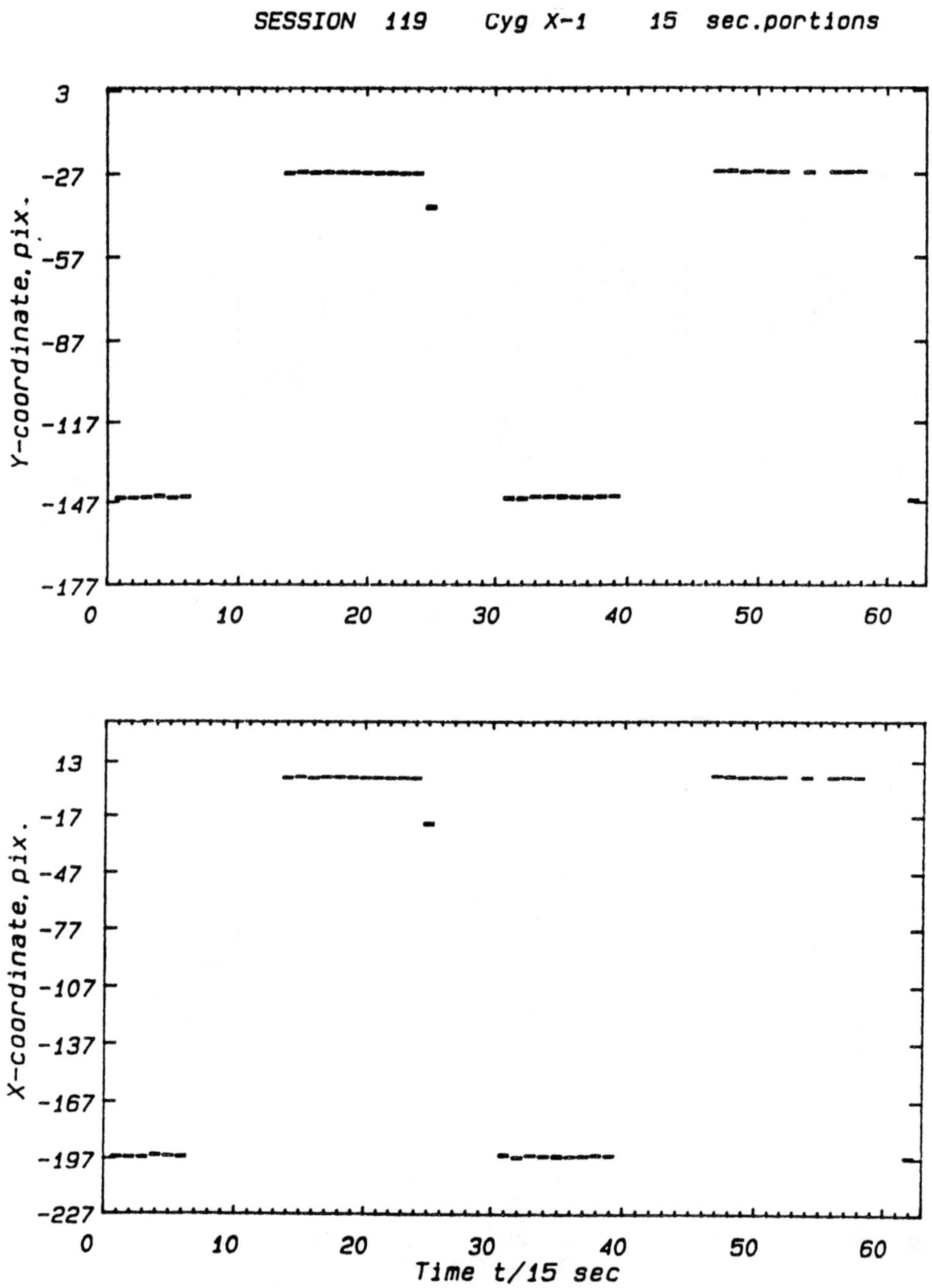

Fig. 2. The session of the Cyg X-1 observation on August 17, 1987. During the session two cycles of the complex reorientation were made from the background area to the source area. Errors in the source position determination cannot be seen because of a large scale along the ordinate axis [1].

Supplement: counts/s. The day after explosion.

15 - keV;

45 - 105 keV;

105 - 200 keV.

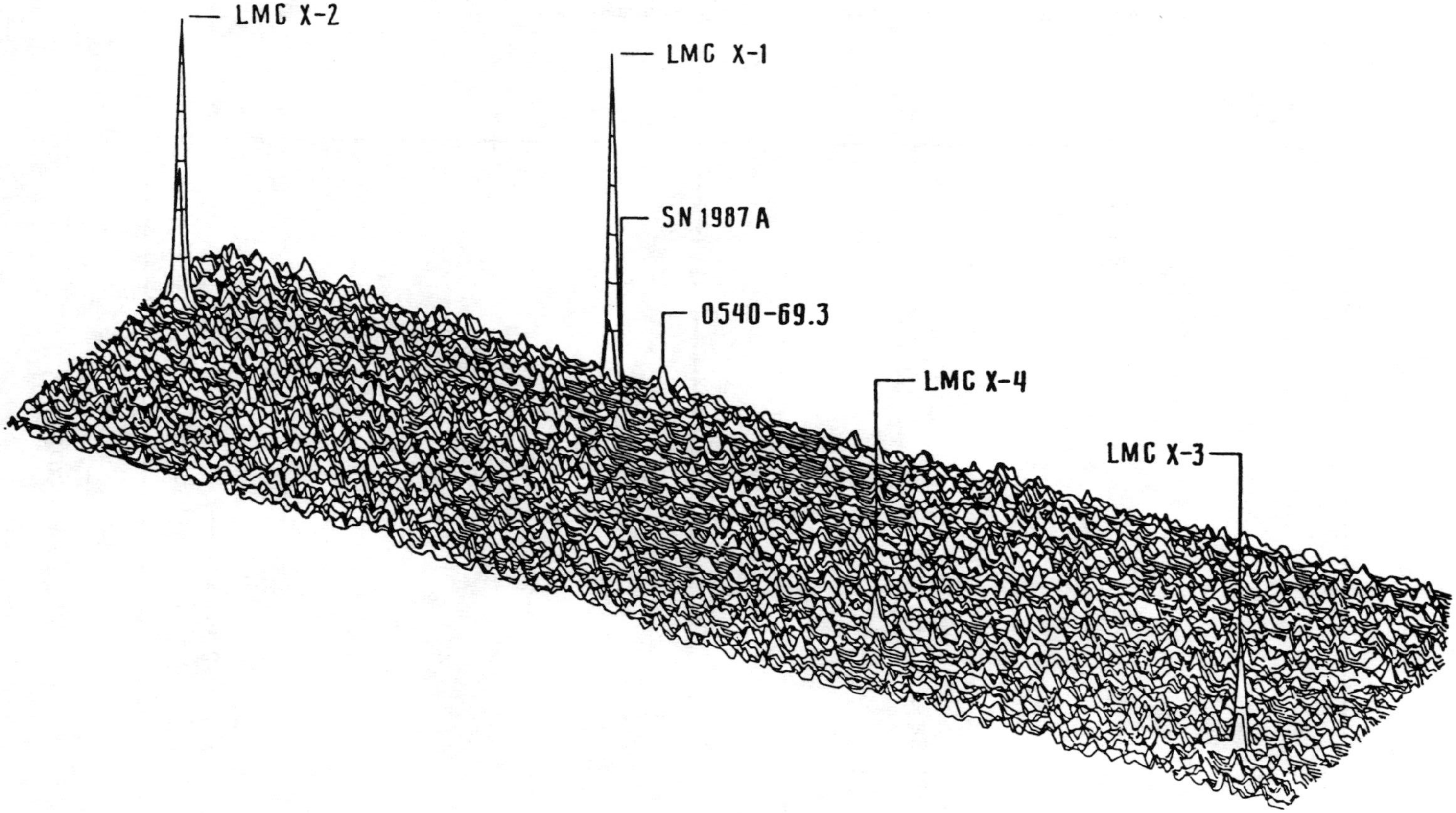

Fig. 3. Image of the Large Magellanic Cloud in the range 2 - 32 keV (2.5° · 8.5°) recorded by the TTM instrument from August 10 to 23, 1987 during 40 000-s observations. The height of the peaks corresponds to the importance of the source detections [2].

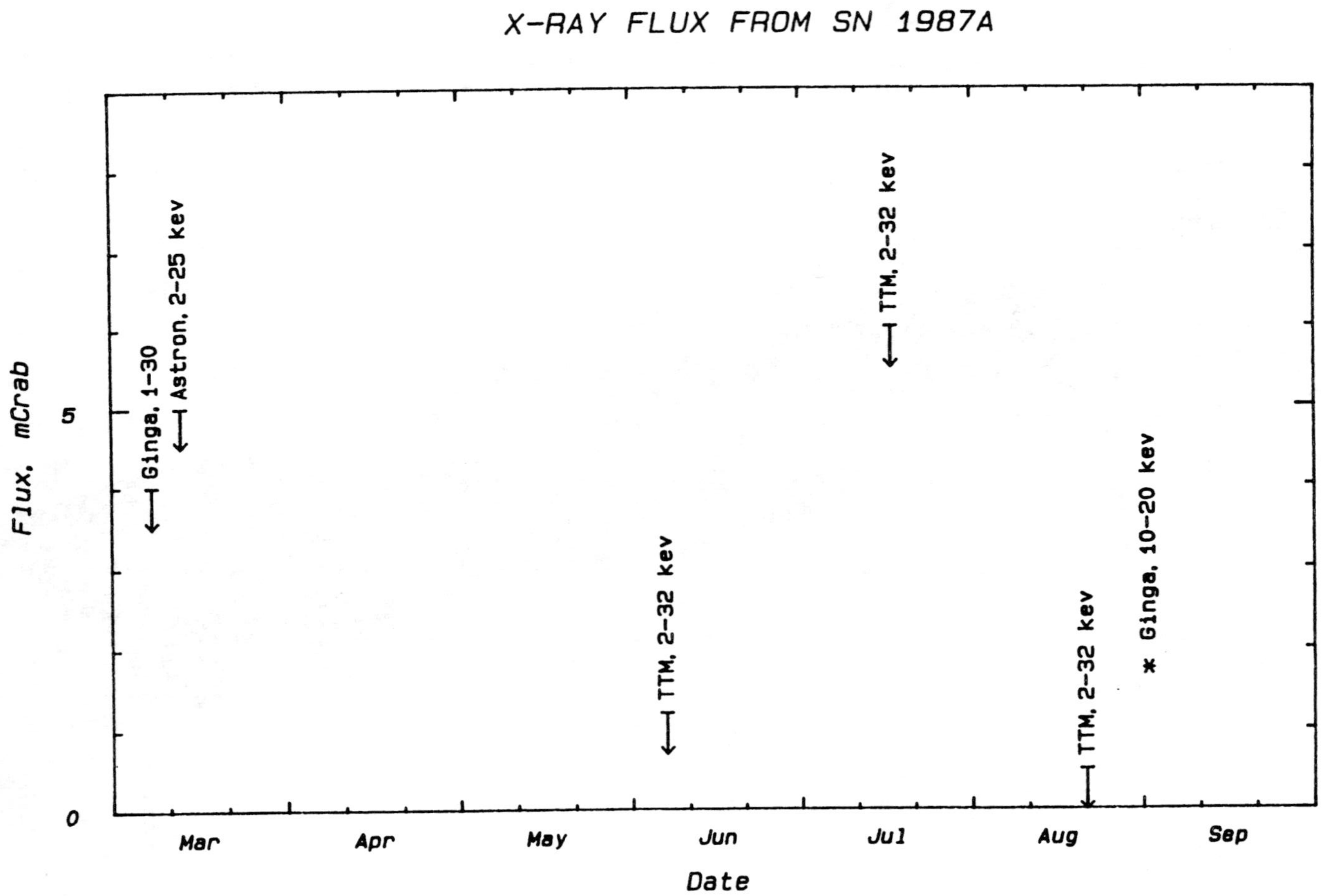

Fig. 4. Upper limits (3 σ) for X-ray emission from SN1987A, assuming its spectrum is similar to the Crab Nebula spectrum, recorded by different instruments in February to September 1987 [14, 15, 2]. The asterisk shows the detection of SN1987A soft X-ray radiation by GINGA [16] (the flux values were transformed into mCrab units by the authors [17]).

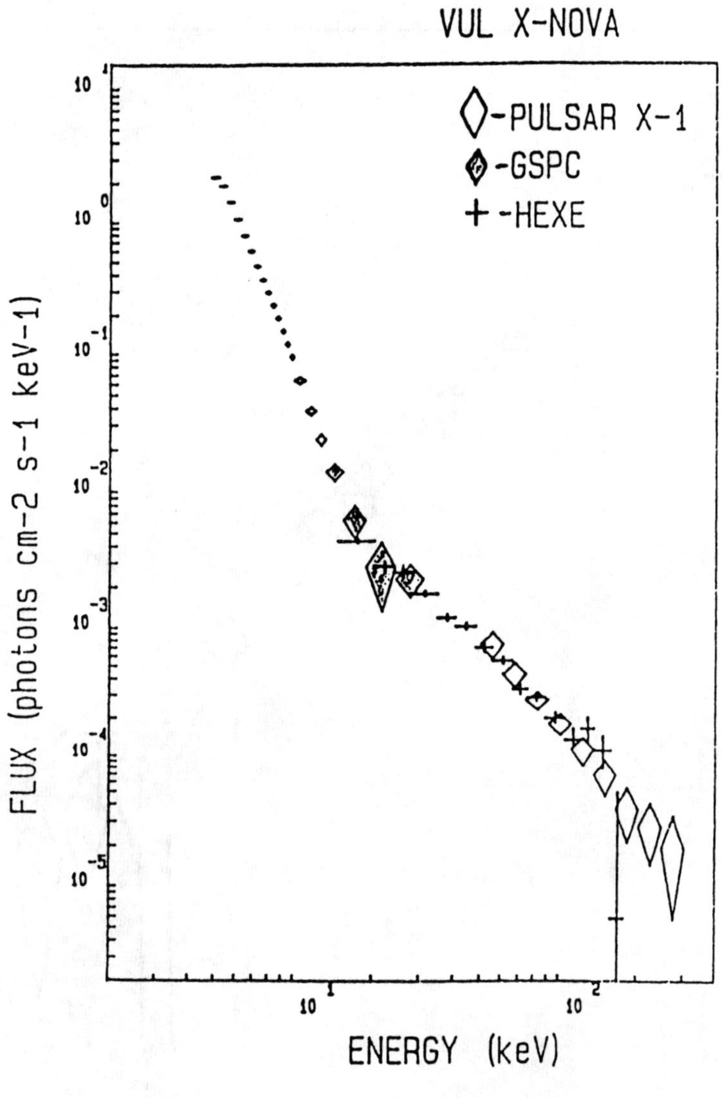

Fig. 5. X-ray spectrum of ASM 2000+25 source according to the module KVANT data: GSPC (17. 05. 88, exposure 240 s), two HEXE detectors (15. 05. 88 and 4, 6, 8. 06. 88, exposure 5400 s), one detector of the Pulsar X-1 instrument (15. 05. 88 and 4, 8. 06. 88, exposure 1300 s). The observed spectrum consists of two components: a soft component with a temperature of $20 \cdot 10^6$ °C and a hard one with a temperature of about 10^9 °C [6].

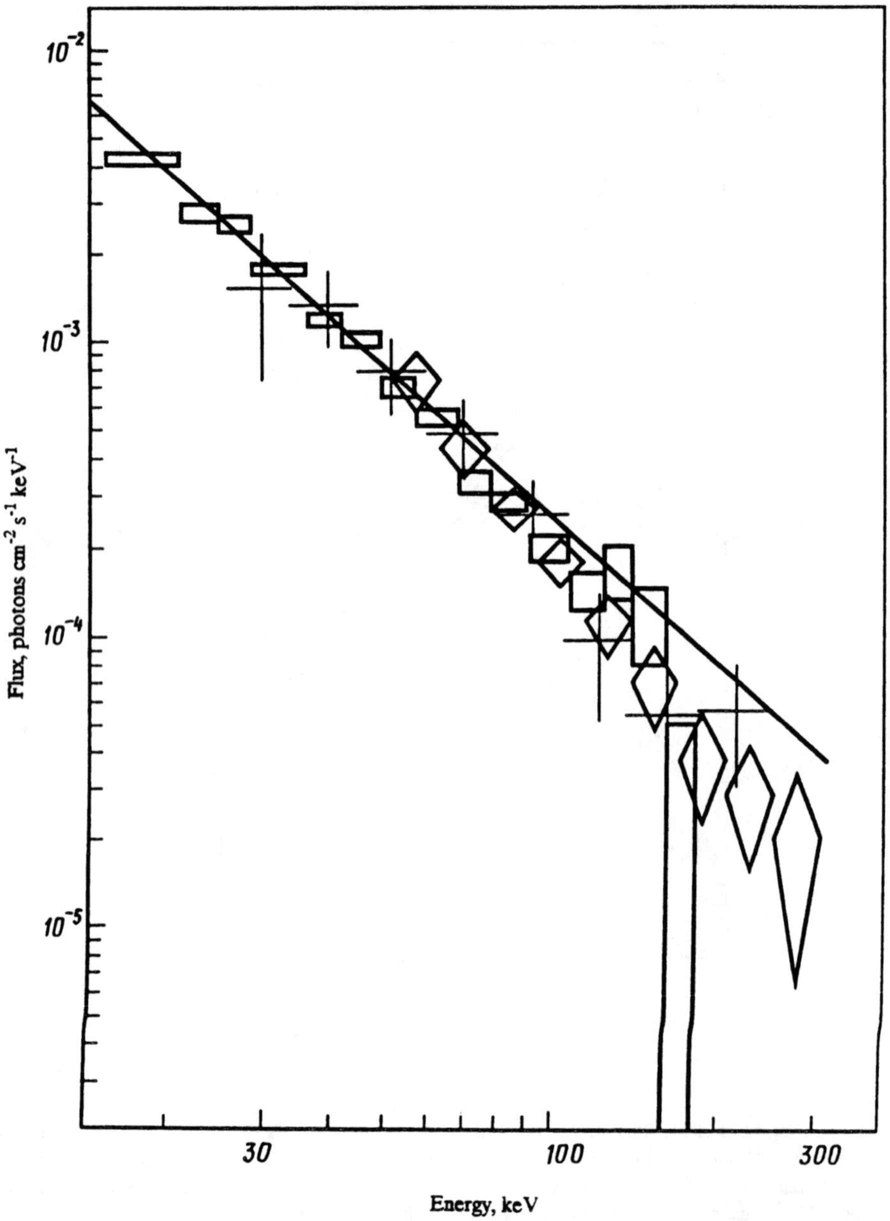

Fig. 6. X-ray radiation of ASM 2000+25 source according to the data of two detectors of the Pulsar X-1 instrument, an IRA analyzer (crosses show the results of the first detector and rhombi, of the second detector). For comparison also given are HEXE data (squares) and a power approximation (a spectral index α = 1.7) [6].

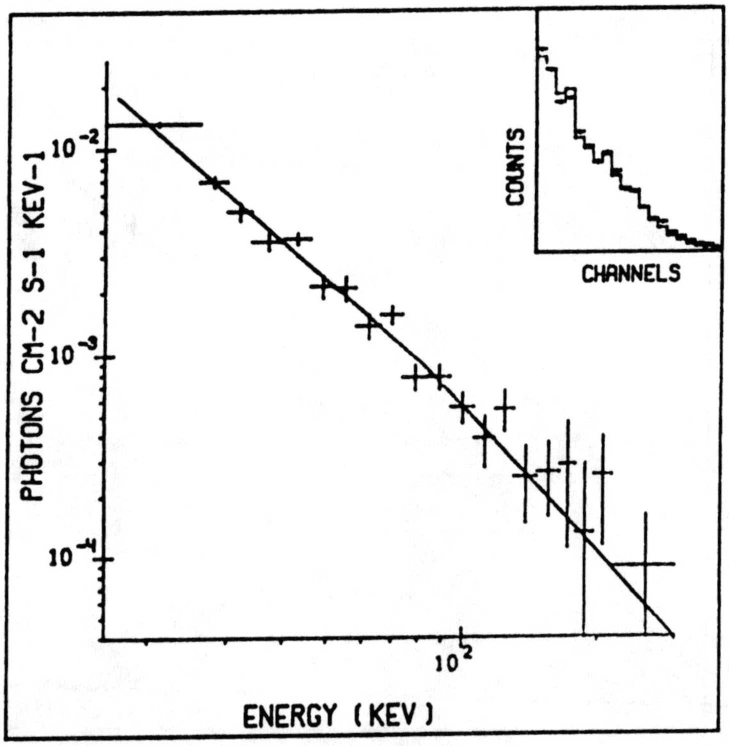

Fig. 7. Spectrum of graduation source TAU X-1 according to HEXE observations [10].

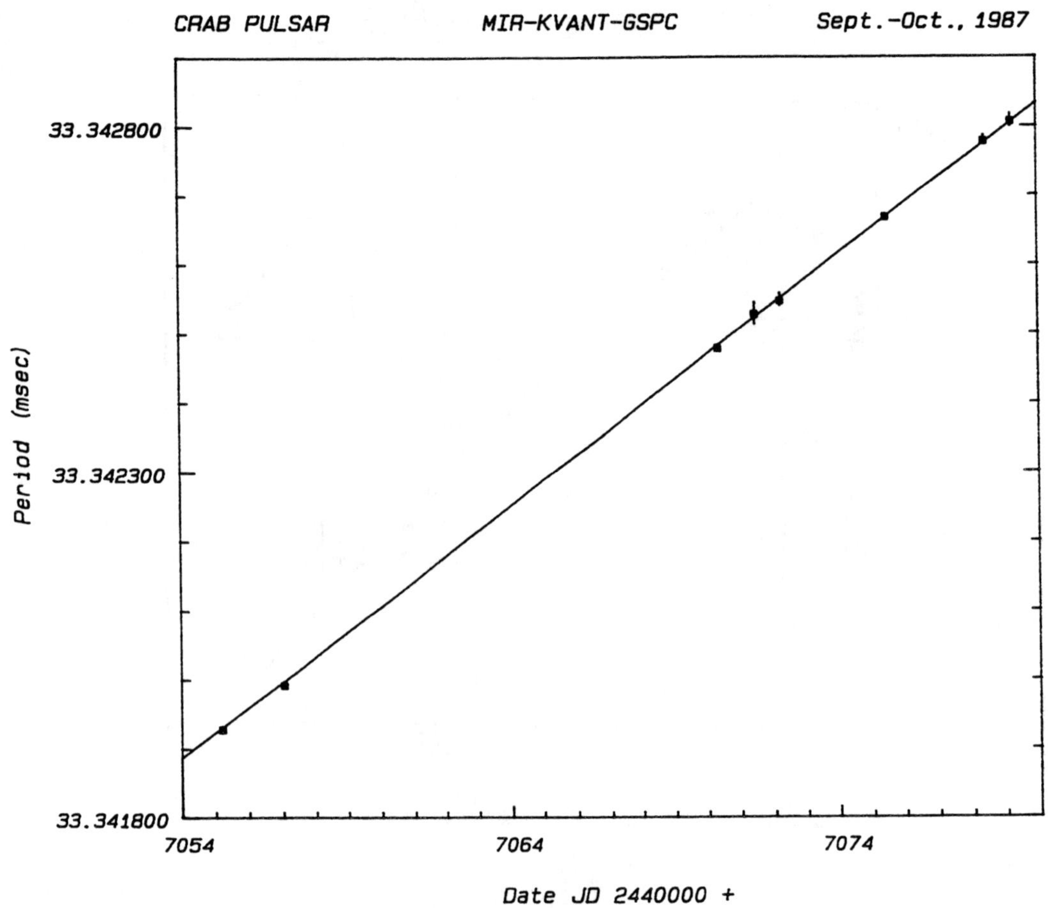

Fig. 8. X-ray pulse period of TAU X-1 according to GSPC data. The solid line shows extrapolation of radio observation data.

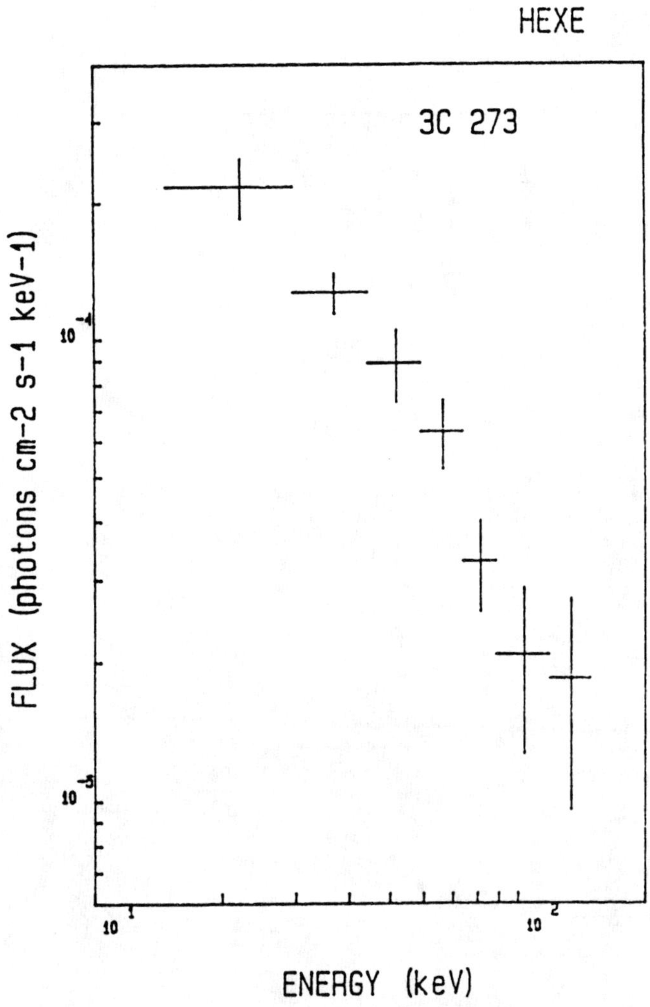

Fig. 9. Spectrum of 3C 273 quasar recorded by HEXE in December 1987.

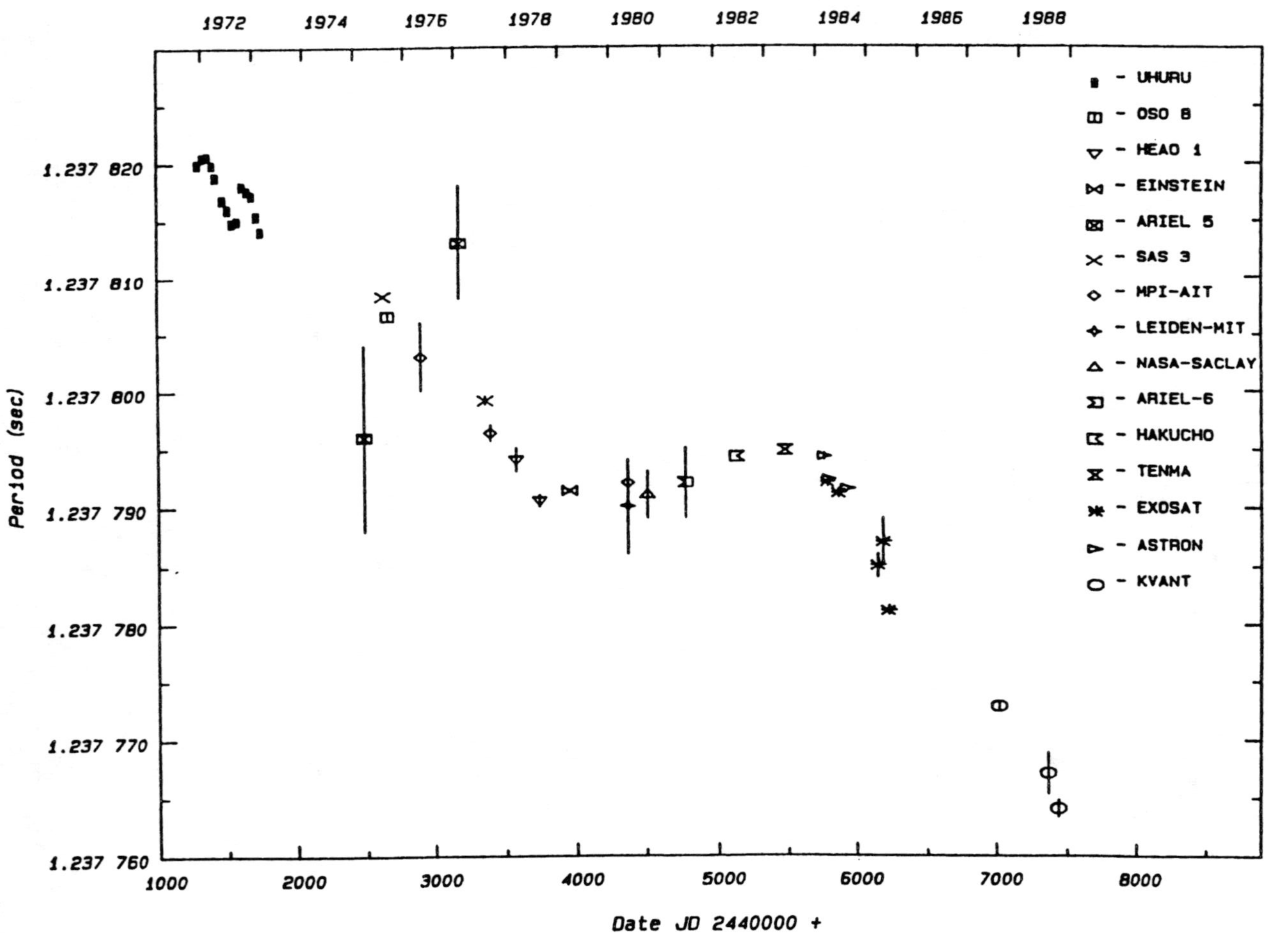

Fig. 10. Measurements of Hercules X-1 pulsar period made in 1972 - 1988 [12].

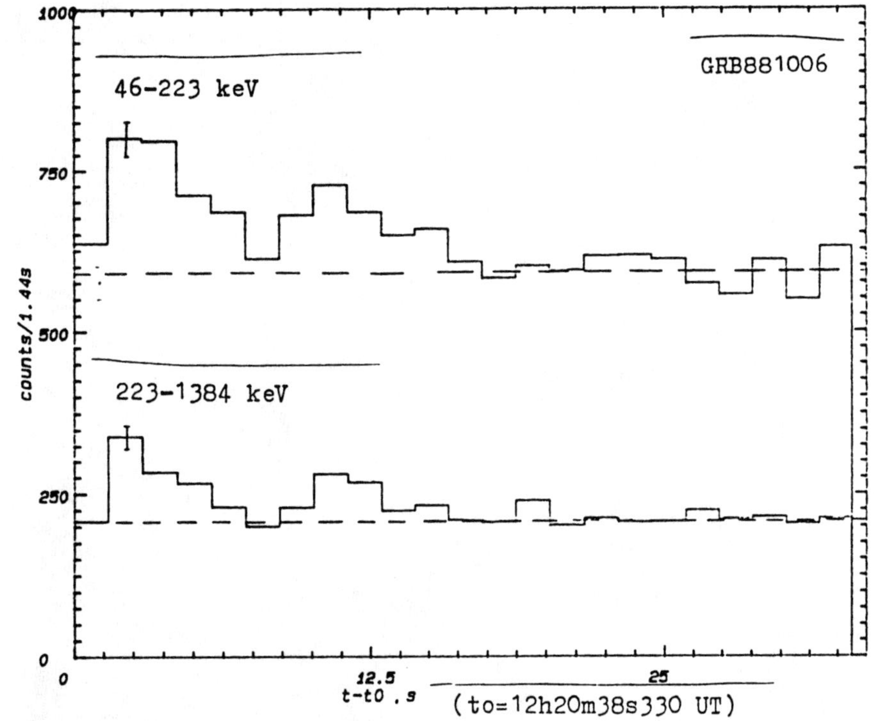

Fig. 11. Time history of the GRB 881006 burst detected by the IRA instrument aboard the module KVANT. Data are given for two spectral regions. The total flux is $3 \cdot 10^{-6}$ erg/cm^2.

REFERENCES

1. R.A. Sunyaev, M.R. Gil'fanov, E.M. Churazov, et al., Pis'ma Astron. Zh., 1988b, v. 14, p. 695.
2. R.A. Sunyaev, A.S. Kaniovsky, V.V. Efremov, et al., Nature, 1987a, v. 330, p. 227.
3. R.A. Sunyaev, A.S. Kaniovsky, V.V. Efremov, et al., Pis'ma Astron. Zh., 1987b, v. 13, p. 1027.
4. F. Makino et al., IAU Circ., no. 4587, 1988.
5. S. Okamura and T. Noguchi, IAU Circ., no. 4589, 1988.
6. R.A. Sunyaev, I.Yu. Lapshov, S.A. Grebenev, et al., Pis'ma Astron. Zh., 1988, v. 14, p. 771.
7. R.A. Sunyaev et al., IAU Circ., no. 4606, 1988e.
8. J.E. McClintock and R.A. Remillard, Astrophys. J., 1986, v. 208, p. 110.
9. N.I. Shakura and R.A. Sunyaev, Astron. and Astrophys., 1973, v. 24, p. 337.
10. R.A. Sunyaev, V.V. Efremov, A.S. Kaniovsky, et al., Pis'ma Astron. Zh., 1988a, v. 14, p. 579.
11. W. Voges et al., Astron. and Astrophys., 1988.
12. R.A. Sunyaev, M.R. Gil'fanov, E.M. Churazov, et al., Pis'ma Astron. Zh., 1988e, v. 14, p. 979.
13. H. Oegelman et al., Space Sci. Rev., 1985, v. 40, p. 347.
14. F. Giovanelly et al., IAU Circ., no. 4367, 1987.
15. F. Makino et al., IAU Circ., no. 4336, 1987a.
16. F. Makino et al., IAU Circ, no. 4447, 1987b.
17. R.A. Sunyaev, M.R. Gil'fanov, E.M. Churazov, et al., Pis'ma Astron. Zh., 1988b, v. 14, p. 591.

Physics of Neutron Stars
Edited by A. M. Kaminker

ISBN 1-56072-164-2

TO THE MECHANISM OF THE SN 1987A BURST

E. K. Grasberg and D. K. Nadyozhin
Radioastrophysical Observatory of the Academy of Sciences
of the Latvian SSR, 226524, Riga

It has been shown that if both neutrino pulses from the SN 1987A were real at the moment of the first pulse (the Mont Blanc) emission, the shell could not receive energy higher than ~ $5 \cdot 10^{49}$ erg.

The burst of the SN 1987A supernova in the Large Magellanic Cloud (LMC) provides a unique opportunity to impose some constraints onto the character of the explosion-like energy release in such bursts of type-II supernovae (SNII) which are associated with the collapse of the star core. Indeed, it was just for the SN 1987A that (a) the earliest stage of the optical burst development was first followed in most detail (literally speaking, hour-by-hour), (b) the real presupernova was first observed, (c) the neutrino radiation from the collapsing star core was first recorded. The data on the character of energy release are very important because up to now there is no clear understanding of all the details of the mechanism of the shell jet in the gravitational collapse of the star core.

The initial phase of the burst development is determined, in fact, by the structure of the presupernova just before the explosion and the character of energy release understood here as a quantity of energy ε transferred to the thrown-off shell, and the characteristic time Δt_e of this energy transfer. The modern concept of the explosion mechanism allows both instantaneous ($\Delta t_e \approx 0$) and continuous energy release (Δt_e is comparable with or larger than the characteristic hydrodynamic time of the star) [1]. Two recorded neutrino pulses from the SN 1987A lead to the idea about a two-stage collapse. Or else one may think of a "double explosion", i.e., of a twice energy injection into the thrown-off shell of the supernova. Note that a possible "double explosion" was indicated earlier for the SN 1983K in NGC 4699 [2].

The star burst was simulated by the numerical solution of the gas dynamics equations in the approximation of the radiative thermal conductivity. The real presupernova of the CN 1987A was a hot supergiant Sk-69^0 202 of the spectral class B3 1a. The ambiguity of the distance modulus (m - M) for the LMC, which may reach the value ~ 0^m, 5, and the errors in the determination of the mass of the supergiant B3 1a lead to the following intervals of the admissible changes in the presupernova characteristics : R_o ~ (30-60)$R_\odot$, M_o ~ (10-35)$M_\odot$. As an initial model for the calculations, a simple hydrostatically equilibrium configuration in the form of a polytrope of index n = 3 was chosen that imitated the density distribution in massive stars (with the exception of its dense core). In all the variants, the mass of the thrown-off shell was M_o = 16$M_\odot$. For variant 1, R_o = 60 $R_\odot$, for all the other variants R_o = 30 $R_\odot$. The other characteristics of different variants are listed in Table 1.

Variant	1	2	3	4	5	6	7
$\varepsilon_{oo}/10^{51}$ erg	-	-	0.01	0.1	0.1	3.0	3.0
Δt_e, hour	-	-	-	-	-	3.0	10.0
$\varepsilon_o/10^{51}$ erg	1.5	3.0	3.0	3.0	1.5	-	-

The following cases were considered:

1. Variants 1 and 2 - a single "instantaneous explosion". Energy ε_o is released at the moment $t_{\nu 2}$.
2. Variants, 3, 4, 5 - a"double explosion". At the moment $t_{\nu 1}$ the energy ε_{oo} is released, then at the moment $t_{\nu 2}$ the energy ε_o is released.
3. Variants 6 and 7 - "prolonged explosion". The energy ε_{oo} is uniformly released for the time Δt_e . The energy starts releasing at the moment $t_{\nu 1}$.

The moments $t_{\nu 1}$ and $t_{\nu 2}$ corresponds to the first and the second neutrino pulses recorded on February 23.124UT [3] and 23.316UT [4-6], respectively. Figure 1 illustrates the initial rise of luminosity in the photometric band V. For the modulus of the LMC the value 18, 5 (50 ?) has been taken and for the absorption $A_v = 0^m,6$. Figure 2 shows the variation in the matter velocity at the level of the photosphere, U_{ph}, for the first 120 hours of the burst development. The consideration of these illustrations leads to the following conclusions.

1. The "simultaneous explosion" with the energy $\varepsilon = (1.5\text{-}3)10^{51}$ erg satisfactorily describes the initial phase of the brilliance increase and the rate of the shell fragment separation of the SN 1987A in the case where the energy release occured at the moment of the second neutrino pulse $t_{\nu 2}$. The "simulataneous explosion" cannot be associated with the moment $t_{\nu 1}$ since in this case an increase of the brilliance should have been started several hours earlier than it was really observed.

2. The "double explosion" can be associated with the observations of CN 1987A only if at the moment $t_{\nu 1}$ the energy release was not too intense, $\varepsilon_{oo} \leq 5\text{-}10^{49}$ erg. For larger values of ε_{oo} the rates of the shell fragment separation turn out to be lower than those really observed for SN 1987A (up to ~ 3-10 kn/s in the first days). This is associated with the fact that if a relatively weak "first explosion" provides the shell fragment separation prior to the arrival to the surface of a shock wave from the "second (main) explosion", then the second shock wave propagates along the moving matter under the conditions unfavorable for the effect of its cumulative enhancement. Therefore it cannot substantially accelerate the shell edge.

3. The "prolonged explosion" is the energy release for a time longer than thee characteristic hydrodynamic time of the presupernova and therefore cannot explain the initial steep rise of the SN 1987A brilliance.

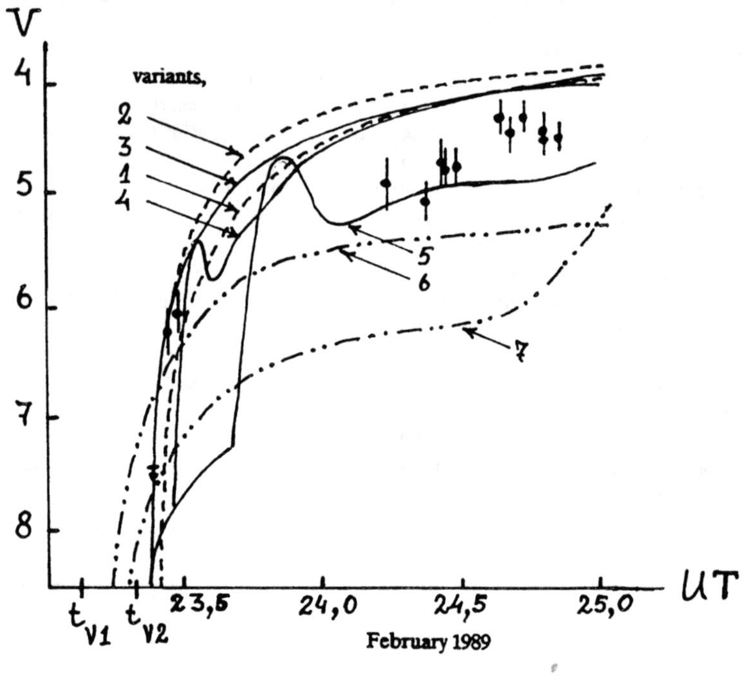

Fig.1.

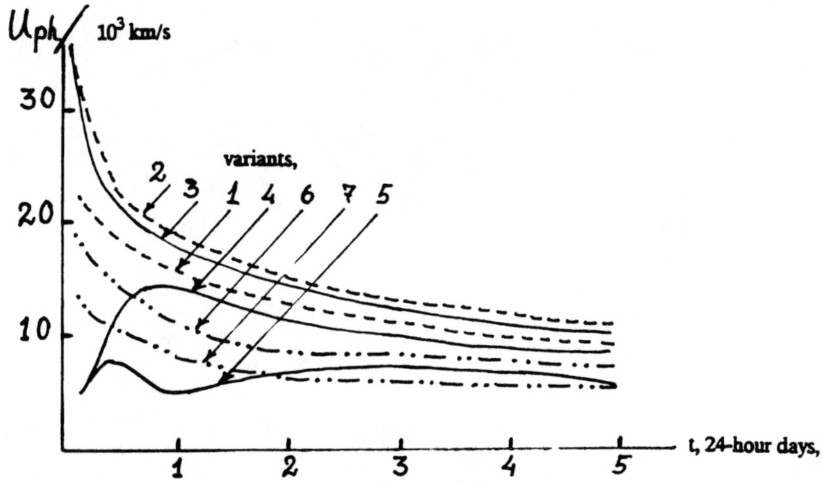

Fig.2.

References
1. E.K. Grasberg, V.S. Imshennik, D.K. Nadezhin, and V.P. Utrobin. Pis'ma v Astron. Zh., 1987, 13, 547.
2. E.K. Grasberg and D.K. Nadezhin, Pis'ma v Astron. Zh., 1987, 13, 547.
3. E.K. Grasberg and D.K. Nadezhin, Pis'ma v Astron. Zh., 1986, 12, 168.
4. B.L. Dadykin, G.T. Zatsepin, V.V. Korchagin, et al. Pis'ma v ZhETF, 1987, 45, 464.
5. E.N. Alekseev, L.N. Alekseeva, V.I. Volchenko, and I.V. Krivosheina, Pis'ma v ZhETF, 1987, 45, 461.
6. R.M. Bionta, G. Blewitt, G.B. Bratton, et al., Phys. Rev. Letters, 1987, 58, 1494.
7. K. Hirata,, T. Kajita, M. Koshiba, et al., Phys. Rev. Letters, 1987, 58, 1490.

Physics of Neutron Stars
Edited by A. M. Kaminker

ISBN 1-56072-164-2

ON POSSIBLE DETECTION OF PULSED HIGH FREQUENCY RADIATION FROM SN 1987A

A. Z. Kazbegi, G. Z. Machabeli and G. I. Melikidze
Abastumani Astrophysical Observatory of the Academy of Sciences of the Georgian SSR

It has been shown that if in the SN 1897A a pulsar is formed, it would emit high-frequency radiation with the spectrum similar to that of the PSR 0531+21 but shifted towards high frequencies.

In February 1987 a burst of a supernova took place in the LMC. According to the theory of star evolution, during an explosion of a supernova the compression of its central part may give rise to the formation of a neutron star - a pulsar. Then, since the compression process proceeds with the conservation of the angular rotation momentum and the magnetic flux, a newly born pulsar may have a revolution period of the order of $P \sim$ 1-2 ms, and the deceleration rate is $\dot{P} \sim 10^{-13}$ [1,2]. As was indicated in [3], the presence of a pulsar in the SN shell is indicated by the maintenance of the optical brightness at a high level for several weeks upon the explosion. The aim of the present study is to consider a possible mechanism of radiation emission and determine the frequency range in which such a pulsar (if it is really formed in the SN 1987A) may emit a radiation.

A pulsar radiation is generated in the manetospheric electron-positron plasma where two types of waves are possible: a "potential-potential" (lt-wave) and a pure transversal (t-wave) [4]. Note that electromagnetic t-waves can leave the magnetosphere without any transformation. In a uniform plasma, t-waves are generated due to asymmetry of the distribution function in the resonance at the anomalous Doppler effect [5],

$$\omega - k_z v_z + \omega_B / \gamma = 0 \qquad (1)$$

Here ω and $\vec{k}$ are the frequency and the wavevector of excited perturbations, V_Z is the particle velocity along the magnetic field $\vec{B} \| \vec{Z}$, $\omega_B = eB / mc$, e and m are the charge and the mass of particles, respectively, and $\gamma = (1 + p^2/m^2c^2)^{1/2}$ is the Lorentz factor of particles. Considering the perturbation propagating along the magnetic field $K \| B$, we have

$$1 = \frac{k^2c^2}{\omega^2} + \sum_a \frac{4\pi e^2 a}{m_a} \int \frac{f_{||}(1 - k_z V_z/\omega)}{\gamma(\omega - k_z V_z \pm \omega_B/\gamma)} d p_z \qquad (2)$$

For the parameters of magnetospheric plasma the following condition is always valid

$$\delta \equiv \frac{\omega_p^2}{\omega_B^2} <\gamma> \ll 1 \qquad (3)$$

where $<\gamma> = \int \gamma f_{||} dp_z, f_{||} = \int f_0 p_\perp dp_\perp$, $\omega_p^2 = 4\pi e^2 n/m$, f_0 is the distribution function for plasma particles. This condition excluydes a development of hydrodynamic (hose) instability.

For real part of the frequency, $\omega = \mathrm{Re}\,\omega \ll \omega_B$, and the quantity $\Gamma = \mathrm{Im}\,\omega$ we have

$$\omega = k_z c \left(1 - 2\frac{\omega_p^2}{\omega_B^2} <\gamma>\right) \qquad (4)$$

$$\Gamma = \frac{\pi^2 e^2}{k_z} \left\{ f_{||}\left(\frac{m\omega_B^3}{2k_z \omega_p^2 <\gamma>}\right) - f_{||}\left(-\frac{m\omega_B}{2k_z}\right) \right\} \qquad (5)$$

The second term in (5) is responsible for the perturbation attenuation at plasma particles. For a symmetric distribution function the attenuation it turned out to be more intense than the excitation. But distribution function of the magnetospheric plasma of the pulsar has a tail elongated towards positive pulses, and the attenuation takes place only at particles with negative pulses. Hence, under the condition

$$k_z < \omega_B / 2c\,\gamma_p \qquad (6)$$

the attenuation is small due to absense of particles that satisfy the resonance interaction $\omega - k_z v_z - \omega_B/\gamma = 0$. Formulae (1), (6), and (3) reduce to the following inequalities:

$$\frac{\gamma_p^2}{\gamma_t} \leq \left(\frac{\omega_p}{\omega_B}\right)^2 < 1 \qquad (7)$$

According to [6], the increment in the instability of t-waves is

$$\Gamma = \frac{\pi}{8}\frac{n_t}{n_p}\gamma_t <\gamma> \frac{\omega_{p0}^4}{\omega_{B0}^3}\left(\frac{R}{R_0}\right)^3 \qquad (8)$$

The calculations performed in the system at rest of the main plasma mass, the subscript 0 indicates the values close to the star surface, the subscripts tl and p indicate the values for a tail and the main mass of the plasma, respectively, n is the particle density, $R = 10^6$ cm is the neutron star radius, and R is the distance to the point under consideration. It was estimated that this instability takes place in magnetospheres of young pulsars. With time, the revolution of pulsars decelerates: a decrease in the rotation frequency results in the changes of the plasma parameters and, for 10^4 years, also in the violation of the excitation conditions.

In order to apply the obtained results to the pulsar in the SN 1987A one should know the plasma parameters. We take that the magnetic field at the star surface is $\sim 10^{12}$ G and the revolution period is $\sim$ 1 ms. Then the Goldreich-Julian density is $n_b \sim 7{\cdot}10^{14}$ cm^{-3} . Not going into the details of the magnetosphere structure, we should like to note here that in the existing models have a polar gap in the vicinity of the magnetic poles of the star where the particles of the primary beam are accelerated up to energies $\sim ehE_{||}$ by the electric-field component $\vec{E}$ directed along $\vec{B}$, where h is the gap height ($E_{||} \sim 10^9$ CGSE). To these energies there correspond the Lorentz-factor $\gamma_b = ehE_{||}/mc^2$. We shall take that $h = 5{\cdot}10^4$ cm and $\gamma \sim 10^9$ [7,8]. Assuming that the energy of the beam particles is of the order of the energy of plasma particles ($n_b\,\gamma_\beta = n_b\,\gamma_b$) and using (7), we have $\gamma_p \leq 30$. Therefore $n_p \sim 2{\cdot}10^{22}$cm^{-3} . Substituting these parameters into (8), we have

$$\Gamma \approx \frac{\pi}{8}\left(\frac{R}{R_0}\right)^3 10^3$$

In order to provide a sufficient time for the development of the instability under consideration, it is necessary that the condition $\Gamma \gg 1/\tau$ was fulfilled, where $\tau = P/2\pi = 1.7{\cdot}10^{-4}$ s is the time of the plasma transfer outside the light cylinder. Whence $\Gamma \gg 10^4$. Taking the maximum possible R (on the light cylinder), we have $\Gamma \sim 10^5$. Thus, the instability has a sufficient time to be developed in the magnetosphere of the pulsar in the SN 1987A. As has already been noted, it is developed at the tail of the distribution function. Taking the estimate $\gamma_t \sim 10^7$, it is possible to determine the resonance frequency. Using (1) and (4), we obtain that $\omega \sim \omega_B/\delta\gamma_t \sim 10^{17}$. Converting this value into the system of observer we obtain that the frequencies $\nu \sim 10^{17}$ would be observed. It should be noted that the frequencies exceeding 10^{19} Hz cannot be excited at this resonance since their wavelength $\lambda \equiv \frac{c}{\nu} < n^{-1/3}$, and therefore they are not plasma waves and cannot be described in the given approximation.

Thus if the distribution function has a long asymmetric tail, the cyclotron excitations are generated whose quasilinear interaction with the beam particles results in the particle diffusion in the pulse space across the magnetic field. On the other hand, in the particle motion to the light cylinder the force determining the existense of the adiabatic invariant in an inhomogeneous magnetic field "press" them to the force lines. As a result of these competing effects, the particles acquire an effective pitch-angle of the order of [6]

$$\Psi = 10^{-5}\left(\frac{R_B}{10^7\,cm}\right)^2$$

where R_B is the curvature radius of the force lines of the magnetic field. If $R_B \sim 10^7$ cm, the pitch-angle is

$$\Psi \sim 10^{-4} - 10^{-5}$$

In this case, the inequality $\Psi < 1/\gamma_t$ is valid for the tail particles and the maximum of the synchrotron radiation occurs at the frequencies

$$\nu_{max} = \frac{\omega_B}{\pi} \Psi \gamma_t^2 \sim 3 \cdot (10^{25} - 10^{26}) \text{ Hz}$$

Recently an X-ray radiation from the SN 1897A was recorded in the frequency range 10^{19} -10^{20} Hz [9]. One of possible explanations of the origin of this radiation is the presence of a central pulsar. But the observed radiation is not pulsed and therefore may also be explained by the occurrence of the radioactive $^{56}C_o$ decay [10,11].

The results can be summarized up as follows. In the formation of a pulsar at the center of the SN 1987A with the magnetosphere parameters close to the above considered the presence of a long asymmetric tail of the distribution function should result in the generation of t-waves that would be observed in the frequency range $\nu \sim 10^{17} - 10^{18}$ Hz. The t-waves, interacting with particles, give rise to quasilinear diffusion in the direction normal to the magnetic field. In this case, particles acquire a certain effective pitch-angle. The distribution function stops being one-dimensional and, hence, the mechanism of synchrotron radiation starts workingin the frequency range $\nu \sim 3 \cdot (10^{25} - 10^{27})$ Hz with the spectrum being close to that of the high-frequency radiation of the pulsar from the Crab Nebula.

References

1. C. Cutlaer and L. Lindblom, Ap. J., 1987, 314, 234.
2. J. Friedman, J./ Isper, and L. Parker, Ap. J., 1986, 304, 234.
3. J.P. Ostriker, Nature, 1987, 327, 287.
4. Dzh.G. Lominadze, G.Z. Machabeli, G.I. Melikidze, and A. Pataraya, Fizika Plazmy, 1986, 12, 1233.
5. Dzh.G. Lominadze, G.Z. Machabeli, and A.B. Mikhailovskii, Fizika Plazmy, 1979, 5, 1337.
6. Dzh.G. Lominadze, G.Z. Machabeli, and V.V. Usov, Astrophys. and Space Sci., 1983, 90, 19.
7. E. Tademaru, Ap. J., 1973, 183, 625.
8. A. Ray and G. Benford, Phys. Rev. D., 1981, 23, 2142.
9. R. Sunaev etc al., Nature, 1987, 330, 227.
10. S.A. Grebenev and R.A. Syunyaev., Pis'ma v AZh., 1987, 13, 945.
11. M. Itch, S. Kumagai, T. Shigeyama, K. Nomoto, and J. Nishimura., Nature, 1987, 330, 233.

Physics of Neutron Stars
Edited by A. M. Kaminker

ISBN 1-56072-164-2

NONEQUILIBRIUM PROPERTIES OF NEUTRON STARS

G. S. Bisnovatyi-Kogan
Institute of Space Research of the USSR Academy of Sciences, 117810 Moscow

The properties has been considered of the neutron star matter in the state of the minimum energy as a result of the formation and evolution of a neutron star. In real cases the composition of the matter building the solid crust turns out to be nonequilibrium. Nonequilibrium may influence the rate of neutron star cooling and result in explosions close to their surfaces that remind the observable gamma-bursts.

1. Equilibrium Matter at the Energy Minimum

We shall take that under the conditions of high density and strong degeneracy of electrons and nuclons the matter is cold at $T = 0$. At a density of $\rho < 8.1 \cdot 10^6$ g/cm^3 the matter consists of iron, ^{56}Fe, with maximally stable nuclei. If the Fermi energy of electrons, ε_{Fe}, exceeds the value ε_β,

$$\varepsilon_{Fe} = (m_e^2 c^4 + p_{Fe}{}^2 c^2)^{1/2} - m_e c^2, \quad \varepsilon_\beta(A,Z) = (m_{A,Z-1} - m_{AZ}) c^2 - m_e c^2,$$

$$\frac{p_{Fe}}{m_e c} = \left(\frac{3\pi^2 h^3 \rho}{m_e^3 c^3 m_p \mu_z}\right)^{1/3} \approx \left(\frac{\rho}{10^6 \mu_z}\right)^{1/3}, \quad \mu_z = \frac{A}{Z} \tag{1}$$

the electron capture by a stable nucleus (A, Z) becomes energetically favorable.

At $T = 0$, a thermodynamical equilibrium corresponds to the minimum of the total energy per unit volume, E_{tot}, which is a function of A and Z at the given nuclon concentration n_b. With an increase in the density the equilibrium shifts towards larger A and Z, and at $\rho > \rho_{nd} = 4 \cdot 10^{11}$ g/cm free neutrons arise in the equilibrium. At $\rho < \rho_{nd}$ we obtain with due regard for the rest energy [1]:

$$E_{tot}(A, Z, n_b) = n_{AZ}(m_{AZ} c^2 + W_L) + E_e(n_e), \tag{2}$$

$$n_e = Z n_{AZ}, \quad n_b = A n_{AZ}$$

The masses of stable nuclei are well known [2]. For nuclei far from the stability region there is no experimental data and therefore semiempirical formulae for the binding energy $B_{A,Z}$ are used of which the simplest is the Weizsacker formula [3]:

$$B_{AZ} = 15.568 - 17.226 A^{2/3} - 0.698 \frac{Z^2}{A^{1/3}} - 23.379 \frac{(A - 2Z)^2}{A} + \frac{34}{A^{3/4}} \delta \ (MeV), \tag{3}$$

$$B_{AZ} = Z m_p + (A - Z) m_n - m_{AZ}$$

$\delta = 1$ for even A and Z, $\delta = 0$ for odd A and any Z, and $\delta = -1$ for even A and odd Z. More exact are the Meyers and Swiatecki formula [4] and also the tables from Ref. [5]. The energy of cold electrons is

$$E_e = \frac{m_e^4 c^5}{24 \pi^2 h^3} \left[3y(2y^2 + 1)\sqrt{y^2 + 1} - 3 Arsh y\right], \quad y = \frac{p_{Fe}}{m_e c} \tag{4}$$

The quantity W_L is the electrostatic-interaction energy whose minimum is attained for a body-centered nucleis lattice [1]:

$$W_L = -1.819620 \frac{Z^2 e^2}{b} \quad n_{AZ} b^3 = 2(5)$$

The dependence of the equilibrium composition on density obtained in [1] is given in Table 1.

At $\rho > \rho_{nd}$, all the bound neutron states in nuclei turn out to be filled, and a further increase in the density results in the formation of free neutrons. A reliable approach to the problem of the determination of the matter properties in the presence of free neutrons has been developed elsewhere [6]. This approach is based on the following principles.

1. The consistent description of the energy of nuclon interactions both inside and outside the nuclei.
2. The allowance for the effect of surrounding neutrons on the surface nucleus energy that goes to zero for a single phase matter.
3. The allowance for the energy of Coulomb interaction of lattice electrons and nuclei and also of protons inside nuclei.

If $n_{A,Z}$ and n_n are the concentrations of nuclei and free neutrons outside nuclei and $V_{A,Z}$ is the nucleus volume, then E_{tot} is written as

$$E_{tot}(A, Z, n_n, n_{A,Z}, V_{A,Z}) = n_{A,Z}(W_{A,Z} + W_L) + (1 - V_{A,Z}n_{A,Z})E_n(n_n) + E_e(n_e) \quad (6)$$

The minimization of E_{tot} with respect to the arguments at a constant nuclon concentration n_b,

$$n_b = A\,n_{A,Z} + (1 - V_{A,Z}n_{A,Z})\,n_n \quad (7)$$

defines the parameters of the equilibrium state. The nucleus energy $W_{A,Z}$ in the drop nuclear model has the form

$$W_{A,Z}(A, Z, V_{A,Z}, n_n) = [(1-x)\,m_n c^2 + x\,m_p c^2 + W(k,x)]A + W_{coul} + W_{surf}, \quad x = \frac{Z}{A} \quad (8)$$

where $W(k,x)$ is the energy per baryon in a homogeneous nuclear matter with the concentration $n = k^3/1.5\pi^2 = A/V_{A,Z}$, W_{coul} is the Coulomb energy of the proton interactions in the nucleus, and W_{surf} is the surface energy of the nucleus. The energy of a neutron gas is

$$E_n(n_n) = n_n\,W(k_n,0) + m_n c^2, \quad n_n = \frac{k_n^{\,3}}{1.5\pi^2} \quad (9)$$

The minimization of E_{tot} with respect to its arguments at n_b = const reduces to four equations defining the equilibrium with respect to the following processes. 1/ Compression and expansion of a nucleus under the pressure of external neutrons. 2/ The neutron exchange between the nucleus and the neutron gas. 3/ The transformation of neutrons and protons into one another inside the nucleus. 4/ The separation of the nucleus mass minimizing the total energy. The results calculated elsewhere [1, 6] are given in Fig.1. With an increase of the density, the mass and the volume of nuclei increase (up to $A \sim 2500$) until they start touching one another at $\rho = \rho_{nn} = 2.4 \cdot 10^{14}$ g/cm^3. In [6] it is assumed that at $\rho = \rho_{nn}$ a first-order phase transition results in the formation of a homogeneous nuclear matter.

At a density exceeding the nuclear density the strong interaction becomes essential. Under such conditions, the calculations become very complicated even if the known interaction laws are known. In the case under consideration the ambiguity becomes even greater because there is no rigorous theory of strong interactions. One of the most realistic variants of the equations of state calculated in [7] is given in Fig.1.

2. Nonequilibrium Neutronization with an Increase of Density in Cold Matter

An equilibrium for a nuclear composition is attained in a matter at high temperatures when all the reaction channels are open. With cooling, most channels start closing, and therefore the attainment of the complete equilibrium in a cold matter, strictly speaking, is impossible. At low temperatures the matter is always in a nonequilibrium state, but the degree of nonequilibrium and its character depend on the way by which the matter attained the state with the given ρ and T. In [8,9] two possible ways for the attainment of such a state are considered - the compression of a cold matter and cooling matter at the given density.

The compression of a cold matter can occur in the gas accretion onto a neutron star in a binary system where heating of deep layers is inessential. Let a matter at low density ρ consist of the most stable element ^{56}Fe and is slowly compressed at a temperature close to zero. At $\rho = 1.15 \cdot 10^9$ g/cm^3 and $\varepsilon_{Fe} = 3.7$ MeV the electron capture by a ^{56}Fe nucleus becomes energetically advantageous. Due to a lower stability of even-odd nuclei, the first capture is immediately followed by the second one, and the following chain of reactions occurs:

$$^{56}Fe + e^- \rightarrow {}^{56}Mn + \nu_e$$
$$^{56}Mn + e^- \rightarrow {}^{56}Cr + \nu_e \quad (10)$$

For the second reaction ε_β = 1.6 MeV; at ε_{Fe} = 3.7 MeV this reaction proceeeds in a nonequilibrium way and is accompanied by heating [10] which, nevertheless, makes no effect on the chemical composition. In the course of a further compression, the transformations $^{56}Cr \rightarrow {}^{56}K \rightarrow {}^{56}Ar$ etc. become energetically advantageous until the formation of a nucleus with the detachment energy of the last neutron $Q_n \approx 0$. Then, a further increase of the density and electron captures are accompanied by cold neutron evaporation Wneutrons from the nuclei and a decrease in both A and Z.

In the process of nonequilibrium neutronization, the number of nuclei per baryon remains unchanged. If A_0 is the atomic number of an initial nucleus, then the concentration of nuclei and electrons are always equilibrium

$$n_{A,Z} \approx \frac{\rho}{A_0 m_p} \qquad n_e = \frac{Z\rho}{A_0 m_p} \tag{11}$$

where $m_{A,Z} = Am_p$. Prior to te neutron evaporation at $A = A_0$, the charge of the nucleus Z is constant in the density range where the following inequality is valid

$$\varepsilon_\beta\,(A_0,Z+1) < \varepsilon_{Fe}\,(n_e) < \varepsilon_\beta\,(A_0,Z\,) \tag{12}$$

When the right-hand relation in (12) turns into an equality, a new electron capture begins in the range $\rho_Z < \rho < \rho_{Z-2}$ at constant $\varepsilon_{Fe}$and P_e :

$$\varepsilon_{Fe}\,(n_e) = \varepsilon_\beta\,(A_0,Z\,) = const$$
$$\frac{\rho_{Z-2}}{\rho_Z} = \frac{Z}{Z-2} \tag{13}$$

The properties of a matter in cold neutronization prior to neutron evaporation are given in Table 2 where the nuclear data [5] are also used. Upon the beginning of neutron evaporation, their concentration is given by the relationship

$$n_n = \frac{\rho}{m_p}\left(1 - \frac{A}{A_0}\right) \tag{14}$$

The evaporation condition in a cold matter after a sequential electron capture has the form:

$$Q_n\,(A,Z\,) = B_{A,Z} - B_{A-1,Z} \leq -\varepsilon_{Fn}\,(n_n) \tag{15}$$

where ε_{Fn} is the energy of Fermi neutrons.

Because of oscillations of the Q_n (A,Z) function depending on the A and Z parity,the whole chain of type-(10) reactions occurs with the evaporation of six to eight neutrons. The calculation of nonequilibrium neutronization with neutron evaporation by (11)-(15) was carried out in [11] with the use of (3) for $B_{A,Z}$ and in [12] with allowance for the nuclear interaction of nuclons by the method suggested in [6]. The results obtained in [12] are also listed in Table 2. The allowance for pycnonuclear reactions in [12] resulted in that at $\rho = 1.4 \cdot 10^{12}$ g/cm^3 the nuclei (A,Z) = (32,9) formed from ^{56}Fe merge together, and, upon the capture of two electrons and evaporation of eight neutrons, the (56, 16) nuclei are formed. The calculation was carried out up to $\rho = 5.13 \cdot 10^{13}$ g/cm^3 with the formation of the (99, 19) nucleus. According to [6,13], in equilibrium A = 375 and Z = 30-40 at such a density.

For a less detailed consideration of the nonequilibrium neutronization one may use the approximation Nemirovskii results for the detachment energy of the last proton, Q_p, at the evaporation boundary with Q_n = 0 [9]:

$$Q_p\,(A,Z\,) = B_{A,Z} - B_{A-1,Z-1} = \varepsilon_\beta\,(A,Z\,) + Q_n\,(A,Z-1) -$$
$$- (m_n - m_p)\,c^2 + m_e\,c^2 \approx 33 - Z/7 \ MeV \tag{16}$$

at $Z \geq 6$ and $A = 47$.

Neglecting the Fermi-energy of free neutrons and considering the evaporation along the line Q_n = 0, we obtain from (16) (with due regard of (1) and (11)-(14)) the relationship for Z(ρ) and A(ρ):

$$\left[1 + (Z\rho / 10^6 A_0\,)^{2/3}\right]^{1/2} = 1.96\,(33 - Z/7) + 2.53 \tag{17}$$
$$A = 4Z$$

This formula takes account of the fact that $m_e c^2 = 0.511$ MeV, $(m_n - m_p)c^2 = 1.239$ MeV. The results of the calculations by (17) are listed in Table 2. The comparison of the results with Fig.1 shows that the nonequilibrium pressure exceeds by ~ 50% the equilibrium pressure at the same density.

3. Nonequilibrium Composition During Cooling of a Hot Dense Matter

A netron star formed as a result of a gravitational collapse is very hot and cools rapidly due to neutrino cooling. When the temperature of the cooling matter becomes lower than $(4\text{-}5) \cdot 10^9$ K, the reaction between the charged particles drastically decelerate and the nucleus concentration is "frozen". Under such conditions, the occurrence of neutron reactions becomes possible and also neutron photodetachment and capture, e- decay at $\varepsilon_\beta > \varepsilon_{Fe} + m_e c^2$ and e^- captures at $\varepsilon_\beta < \varepsilon_{Fe} + m_e c^2$. At a finite temperature the nuclei may attach and detach neutrons if

$$-\varepsilon_{Fn} < Q_n < Q_{nb} \approx 20\,kT - \varepsilon_{Fn} \qquad (18)$$

At high temperatures, the matter has numerous free neutrons. Under the conditions of neutron excess, the formation of a chemical composition during the rapid cooling of the matter is illustrated by Fig.2 [9]. The (A,Z) plane for nuclei is divided into three regions:

region I with $Q_n > Q_{nb}$;
region II with $-\varepsilon_{Fn} < Q_n < Q_{nb}$, $\varepsilon_\beta > \varepsilon_{Fe} + m_e c^2$;
region III with $-\varepsilon_{Fn} < Q_n < Q_{nb}$, $\varepsilon_\beta < \varepsilon_{Fe} + m_e c^2$.

At high neutron concentration the existing nuclei rapidly pass from region I to regions II and III because of the neutron capture. In regions II and III there exists an equilibrium with respect to the capture and neutron detachment. In region II, the beta-decay results in an increase of Z, and the nuclei pass into the region above the cd line. In region III, the beta-capture decrease Z and "transfer" the nuclei to the region below the ab line. Thus, under the conditions of a limited equilibrium at $\varepsilon_{Fe} \gg Q_{nb} \gg kT$ where the temperature effects do not influence the beta-processes, the nuclear composition is determined by a narrow abcd region on the (A,Z) plane (Fig.2). The composition remains within this regionbecause there are no admissible beta-processes and neutron photodetachment. At $T < 5 \cdot 10^8$ K only one nucleus is remained at the boundary (15).

If to neglect ε_{Fn} and take into account (16), then the nuclear composition can be roughly determined from the relationships analogous to (17):

$$Z = 7\left\{33 - 0.511\left[\left(\frac{\rho}{\mu_z 10^6}\right) + 1\right]^{1/2} + 1.293\right\} \quad A = 4Z \qquad (19)$$

Here μ_z is the number of nuclons per electron. The $Z(\rho)$ dependence for the weight neutron concentration $x_n = 2$ is shown in Fig.3. At $\rho = \rho_1 = 2.24 \cdot 10^{12}$ g/cm^3 a nucleus with Z = 6 is formed with $\varepsilon_\beta = \varepsilon_{\beta max}$, and at $\rho = \rho_2 = 1.2 \cdot 10^{12}$ g/cm^3 a nucleus with Z = 150 is formed with $Z^2/A = Z/4 = 37.5$,and the fission time is $\tau_f = 3 \cdot 10^7$ years [14]. At $\rho < \rho_1$ the quantity Z^2/A increases, whereas the fission time decreases. Fission and alfa-decay result in the growth of the number of nucleating nuclei, i.e., at $\rho > \rho_2$ or $\rho < \rho_1$ the chemical composition can approach the equilibrium composition. An essential nonequilibrium is possible only if $\rho_1 < \rho < \rho_2$.

The difference of the nonequilibrium composition in cooling from the composition in compression arises already at low densities (where for the hot case the superheavy nuclei are formed). In both variants, there are some principled differences from the equilibrium where $Z(\rho)$ increases with growth in ρ, since in both nonequilibrium compositions $Z(\rho)$ drops with an increase in density.

4. Models of Neutron Stars

Models of neutron stars are based on solution of the equilibrium equations of the general relativity theory (GRT)

$$\frac{dP}{dr} = -\frac{G(\rho + P/c^2)(m + 4\pi\rho^3 P/c^2)}{r^2(1 - 2Gm/rc^2)} \qquad (20)$$

$$\frac{dm}{dr} = 4\pi\rho r^2 \qquad (21)$$

Here the density ρ includes in addition to the rest mass also the kinetic energy and the energy of nuclon interaction. Becuase of the difficulties encountered in the determination of thermodynamic functions for supernuclear densities, the models of neutron stars include some ambiguities. The dependence of the mass of a neutron star, M_n, on the central density of the rest mass of baryons, ρ_{c0}, from [15] is shown in Fig.4 for different equations of state at supernuclear densities from [7]. As a rule a neutron star has the following zones (see, e.g., [16]):

1. A surface zone ($\rho \leq 10^6$ g/cm^3). In this zone the temperature and the magnetic field can influence the equation of state and the shell structure. The equilibrium composition is iron, ^{56}Fe.
2. The external crust ($10^6 \leq \rho \leq 4.3 \cdot 10^{11}$ g/cm^3) is the zone of the existence of a Coulomb crystal, and in the equilibrium it consists of nuclei from Table 1. The upper density boundary corresponds to the beginning of neutron evaporation in the equilibrium.
3. The internal crust ($4.3 \cdot 10^{11} \leq \rho \leq (2.0+2.4) \cdot 10^{14}$g/cm^3) is the zone of the existence of a crystal lattice of nuclei in which neutrons are close to the evaporation boundary together with free neutrons which may be superfluid. Both density boundaries are given for the equilibrium conposition. A layer with strong nonequilibrium of the chemical composition should occupy a narrow region between the external and internal crust with almost equilibrium composition. Nuclei there also form a crystal lattice and its effect on the integral parameters of a neutron star is inessential.
4. Neuton liquid ($(2.0 + 2.4) \,.\, 10^{14} \leq \rho < \rho_{core}$) In this zone there exist superfluid neutrons, protons, and normal electrons.
5. The nucleus zone ($\rho > \rho_{core}$) has not been studied as yet; here pion condensation is possible, a solid neutron matter, the generation of hyperons, quarks, or any other exotic phase different from a neutron liquid. Here $\rho_{core} \geq 6 \cdot 10^{14}$ g/cm^3 . For different equations of state the maximum mass values in Fig.4 vary from 1.65 to 3.02 M$\odot$. Τηε λοωερ λιμιτ M_{max} seems to be more realistic than the upper one. The most accurate value of the mass of a neutron star is known for a binary pulsar PSR 1913 + 16 and is equal to 1.41 ± 0.06 M$\odot$.

In the formation of a neutron star, it reaches enormously high temperatures - $10^{11} \div 10^{12}$ K. The limit mass of a star increases with the temperature rise, and the mass maximum is shifted towards the region of lower densities. The calculations of the models of hot neutron stars with adiabatic temperature distributions were performed in [17]. The maximum mass for a hot superdense star is $M_{max} \approx 70$ M$\odot$, in this case the star center is built by of approximately equal numbers of protons and neutrons. Some of the above discussed problems are considered in more detail elsewhere [18].

5. Possible Observational Manifestations of Nonequilibrium

a/ Cooling of a Neutron Star

An excess of free neutrons in a nonequilibrium layer results in inward neutrons diffusion under the action of gravitation (barodiffusion) which, in turn, results in energy release. The calculation of the barodiffusion process has shown [19] that at the initial weight concentration of neutrons in a nonequilibrium layer $x_{n0} = 1/2$ the energy release due to barodiffusion for $t \approx 10^4$ years amounts to more than 10^{35} erg/s, then it rapidly drops and at $t \approx 1.5 \cdot 10^4$ years is ~ 10^{33} erg/s. The neutron barodiffusion substantially increases the duration of the neutron star phase with a luminosity of 10 erg/s in comparison with the equilibrium case where the energy emission occurs due to neutron star cooling. The duration of the stage with luminosity $L > 10^{35}$ erg/s due to cooling does not exceed 10 years [20].

b/ Accumulation of Deformation and Crust Fracture

Neutrons of the nonequilibrium layer give a small contribution into pressure in comparison with that of electrons, but their mass is comparable with the mass of nuclei. In a thin star shell with $\Delta R \ll R_H$ the pressure P is uniquely related to the mass m of a matter lying at the surface

$$P = \frac{G M_H m}{4\pi R_H{}^4} \qquad (22)$$

where R_H is the neutron-star radius . In the inward neutron diffusion a decrease in m results in a decrease in P occurring because of the matter expansion. For 10^4 years, the thickness of the nonequilibrium layer may increase by a factor exceeding 1.5 [19]. Such a large extension of the crystal shell should give rise to cracks. Because of this the expansion should proceed not in s smooth way but in a jumpwise manner and should be accompanied by a small increase of the period. Such jumps in the period increase with $\Delta P/P \approx 10^{-9}$ were observed for Crab pulsar. For Vela and some older pulsers only the jumps with a decrease of the period were observed, which is explained by a loss of a rotational moment.

c/ Nuclear Explosions and Gamma-bursts

Starquakes accompanied by the convective-type motion result to the transfer of the matter from the nonequilibrium layer to the low-density regions (with $\rho \ll \rho_1$) where it is unstable relative to the beta-decay. An increase of Z in the beta-decay make the nuclei unstable with respect to the fission, and a nuclear explosion occurs as a result of the chain reaction. The emerge of the shock wave formed in the explosion to the neutron star surface is accompanied by the temperature rise up to 10^9 K and can result in observable gamma-bursts [21-23]. The matter jet in the explosion by a shock wave and a strong star wind (due to supercritical luminosity) can result in an optical burst [24] and the formation of observable absorption lines due to the presence of Ba- and J-type heavy elements which are the decomposition products of the chain fission reaction [25].

d/ A matter outbursted from a neutron star as a result of an explosion contains an impurity of superheavy nuclei with A ≥ 290 whose formation in r-processes during supernova explosions requires unrealisically intense inward neutron fluxes. The neutrons outburst in the explosion interact with the medium protons and are transformed into deuterium. Thus in an interstellar matter the existence of excessive deuterium around the neutron stars is possible; the deuterium 2.2 MeV line is observed in the spectra of some gamma-bursts; the origin of transuranium elements seems to be associated with neutron star outbursts [9, 19]. All these effects are caused by nonequilibrium properties of neutron stars.

Table 1. Equilibrium nuclear composition at densities lower than the density of the neutron evaporation from [1]

Nucleus	ρ_{max}, g/cm^3	μ_{tl}, MeV	$\Delta\rho/\rho$,%
26 ^{56}Fe	8.1(6)	0.95	2.9
28 ^{62}Ni	2.7(8)	2.6	3.1
28 ^{64}Ni	1.2(9)	4.2	7.9
34 ^{84}Se	8.2(9)	7.7	3.5
32 ^{82}Ge	2.2(10)	10.6	3.8
30 ^{80}Zn	4.8(10)	13.6	4.1
28 ^{78}Ni	1.6(11)	20.0	4.6
26 ^{76}Fe	1.8(11)	20.2	2.2
42 ^{124}Mo	1.9(11)	20.5	3.1
40 ^{122}Zr	2.7(11)	22.9	3.3
38 ^{120}Sr	3.7(11)	25.2	3.5
36 ^{118}Kr	4.3(11)	26.2	

μ_{tl} is the chemical potential of electrons at ρ_{max} - the maximum density at which the given nuclide exists, $\Delta\rho/\rho$ is a relative increase of the density in the transition to a next nuclide.

Table 2. Nonequilibriium properties of a matter in cold compression

$\rho \cdot 10^{-11}$, g/cm^3		A, Z		ε_{Fe}, MeV		ε_{Fn}, MeV		$P \cdot 10^{-30}$, dyne/cm^2	
0.012		5624		37		0		55 10^{-4}	
014		5622		883		0		14 10^{-2}	
057		5620		140		0		80 10^{-2}	
16		5618		193		0		028	
39		5616		252		0		073	
	501		5616		2698		0		09333
	617		5416		2885		009		120
	724		4614		2919		026		129
101		5614		3178		0		197	
	1023		4012		3100		044		166
121		4812		3206		043		207	
	1288		3610		3160		058		186
	1413		329		3147		065		186
150		4010		3235		079		225	
168		369		3250		099		238	
192		328		3264		122		259	
222		287		3278		149		288	
263		246		3292		182		335	

The right-hand columns are the results taken from [12], the left-hand colums are the results calculated by formulae of Sect.2.

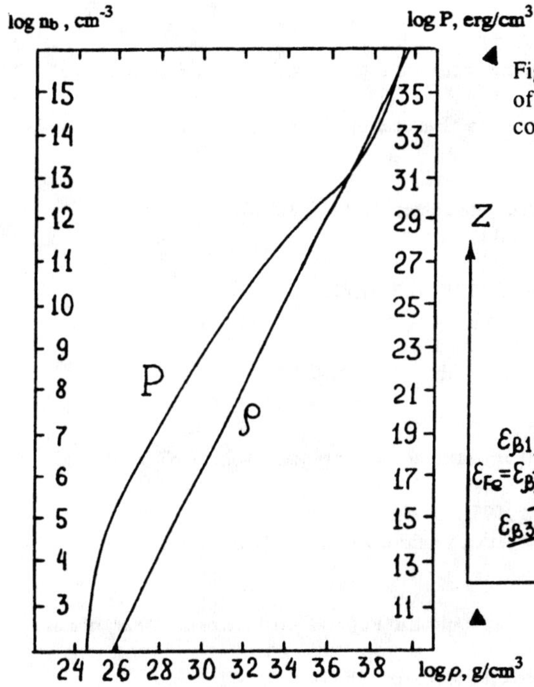

◀ Figure 1. Pressure P and the total density ρ of the mass as functions of the baryon concentration n_b.

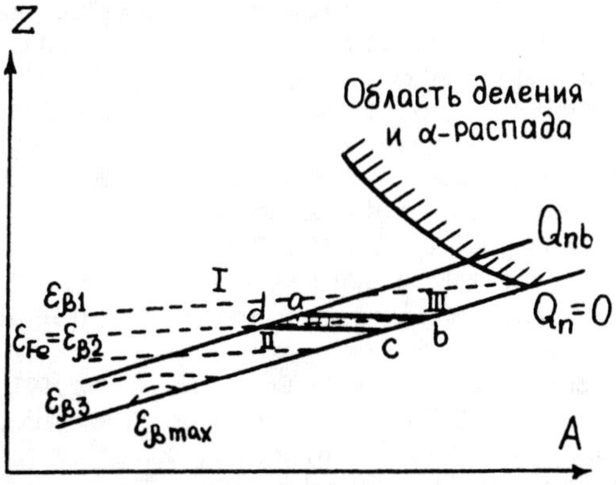

▲ Figure 2. Formation of nonequilibrium composition in cooling of a hot neutronn star. The region of fission and α-decay

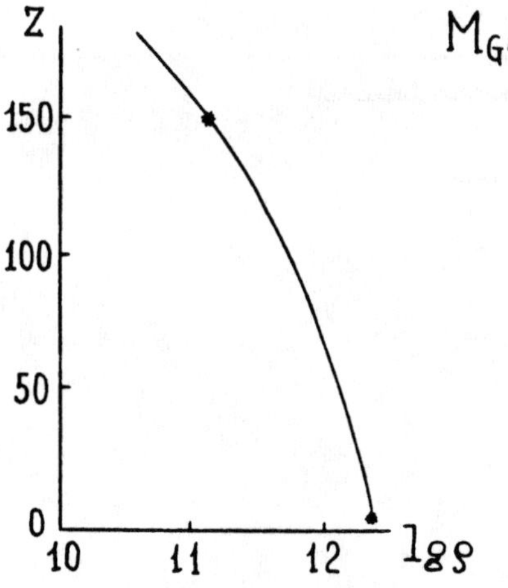

Figure 3. Composition of the nuclear of a neutron star after cooling at X_n=1/2; a nonuniform layer is located between the stars.

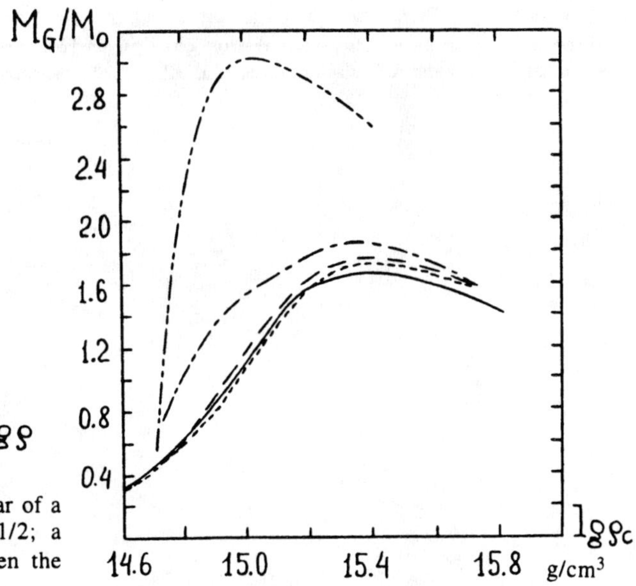

Figure 4. Dependence of the mass of a neutron star for various equations of state. The solid line indicates the equation of state from figure 1.

References

1. G. Baym, Ch.Pethick, and P. Sutherland. Astrophys. J. , 1971, 170, 306.
2. V.A. Kravtsov. Massy atomov i energii svyasi yader (Atomic masses and the binding energies of nuclei)1974 (Moscow: atomizdat) (in Russian).
3. G. Bethe and F. Morrison. Elementarnaya teoriya yadra (Elementary theory of a nuclei) 1958 (Moscow: IL) (in Russian).
4. W. Meyers and W. Swiatecki. Nucl. Phys., 1966, 81, 1.
5. H. Groote, E.R. Half, and R. Takahashi. At. Data Nucl. Data Tables., 1976, 17, 418, 476.
6. G. Baym, H. Bethe, and Ch.Pethick. Nucl. Phys., 1971, A175, 225.
7. H.Bethe and M. Johnson., Nucl. Phys., 1974, A230, 1.
8. G.S. Bisnovatyi-Kogan and V.M. Chechetkina. Pis'ma v ZhETF, 1973, 17, 622.
9. G.S. Bisnovatyi-Kogan and V.M. Chechetkina. UFN, 1979 127 263.
10. G.S. Bisnovatyi-Kogan and E.F. Seidov, Astron. Zh., 1970, 47, 139.
11. Yu.L. Vartanyan and N.K. Ovakimova. Soobshch. Barakan. Obs., 1976, No.49, 87.
12. K. Sato. Progr. Theor. Phys., 1979, 62, 957.
13. G. Baym and Ch. Pethick. Ann. Rev. Nucl. Sci., 1975, 25, 27.
14. K.N. Mukhin. Eksperimentalnaya yadernaya fizika (Experimental nuclear physics) vol.1, 1983 (Moscow: Energoizdat) (in Russian).
15. R. Malone, M. Johnson, and M. Bethe. Astrophys. J.,, 1975, 199, 741.
16. S. Shapiro and S. T'yukol'skii. Chernye dyry, belye karliki i neitronnye zvezdy (Black holes, white dwarf stars, and neutron stars) 1985 (Moscow: Mir) (in Russian).
17. G.S. Bisonovatyi-Kogan, Astrofizika, 1968, 4, 221.
18. G.S. Bisnovatyi-Kogan Fizicheskie voprosy teorii zvesdnoi evolyutsii (Physical problems of the theory of star evolution) 1988 (Moscow: Nauka) (in Russian).
19.G.S. Bisnovatyi-Kogan, Yu.N. Kulikov, and V.M. Chechetkina. Astron. Zh., 1976, 53, 975.
20. K. Nomoto and S. Tsuruta. Astrophys. J., 1987, 312, 711.
21. G.S. Bisnovatyi-Kogan, V.S. Imshennik, D.K. Nadezhdin, and V.M. Chechetkina. Astrophys. Space. Sci., 1979, 35, 3.
22. G.S. Bisnovatyi-Kogan, C.I. Blinnikov, and A.F. Zakharov. Astron. Zh., 1984, 61, 104.
23. G.S. Bisnovatyi-Kogan. Proc. Symp. The Origin and Evolution of Neutron Stars, 1987, 501, D. Reidel.
24. G.S. Bisnovatyi-Kogan and A.F. Illarionov, Astron. Zh., 1986, 63, 984.
25. G.S. Bisnovatyi-Kogan and A.F. Illarionov. Preprint IKI, Pr-1292, 1987 (in Russian).

Physics of Neutron Stars
Edited by A. M. Kaminker

ISBN 1-56072-164-2

NEUTRON STARS CONTAINING PION CONDENSATE

Yu. L. Vartanyan and G. B. Alaverdyan
Erevan State University, 375049 Erevan

Neutron stars are investigated in which the presence of a pion condensate causes a second-order phase transition. The calculations have been carried out for two equations of state for the nuclon component. For the pion condensate, a model of developed condensate taking account of degenerate electrons and μ^--mesons was used . The calculations were carried out for various sets of nuclear constants. For three series of neutron stars with the jumps in density the following quantities were calculated as functions of the central pressure:mass, the total number of baryons, the binding energy, the relativistic moment of inertia, and the red shift. The models were studied in detail in the region of the phase transition, the energy of the transition onto a stable branch of neutron stars was calculated. The models of self-nucleus-like superdense bodies - pion stars - have also been considered.

1. Introduction

In recent ten-fifteen years one of the most crucial problem of the theory of nuclear matter was that of possible formation in there of a pion condensation the idea of which was put forward by Migdal, Sawyer, and Scalapino [1-3]. Theoretical calculations of the threshold for generation of a pion condensate lead to a conclusion that such a quantity should exceed the nuclear density [4]. This is also confirmed by the fact that the experiments on detection of pion condensation in conventional nuclei gave negative result.

Since the densities exceeding nuclear may be formed in the interior of neutron stars, the pion condensation in such a matter acquires great importance in the consideration of the internal structure and observed manifestations of superdense stars. The actuality of this problem is caused, to a large degree, also by the fact that in recent decades the progress in astronomy substantially increased the accuracy of the determination of such integral parameters of stars as their mass, radius, and the moment of inertia [5,6]. Their comparison with the theoretically calculated data allows one to determine which of the theoretical models may really be implemented in nature.

The existence of pion condensation in neutron stars drastically "softens" the equation of state - diminishes nuclear repulsion forces. Unfortunately, at present the difficulties encountered by the theory of strong multiparticle interaction do not allow the unique construction of a theoretical model. Depending of various insufficiently accurate parameters several variants can be chosen that can lead to both the first- and the second-order phase transitions in a dense matteer. This, in turn, can lead to a series of interesting observational manifestations considered below.

In connection with the phase transition, we should like to note that recently a series of theoretical studies [7-9] on phase transitions to the quark phase in a nuclear plasma with the formation of substances having a strangeness. Such trnasitions seem to occur at densities higher than the density of the pion condensate formation. But this problem is beyond the interests of the present article.

In distinction from the conventional theory of a dense condensate [1-3], the authors of [10] have developed a"phenomenological" (their own term) approach to explain the existense of π^--mesons in heavy neclei. It reduces to the refinement (by adding some new terms) of the Weizsaker formula, which for nuclei with A > 208 yields the binding energy per nuclon lower by 0.01 MeV than the experimental values. In the following publications of the same authors (see review artiicles [11]) these results were extended to a solid nuclear matter, and far-reaching astrophysical conclusions were made. Thus, in the equation of state, a density jump from 10^{11} to 10^{14} g/cm^3 was obtained. This makes the absense of the region where the matter consists of anomalously heavy atomic nuclei to be overloaded with neutrons that are in a degenerate electron and neutron gas (the matter forming external layers of neutron stars) [12, 13]. Not going into details, we should like to note that we do not share this standpoint and consider it to be erroneous. Note here only that the Weizsaker formula does not pretend for such a high accuracy (0.01 MeV is of about 0.1% of the mean binding energy of a nuclon in a nucleus). Such an accuracy is attained only for heavy nuclei, i.e., just there where the formula is modified, whereas for moderately heavy nuclei the deviation of the binding energy per nuclon from its experimental values is substantially larger and is of the oscillating nature. Moreover, as has already been noted, at present the absence of π-mesons in nuclei have been proved experimentally. Thus there is no need in their theoretical"detection" and any astrophysical conclusions on this basis.

In the investigation of the stability problem for superdense stars one usually considers the dependence of the mass on the central density $M(\rho_c)$ and, in the case of the density jump, also $M(P_c)$ (P_c is the central pressure). The phase transition with the density jump on the stable branch of this curve can result in the formation of an additional sharp maximum. It is the configurations located in the vicinity of this maximum that can pass from the unstable branch to the stable one with simultaneous the release of a tremendous energy. Whereas the calculation of $M(\rho_c)$ and other integral parameters of conventional neutron stars (without the jump in density) are considered in numerous investigations, it is not so with the configurations having a jump in density. Such superdense stars were studied earlier [14-16]. In [15] only the $M(\rho_c)$ curve is given, but there are no tabulated data on the binding energy. Such tabulated data are given in [14] but they show that all the configurations located in the region of the phase transition have positive binding energy (anomalous mass defect), i.e., cannot be implemented in reality. In [16] the authors studied how different variants of the equation of state influence the integral parameters of neutron stars with the density jump at the maximum mass; the model with the fixed value of the total baryon number was also considered there. Such models were also studied in the phase transition region.

Below neutron stars with the first-order phase transition are studied in detail as a function of the central pressure. The calculations are performed for two equations of state of the nuclon component. For a pion condensate, the model of the developed condensate is considered. In distinction from [16], in this case the presence of degenerate electrons and μ^--mesons has been taken into account. The calculations were performed for three different sets of nuclear constants.

One of interesting consequences of the pion condensation theory is possible arising of self-holding nuclear drops whose mass can take any low values whereas from above it is limited only by the effects of the general theory of relativity. In distinction from neutron stars the existense of such drops is due to nuclear forces alone. They are similar to gigantic atomic muclei with the density jump on the surface varying from supernuclear to zero values. Such models were first studied in [14] and in more detail in [17]. The latter data will be considered later on.

2. Equation of State

The equation of state of superdense stars, $P(\rho)$, is usually written in the parametric form depending on the total baryon concentration n (the number of baryon per unit volume):

$$P = n^2 \, d\varepsilon(n) / dn \qquad (1)$$

$$\rho = \rho_0 (1 + \varepsilon(n) / m_n) \qquad (2)$$

$$\rho_0 = m_n n \qquad (3)$$

Here P is the pressure, ρ is the density of the total energy, ρ_0 is the density of the rest mass, and $\varepsilon(n)$ is the energy per baryon. With due regard for the pion condensation, ε is respresented as a the sum of two terms

$$\varepsilon(n) = \varepsilon_0(n) + \varepsilon_\pi(n) \qquad (4)$$

where $\varepsilon_0(n)$ is the energy of the normal component and $\varepsilon_\pi(n)$ is the energy of the pion condensate. Each of these quantities is determined from the condition of its minimum. The σ-model of pion condensate uses not quite correct step: one minimizes not the total energy ε but its individual terms which is associated with the difficulties encountered by the multiparticle theory of strong interactions. If $n < n_0 = 0.17$ $Ф^3$ (nuclear density), one can use for e the results obtained by many authors and tabulated in [18]. In the region $n > n_0$ the data about the nuclear interactions are inconsistent. Therefore ε_0 is found, as a rule, from several models that differ from one another by the steepness of the $P(\rho)$ curve. The steeper the curve, the more rigid the equation of state. An example of a soft equation of state is the Pandharipande (PN) model [19], and example of a more rigid equation is the first Bethe-Johnson-Malone model (BJ) [20].

The difficulties encountered by the theory dictate the determination of $\varepsilon\pi$ also from the model representations. We use here the results obtained in [21] for a model of developed pion condensate, which, generally speaking, is valid for rather high densities. The results obtained in the above cited study with due regard for the electric charge of the condensate were modified by A.B. Migdal [22] to the form

$$\varepsilon_\pi = \alpha_\pi n(1 - 2\nu)^2 - 81 f^2 (1 - \gamma)(n/50) + (\Delta/3) \qquad (5)$$

where $\Delta = m_{N^*} - m_N = 2.1$, $\alpha_\pi = 1/(2F^2)$, F = 1.35 is the constant of pion decay, $f = g_A/F$, $g_A = 1.36$ is the axial constant, γ is the factor taking into account the contribution of nuclon correlations. The constants f and γ cannot be determined uniquely because of the model character of the multiparticle theory of strong interactions. As a rule, models corresponding to different values of these constants are constructed. At this stage, we do not consider their values. Hereinafter the following system of units is used:

$$m_\pi = \hbar = c = 1 \qquad (6)$$

where m_π is the mass of a π-meson, $\hbar$ is the Planck constant, and c is the light velocity.

In (5), ν is an electric charge of the hadron system per a baryon. In concrete calculations, two cases are usually considered [22]: $\nu = 0$ (neutron matter) and $\nu = 1/2$ (nuclear matter). It is correct to consider this quantity as a function of density, $\nu = \nu\nu\nu$ (n) and to take into account that since the pion condensate is charged it is necessary to neutralize it via degenerate electrons and μ^--mesons [23]. In this case

$$\varepsilon_\pi = \varepsilon_\pi^{(0)} + n_e\, \varepsilon_e / n + n_\mu\, \varepsilon_\mu / n \qquad (7)$$

where n_e and n_μ are the densities of electrons and μ^--mesons, respectively, and ε_e and ε_μ are the mean energies of these particles equal to [24]

$$\varepsilon_i = (3\, m_i\, / 8x_i^3) \left[x_i\, (1 + 2x_i^{\,2})\, \sqrt{1 + x_i^{\,2}} - \ln\left(x_i + \sqrt{1 + x_i^{\,2}}\right)\right] \qquad (8)$$

where $x_i = (3\pi^2 n_i)^{1/3} / m_i$, $\quad i = \varepsilon, \mu$.
Substituting (5) and (8) into (7), we arrive at

$$\varepsilon_\pi = \varepsilon_\pi\, (n, n_e, n_\mu, \nu, a) \qquad (9)$$

where $a = f^2\,(1 - \gamma)$. The condition of the ε minimum with respect to n_e , n_μ and ν and an additional condition of electroneutrality

$$n\,\nu = n_e + n_\mu \qquad (10)$$

lead to the following relationship between the chemical potentials

$$\mu_e = \mu_\mu = \mu_\pi \qquad (11)$$

As a rule, for a nuclear plasma condition (11) is complemented with the condition of the β-transitions between different baryon components. hese conditions determine the relative concentrations of the components. But in the model of a developed pion condensate it is assumed that the concentrations of all six baryon quasiparticles (N^{*++} $\cdot$ N^{*+}, p, m, N^{*0}, N) are equal. Therefore the electric charge of the hadron subsystem n is determined by the pion component alone:

$$n\,\nu = n/2 - n_\pi \qquad (12)$$

Substituting the values of chemical potentials, we obtain from (10) and (11) [25]:

$$4\,\alpha_\pi\, n\,(1 - 2\nu) = m_e x_e \qquad (13)$$
$$m_e x_e = m_\mu \sqrt{1 + x_\mu^{\,2}} \qquad (14)$$

$$n\,\nu = \frac{1}{3\pi^2}\left[(m_e x_e)^3 + (m_\mu x_\mu)^3\right] \qquad (15)$$

The solution of the system (13)-(15) yields the dependences n_e (n), n_μ (n) and ν(n) or, which is the same, n_π(n). These dependences and also the dependence μ_π(n) are shown in Fig.1. Substituting n_e(n), n_μ(n) and ν(n) into (7), we arrive at $\varepsilon_\pi = \varepsilon_\pi$ (n,a), and hence, from (4) also at ε(n,a). Taking different values for constant a, it is possible to construct the curves ε(n,a). For small values of the parameter a the energy of the condensate is low, and the curve ε(n,a) has a kink without loosing its monotony. In this case a second-order phase transition occurs.

Let us consider in more detail such value of the constant a that, upon the generation of the pion conndensate, lead to a negative compression modulus. The curves $\varepsilon(n)$ and $P(n)$ in this case are of the Van-der-Waals nature. The first-order phase transition takes place with a jump in density determined from the Gibbs condition if the chemical potentials and pressure are taken to be equal:

$$P_1 = P_2 = P_0, \qquad \mu_1 = \mu_2 \qquad (16)$$

If to take into account that $\mu = (P + \rho)/n$, it is easy to obtain from (16) that at the transition point the following condition should be fulfilled

$$\varepsilon(n_2) - \varepsilon(n_1) = P(1/n_1 - 1/n_2) \qquad (17)$$

The parameters of the transition point can also be determined graphically from the Maxwell construction shown in Fig.2 for one of the variants of the equations of state. Table 1 lists thus determined values of the parameters at the point of the phase transition (jump in density) for three equations of state.

Table 1.

Equation of state	n_1 Φ^{-3}	ρ_1 10^{15} g/cm^3	n_2 Φ^{-3}	ρ_2 10^{15} g/cm^3	ρ_0 10^{34} dyne/cm2	λ
PN, a = 0.475	0.198	0.333	0.985	1.713	0.510	4.98
BJ, a = 0.411	0.471	0.845	0.674	1.280	8.311	1.36
BJ, a = 0.5	0.322	0.568	0.655	1.180	3.166	1.96

If $0.53 < a < 0.8$, then the energy curve not only has the minimum, but also $\varepsilon_{min} < 0$. Thus for $a = 0.55$ $\varepsilon_{min}/m_\pi c^2 = -0.132$, $n_{min}/n_0 = 7.66$. As follows from (17) and Fig.2, in such models the phase transition to the normal component is impossible. The nuclear matter with a pion condensate cannot coexist in an equilibrium state with the matter of conventional neutron stars. In this case the pressure decreases from high values down to zero at the density n_{min}. If self-gravitation was absent, the nuclear plasma could have formed bound systems of constant density n_{min} of any size. Their structure would have reminded the atomic nuclei with the density jump on the surface from n_{min} to zero. But if the mass of such objects is of the order of the star mass, the presence of self-gravitation results in the compression of their inner regions, the densities there would start exceeding n_{min}, and only on the surface the density would be the same as in the absense of gravitation. Such objects received the name of pion stars. They are studied in detail elsewhere [17].

3. Basic Equations

In order to determine the basic parameters of spherically-symmetric superdense stars it is necessary to integrate the relativistic equations of the star equilibrium (the Tallman-Oppenheimer-Volkoff (TOV) equations [26] complemented with the equations for the determination of the relativistic moment of inertia I [27, 18]. The system of such equations has the form

$$dP/dr = -1/2\, r_g (m\rho/r^2)(1 + P/\rho)(1 + b r^3 P/m)/(-r_g m/r) \qquad (18)$$
$$dm/dr = b \rho r^2 \qquad (19)$$
$$dm_0/dr = b \rho_0 r^2/(1 - r_g m/r)^{1/2} \qquad (20)$$
$$dm_p/dr = b \rho r^2/(1 - r_g m/r)1/2 \qquad (21)$$
$$d\nu/dr = (r_g m/r^2)(1 + b r^3 P/m)/(1 - r_g m/r) \qquad (22)$$
$$d\omega/dr = 3 r_g l r^{-4} e^{\nu/2} / (1 - r_g m/r)^{1/2} \qquad (23)$$
$$dl/dr = 2/3\, b\, \omega \rho r^4 (1 + P/\rho) e^{-\nu/2} / (1 - r_g m/r)1/2 \qquad (24)$$

Here P, ρ and ρ_0 are determined by (1-3)-(3), $m(r)$ is a mass accumulated within a sphere of radius r, $m(R) = M$, where M is the total mass of the star, and R is its radius determined by the condition $P(R) = 0$. Here $M_0 \equiv m_0(R) = m_n A$, where A is the total number of baryons in the star, and $M_p \equiv m_p(R)$ is the proper mass. In the Newton limit, this quantity provides the determination of the gravitational energy E_G ($E_G = M - M_p$, $E_G = -G/c^2 \int_0^R m(r)dm/r$) and the total internal energy E_{in} ($E_{in} = Mp - M_0$, $E_{in} = 1/c^2 \int n\varepsilon d\nu$). The metric coefficient $g_{00}(r)$ $dS^2 = g_{ik}\, dx^i dx^k$) is associated with the function $\nu(r)$ defined in [22] by the following relationship

$$g_{00}(r) = (1 - r_g M/R)\exp[\nu(r) - \nu(R)] \qquad (25)$$

Equation (25) provides the calculation of the red shift

$$Z = \Delta\lambda/\lambda = [g_{00}(r)]^{-1/2} - 1 \qquad (26)$$

Equations (23) andd (24) provide the determination of the relativistic moment of inertia I [16]:

$$I = l(R)/\Omega \qquad (27)$$

The parameters of the transition point can also be determined graphically from the Maxwell construction shown in Fig.2 for one of the variants of the equations of state. Table 1 lists thus determined values of the parameters at the point of the phase transition (jump in density) for three equations of state.

Table 1.

Equation of state	n_1 $Ф^{-3}$	ρ_1 10^{15} g/cm^3	n_2 $Ф^{-3}$	ρ_2 10^{15} g/cm^3	P_0 10^{34} dyne/cm2	λ
PN, a = 0.475	0.198	0.333	0.985	1.713	0.510	4.98
BJ, a = 0.411	0.471	0.845	0.674	1.280	8.311	1.36
BJ, a = 0.5	0.322	0.568	0.655	1.180	3.166	1.96

If $0.53 < a < 0.8$, then the energy curve not only has the minimum, but also $\varepsilon_{min} < 0$. Thus for $a = 0.55$ $\varepsilon_{min}/m_\pi c^2 = -0.132$, $n_{min}/n_0 = 7.66$. As follows from (17) and Fig.2, in such models the phase transition to the normal component is impossible. The nuclear matter with a pion condensate cannot coexist in an equilibrium state with the matter of conventional neutron stars. In this case the pressure decreases from high values down to zero at the density n_{min}. If self-gravitation was absent, the nuclear plasma could have formed bound systems of constant density n_{min} of any size. Their structure would have reminded the atomic nuclei with the density jump on the surface from n_{min} to zero. But if the mass of such objects is of the order of the star mass, the presence of self-gravitation results in the compression of their inner regions, the densities there would start exceeding n_{min}, and only on the surface the density would be the same as in the absense of gravitation. Such objects received the name of pion stars. They are studied in detail elsewhere [17].

3. Basic Equations

In order to determine the basic parameters of spherically-symmetric superdense stars it is necessary to integrate the relativistic equations of the star equilibrium (the Tallman-Oppenheimer-Volkoff (TOV) equations [26] complemented with the equations for the determination of the relativistic moment of inertia I [27, 18]. The system of such equations has the form

$$dP/dr = -1/2\, r_g (m\rho/r^2)(1 + P/\rho)(1 + b r^3 P/m)/(- r_g m/r) \qquad (18)$$
$$dm/dr = b\rho r^2 \qquad (19)$$
$$dm_0/dr = b\rho_0 r^2/(1 - r_g m/r)^{1/2} \qquad (20)$$
$$dm_p/dr = b\rho r^2/(1 - r_g m/r)^{1/2} \qquad (21)$$
$$d\nu/dr = (r_g m/r^2)(1 + b r^3 P/m)/(1 - r_g m/r) \qquad (22)$$
$$d\omega/dr = 3 r_g l r^{-4} e^{\nu/2}/(1 - r_g m/r)^{1/2} \qquad (23)$$
$$dl/dr = 2/3\, b\omega\rho r^4 (1 + P/\rho) e^{-\nu/2}/(1 - r_g m/r)^{1/2} \qquad (24)$$

Here P, ρ and ρ_0 are determined by (1-3)-(3), m(r) is a mass accumulated within a sphere of radius r, m(R) = M, where M is the total mass of the star, and R is its radius determined by the condition P(R) = 0. Here $M_0 \equiv m_0(R) = m_n A$, where A is the total number of baryons in the star, and $M_p \equiv m_p(R)$ is the proper mass. In the Newton limit, this quantity provides the determination of the gravitational energy E_G ($E_G = M - M_p$, $E_G = -G/c^2 \int_0 m(r)dm/r$) and the total internal energy E_{in} ($E_{in} = M_p - M_0$, $E_{in} = 1/c^2 \int n\varepsilon dv$). The metric coefficient $g_{00}(r)$ $dS^2 = g_{ik}\, dx^i dx^k$) is associated with the function $\nu(r)$ defined in [22] by the following relationship

$$g_{00}(r) = (1 - r_g M/R)\exp[\nu(r) - \nu(R)] \quad (25)$$

Equation (25) provides the calculation of the red shift

$$Z = \Delta\lambda/\lambda = [g_{00}(r)]^{-1/2} - 1 \quad (26)$$

Equations (23) andd (24) provide the determination of the relativistic moment of inertia I [16]:

$$I = l(R)/\Omega \quad (27)$$

where

$$\Omega = \omega(R)\, e^{-\nu(R)}(1 - r_g M/R)^{1/2} + r_g l(R)/R^3 \quad (28)$$

It should be noted that an error slipped into expression (23) from Ref.[16]: coefficient 3 is mistakenly written as 1/2.
In (18)-(24), P, ρ, and ρ_0 are expressed in the system of units (5), m, m_0, and m_p are expressed in solar masses $M_\odot$, r in km, the moment of inertia I in $M_\odot$ km^2 (i.e. in the units $1.989 \cdot 10^{43}$ g.cm^2). The coefficient $r_g = 2GM_\odot/c^2 = 2.949$ km is the gravitational radius of the Sun. The appearance of this coefficient in (18)-(24) is obvious and requires no additional explanations. The appearance of the coefficient $b = 5.546 \cdot 10^{-4}$ is explained as follows. Consider as an example a cofactor in the expression (18) that includes this coefficient, $1 + br^3P/m$. If to use the CGS system, we have $1 + 4pr^3P/mc^2$. Therefore, $b = 4pr_n^3P_n/M_nc^2$, where r_n, P_n, and M_n are the coefficients for converting the GGS units into the units used in our study. Substituting the values of the conversion coefficients $r_n = 10^5$, $M_n = 1.989 \cdot 10^{33}$, and $P_n = m_\pi c^2/\lambda\!\!\!^-_\pi{}^3 = 7.9 \cdot 10^{34}$ ($\lambda\!\!\!^-_\pi = \hbar/m_\pi c$) into the expression for b, we obtain: $b = 5.546 \cdot 10^{-4}$. In a similar way it is possible to explain the appearance of this coefficient in other equations of system (18)-(24).
Integration is started from the configuration center at r = 0 where the initial condition is given by setting the pressure P(0) and the respective $\rho(0)$ and $\rho_0(0)$ values from the equation of state and also the conditions $m(0) = m_0(0) = m_p(0) = \nu(0) = l(0) = 0$; $\omega(0)$ is set arbitrarily, e.g., $\omega(0) = 1$. The integration is completed at r = R when P(R) = 0.

4. Calculational Results
The calculations were carried out for three series of models by the equations of state whose characteristics at the phase-transition point are given in Table 1. Since for such configurations the density experiences a jump, then it is convenient to consider the models as functions of a continuous parameter - the central pressure P_c. Table 2 lists the values of the total mass M, the rest mass M_0, the proper mass M_p, radius R, and the relativistic moment of inertia I. The last two columns yield the values of mass M_c and radius R_c of the central nucleus of the star where the nuclear matter contains a pion condensate. For stable models we have: $0.25 \le R_c/R \le 0.83$ and $0.26 \le M_c/M \le 0.95$ (the BJ variant with a = 0.5).
From Table 2 it follows that for the variant PN (a = 0.475) the maximum mass is smaller than 1.36 $M_\odot$, the most accuratately determined pulsar mass (PSR 1913 + 16, $M/M_\odot = 1.43 \pm 0.7$). Theefore this series can be considered only as a theoretical illustration.
For the BJ variant with a = 0.411, the maximum mass is $M_{max}/M_\odot = 1.45$. In [16], this set of nuclear constants ($f^2 = 0.822$, $\gamma = 0.5$) is given as $M_{max}/M_\odot = 1.59$ (the BJ1 + b variant, according to the notation accepted in [16]). The difference in these values is explained by the fact that the energy of the pion condensate in [16] does not take account of leptons, indicated in Sect.2. Thus, the allowance for leptons can substantially lower the value of the maximum mass.
Figure 3 shows the dependence of the mass on the central pressure. We should like to note that the shape of these curves in the region of the pion condensate generation depends on the value of the parameter λ which in the Newton approximation is defined as $\lambda = \rho_2/\rho_1$ and in the case of the relativistic models [29] as

$$\lambda = \rho_2/(\rho_1 + P_1) \quad (29)$$

where $\pi 1$, π_2 and P_1 are the values of density and pressure at the phase-transition point. If $\lambda > 3/2$ (in our case it is PN with a = 0.475 and BJ with a = 0.5), then an additional maximum is formed on the stable branch of neutron stars - the instability region. The configurations for which $dM/dP_c < 0$ are unstable. This phenomenon gives rise to rather interesting observational manifestations of the type of supernova outbursts. Indeed, the configuration close to the additional maximum (point A in Fig.3) is unstable with respect to the increase of the general number of baryons which can occur, e.g., due to accretion. Such a star should contract and transit to the stable branch of the curve. During this process an energy is released, which is equal to the difference of the masses of these stars, $\Delta\varepsilon = M_A - M_B$, M = const. This energv is strongly dependent on the form of the equation of state. Thus for the variant BJ with a = 0.5 (curve w in Fig.3) $\Delta\varepsilon \approx 7.4 \cdot 10^{50}$ erg, M0 = 0.85 $M_\odot$. In [30, 31] the authors have considered the dynamics of such a process and calculated the contraction time, which is tuned out to be of the order of the time of free fall, $t_{ff} \sim 10^{-4}$ s.

In order to study the stability of superdense stars it is interesting to consider the dependence of the mass on the radius (Fig.4). It is seen that at the point of the pion condensate formation (jump in density) a sharp peak is formed, and the general run of the M(R) curve is dependent on the value of the parameter λ. For stable neutron stars with the phase transition an increase of the mass leads to smaller radii. As is seen from Table 2, the existense of the pion condensate leads to a substantial decrease in the moment of inertia in comparison with those of conventional neutron stars.

From Table 2 it also follows that for all three above considered series of models the binding energy is negative in the region of the phase transition (condensate nucleation): $M < M_0$. This is emphasized in connection with the fact that in [4] (whose authors were the first to calculate the integral parameters of neutron stars with the density jump depending on the central pressure) the binding energy was shown to have an anomalous sign in the condensate nucleation: $M > M_0$. If this result was correct, then such models would have not existed in nature, and the above considered mechanism of energy release in compression of a limit neutron star (point A in Fig.3) should have been excluded. Our calculation of the binding energy proves that the result obtained in [14] is unrealistic, and, probably, even erroneous.

In the lower left corner of Fig.4 the dependence of the mass on the radius for the nucleus-like objects (pion stars) is shown (according to the data [17]), and on Fig.5 the dependence of the mass and the packing coefficient ($\alpha = 1 - M/M_0$) on the central density. The shape of the M(R) curve leads to a conclusion that, in distinction from stable neutron stars, an increase of the mass is also accompanied by an increase of the radius. Such a behavior of the radius is caused by a very small compressibility of such objects.

The values of the coefficient α (Fig.5) show that pion stars are bound much stronger than neutron stars. Another distinctive feature of such models is the fact that their mass can have infinitesimal values.

It is seen from Fig.5 that the masses of pion stars are smaller than that of the PSR 1913 + 16. The mass of these objects calculated in [14, 16] turn out to be the same. But, if to use the most rigid equation of state ($\partial P/dp = 1$) for the interior of pion stars, then $M_{max}/M_\odot \geq 2$ [32].

In [33] some observational manifestations of pion stars were considered. In particular, it was shown that their cooling proceeds much faster than cooling of neutron stars; it occurs mainly through the photon channel. The occurrence of specific nuclear processes on the surface of such objects during the matter accretion can result in the formation of γ-pulsars.

Table 2.

P_c 10^{35} dyne/cm^2	R km	M M⊖	M_0 M⊖	M_p M⊖	I MQ km^2	M_c M⊖	R_c km
Pandharipande equation, a = 0.475							
0.050	18.374	0.257	0.260	0.263	11.813	–	–
0.055	17.846	0.261	0.264	0.268	11.320	–	–
0.083	18.574	0.252	0.254	0.258	11.749	0.002	0.819
0.198	20.430	0.2112	0.213	0.216	11.698	0.020	1.769
0.317	18.567	0.169	0.170	0.170	6.489	0.046	2.306
0.475	12.632	0.149	0.149	0.153	1.808	0.083	2.794
0.633	9.365	0.168	0.170	0.175	1.256	0.122	3.156
0.791	7.972	0.199	0.202	0.209	1.378	0.160	3.439
1.824	6.378	0.302	0.421	0.439	3.732	0.374	4.445
9.404	6.234	0.991	1.152	1.225	15.798	0.978	5.616
35.974	5.821	1.238	1.519	1.701	20.819	1.231	5.504
68.232	5.503	1.239	1.520	1.768	19.344	1.234	5.246
126.093	5.191	1.196	1.444	1.755	16.751	1.192	4.963
202.839	4.968	1.147	1.357	1.712	14.583	1.144	4.753
Bethe-Johnson equation, a = 0.411							
0.317	17.772	0.803	0.848	0.867	39.298	–	–
0.475	16.156	0.950	1.013	1.042	46.083	–	–
0.791	14.740	1.150	1.247	1.294	56.125	–	–
0.830	14.621	1.171	1.271	1.321	57.174	–	–
0.910	14.560	1.172	1.272	1.323	56.777	0.011	1.584
1.741	13.418	1.175	1.275	1.348	48.355	0.241	4.297
3.363	11.784	1.236	1.346	1.464	42.031	0.568	5.480
6.689	10.261	1.342	1.479	1.665	39.697	0.896	6.060
11.493	9.312	1.412	1.572	1.820	38.424	1.093	6.200
18.162	8.644	1.443	1.615	1.920	36.604	1.201	6.150
23.746	8.295	1.448	1.621	1.961	35.119	1.240	6.071
31.662	7.951	1.443	1.613	1.990	33.249	1.266	5.963
47.493	7.519	1.419	1.577	2.002	30.204	1.276	5.775
55.408	7.370	1.406	1.556	1.999	28.973	1.273	5.697

Bethe-Johnson equation, a = 0.5

0.079	32.862	0.546	0.560	0.567	69.320	–	–
0.261	18.796	0.741	0.779	0.794	36.940	–	–
0.396	17.846	0.780	0.821	0.841	37.143	0.016	1.833
0.554	17.473	0.744	0.781	0.803	32.100	0.068	2.958
0.712	16.768	0.724	0.760	0.785	27.977	0.125	3.590
1.658	12.854	0.775	0.816	0.865	19.935	0.408	5.106
2.260	11.598	0.841	0.892	0.956	20.010	0.537	5.493
3.432	10.336	0.955	1.028	1.116	21.736	0.718	5.888
4.899	9.565	1.059	1.156	1.270	23.913	0.868	6.114
6.153	9.166	1.124	1.238	1.370	25.362	0.957	6.208
8.859	8.633	1.216	1.359	1.524	27.407	1.082	6.286
14.517	8.036	1.308	1.487	1.703	29.001	1.209	6.261
27.971	7.363	1.366	1.571	1.863	28.632	1.299	6.064
47.208	6.897	1.362	1.565	1.918	26.500	1.311	5.823

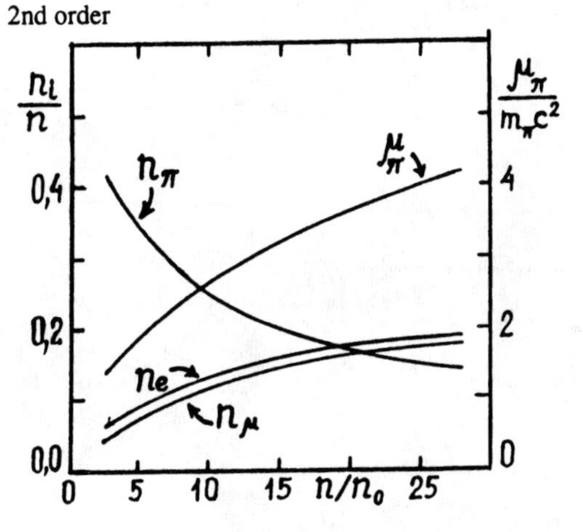

Figure 1

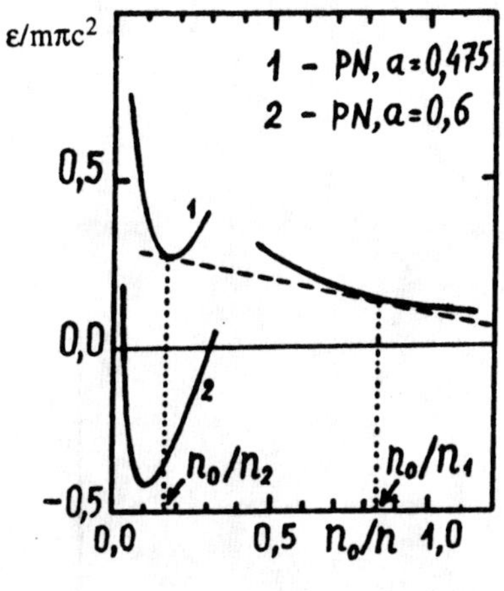

Figure 2

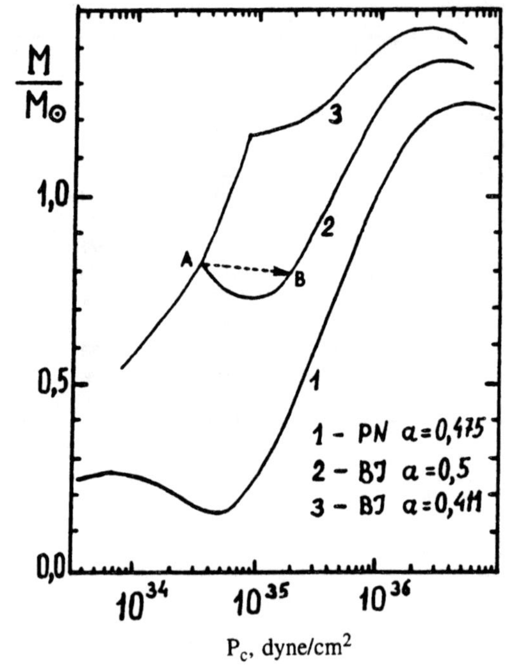

Figure 3

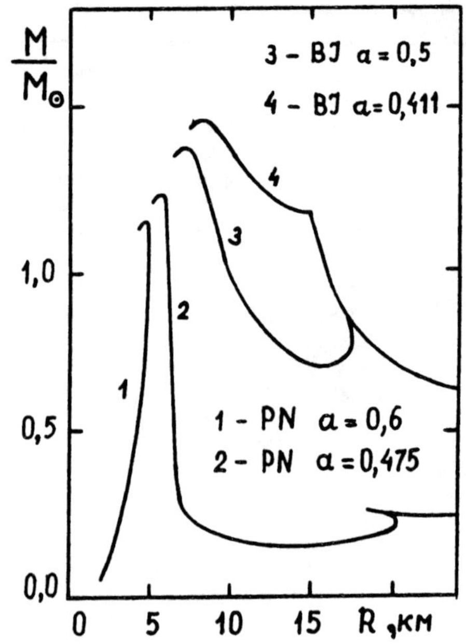

Figure 4

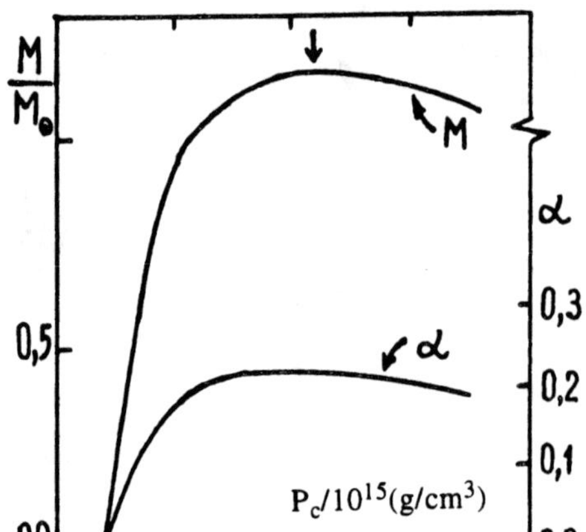

Figure 5

References

1. A.B. Migdal, ZhETF, 1971, 61, 2209.
2. R.F. Sawyer, Phys. Rev. Lett., 1972, 29, 382.
3. R.F. Sawyer and D.J. Scalapino, Phys. Rev., 1972, D.7, 953.
4. S.O. Backman and W. Weise, Mesons in Nuclei, ed. M. Rho and D. Wilkinsona (amsterdam: North Nolland),, 1979, 3, 1097.
5. P. Pains, UFN, 1980, 131, 479.
6. V.S. Beskin, UFN, 1987, 152, 683.
7. C. Alcock, E. Farhi, and A. Olinto, Astrophys. J., 1986, 310, 261.
8. P. Haensel et al., Astron. Astrophys., 1986, 160, 121.
9. H.A. Bethe et al., Nucl. Phys., 1987, A462. 791.
10. L.G. Grigoryan and G.S. Saakyan, Astrofizika, 1977, 13, 463;
Dokl. AN SSSR, 1977, 237, 299.
11. L.G. Grigoryan and G.S. Saakyan, EChAYa, 1979, 10, 1075;
Astrophys. and Space. Sci., 1983, 95, 305.
12. G. Wiler et al., Teoriya gravitatsii i gravitational collaps (Theory of gravitation and gravitational collapse) 1967 (Moscow:: Mir) (in Russian).
13. Yu.L. Vartanyan and N.K. Ovakimova, Astron. Zh., 1972, 49, 306.
14. J.B. Hartle et al., Astrophys. J., 1975, 199, 471.
15. O. Maxwell, Phys. Lett., 1976, 62B, 159.
16. P. Haensel and M. Proszynski, Astrophys. J., 1982. 258, 306.
17. Yu.L. Vartanyan, G.S. Adzhyan, and G.B. Alaverdyan, Astron. Zh., 1984, 61, 677.
18. G. Baum et al., Astrophys. J., 1971, 170, 290.
19. V.R. Pandharipande, Nucl. Phys., 1971, A178, 123.
20. R.C. Malone et al., Astrophys. J., 1975, 199, 741.
21. D. Campbell et al., Phys. Rev., 1975, D12, 679, 1010.
22. A.B. Migdal, Fermiony i bozony v sil'nykh polyakh (Fermions and bosons in strongs fields) 1978 (Moscow: Nauka) (in Russian).
23. D.N. Voskresenskii et al., Pis'ma v ZhETF, 1977, 25, 495.
24. Ya.B. Zel'dovich and I.D. Novikov, Teoriya tyagoteniya i evolutsiya zvezd (Gravitation theory and star evolution), 1971 (Moscow: Nauka) (in Russian).
25. G.S. Adzhyan, A.S. Arutyunyan, and Yu.L. Vartanyan, Uch. Zapiski EGU, 1982, 3, 78.
26. J.R. Oppenheimer and G.M. Volkoff, Phys. Rev., 55, 374.
27. J.B. Hartle, Astrophys. J., 1967, 150, 1005.
28. W.H. Ramsey, M.N., 1950, 110, 325.
29. Z.F. Seidov, Astron. Zh., 1971, 253, 908.
30. A.B. Migdal et al., Phys. Lett., 1979, 83B, 158.
31. P. Haensel and M. Proszynski, Phys. Lett., 1980, 96B, 233.
32. Yu.L. Vartanyan and G.B. Alaverdyan, Astrofizika, 1986, 24, 420.
33. G.S. Adzhyan and Yu.L. Vartanyan, Astrofizika, 1986, 24, 301.

Physics of Neutron Stars
Edited by A. M. Kaminker

ISBN 1-56072-164-2

ON SELF-ABSORPTION OF GAMMA RADIATION OF NEUTRON STARS

A. A. Kozlenkov and I. G. Mitrofanov
Institute of Space Research of the USSR Academy of Sciences, 117810 Moscow

Absorption of a photon the with energy $\geq 2m_e c^2$ occurring in the birth of $e^- e^+$ pairs under the conditions where the magnetic field is so strong and the photon concentration is so high that both one-and two-birth may be essential.

1. Introduction

One- and two-photon birth of $e^{\pm}$ pairs in a magnetic field of a strength close to critical $B_c = m^2c^3/e\hbar = 4.414 \cdot 10$ G can play an important role in the generation of gamma-radiation by neutron stars (see, e.g., [1]). It is well known that one-photon birth [2,3] cannot occur in a weak field $B \ll 1$. Two-photon birth, which is the main process of self-absorption of radiation in the absence of a magnetic field [4], at $B \ll 1$ is quite different from that in the case $B = 0$. This process was considered in detail in the case where the energy of each photon was insufficient for pair genration [5]. The main results of this study are: the existence of singularities at the energies equal to the generation threshold of the pair on the discrete Landau levels; an increase of the threshold of two-phonon process, the establishment of the strong dependence of the process cross-section on photon polarization.

If the photon energy is sufficient for generation of one-photon pair generation, then the total coefficient of radiation absorption is determined by the probabilities of one- and two-phonon processes. Below the conditions of resonance interaction of photons are considered under which the one- and two-pnonon processes are nonadditive, and therefore one can calculate only the total absorption coefficient. For the given flux of photons with energy ω_1 the dependence of the ratio between the indicated processes is studied as a function of the magnetic field strength B and the concentration of photons with energy ω_2.

2. Nonresonance and Resonance Components of Two-Photon Birth

The Johnson-Lippman-type wave functions $C^{\pm}$ used in the previous work [5] yield new results for the cross-section of two-photon processes only under the condition that the summation over the spin states of finite particles is performed. Below the relativistic wave functions obtained in [6] are used in quantum calculations of synchrotron radiation in [7] for the study of arbitrary cases. The wave functions of particles with positive and negative energies have the form:

$$\psi^{(+)}(x) = \frac{1}{2}\sqrt{\frac{E_{n0}+1}{E_n E_{n0}(E_n + E_{n0})}}\, v_n^{(s)} \exp(-iE_n t)$$

$$\psi^{(-)}(x) = \frac{1}{2}\sqrt{\frac{E_{n0}+1}{E_n E_{n0}(E_n + E_{n0})}}\, v_n^{(s)} \exp(+iE_n t) \qquad (1)$$

where x is the 4-coordinate, $E = (1 + p^2 + 2B_n)^{1/2}$ is the energy of the state with the principal quantum number n, $E_{n0} = (1 + 2bn)^{1/2}$, p is the pulse component along the magnetic field directions, and s is the spin projection onto this direction (for an electron $s = +1$ corresponds to the spin orientation along the field). For particles with negative energy, pulse p is positive if its projectiononto OX-axis is negative. The spinors

$$U_n^{(1)} = \begin{pmatrix} (E_n + E_{n0})\chi_{n-1} \\ -pV_{n0}\chi_n \\ p\chi_{n-1} \\ (E_n + E_{n0})V_{n0}\chi_n \end{pmatrix}; \qquad U_n^{(2)} = \begin{pmatrix} pV_{n0}\chi_{n-1} \\ (E_n + E_{n0})\chi_n \\ (E_n + E_{no})V_{n0}\chi_{n-1} \\ -p\chi_n \end{pmatrix};$$

$$v_n^{(1)} = \begin{pmatrix} -p\chi_{n-1} \\ -(E_n + E_{n0})\chi_n \\ (Esubn + E_{n0})\chi_{n-1} \\ -pV_{n0}\chi_n \end{pmatrix}; \qquad v_n^{(2)} = \begin{pmatrix} -(E_n + E_{n0})V_{n0}\chi_{n-1} \\ p\chi_n \\ pV_{n0}\chi_{n-1} \\ (E_n + E_{n0})\chi_n \end{pmatrix} \qquad (2)$$

are expressed in terms of a scalar function χ_n, defined in [5] as

$$\chi_n(x) = \frac{i^n}{L\,(\pi\lambda^2)^{1/2}\,(2^n n!)^{1/2}} \exp\left[-\frac{(x-a)^2}{2\lambda^2} - \frac{iay}{\lambda^2} + ipz\right] H_n\left(\frac{x-a}{\lambda}\right),$$

$$V_{n0} = (2nB)^{1/2}/(E_{n0}+1), \quad \lambda = \lambda_c B^{-1/2}$$

where λ_c is the Compton wavelength. In the argument of the Hermit polynomial H_n a x is the coordinate of the center of electron rotation. The wavefunction $e^{\pm}e$ are normalized proceeding from the "one particlee in the volume L^3 " condition. Similar to [3,5], it is assumed that interacting photons are linearly polarized either in the plane formed by vectors $\vec{K}$ and $\vec{B}$, or normally to it.

For arbitrary photon energies satisfying the conditions of one- and two-phonon generation, the correct statement of the problem requires the knowledge of the total coefficient of photon absorption, ω_1 , with due regard for all possible two-pnoton processes. This coefficient can be represented in the form

$$k(\omega_1)=k^{(1)}+k^{(2)}= \sum_{\zeta} \frac{\gamma_\zeta}{\gamma_\zeta + \Gamma_z(\omega_2)} k^{1\Phi}(\omega_1)+N_2(\omega_2)\sigma(\omega_1,\omega_2)\,(1-\cos(\vec{k}_1,\vec{k}_2)) \quad (3)$$

The first term, $k^{(1)}$, corresponds to one-photon generation by the ω_1 photon of $e^{\pm}$ pairs in all possible states ζ with subsequent emission of a photon by one of particles. These terms include factors which are determined by relationships between the natural width of levels γ_ζ and their collisional broadening $\Gamma_\zeta(\omega_2)$ under the action of impinging photons. Obviously, that at $\Gamma_\zeta(\omega_2) = 0$, the first term in $K^{(1)}$ in (3) coincides with the conventional one-phonon coefficient. The second term $K^{(2)}$ fully describes the two-photon process with absorption of photons ω_1 and photons ω_2 of the radiating field. This term depends on photon concentration $N_2(\omega_2)$ and the cross-section of their interaction $\sigma(\omega_1,\omega_2)$ expressed in terms of squared amplitude.

In order to consider the second term in (3) we apply the method stated in [8]. In studying the evolution of a quantum system by this method one considers insteand of the scattering amplitude describing the probability of the transition from the in state at $t = -\infty$ to the out state at $t = +\infty$ the evolution of the quantum system from the moment $t = 0$ untill $t = +\infty$. In this case one may consider as $t = 0$ the moment of the "beginning of $e^{\pm}$ pair generation". It is also necessary to take into account the nonstationarity of excited quantum states. If the photon concentration ω_2 is low and the corresponding filling number « 1, then the width of the excited levels is deteremined only by spontaneous transitions. The natural widths of the levels obtained in the quantum calculations of these transitions at $B \leq 1$ [7] were used. On the other hand, the width of the ground state is always deteremined by the flux of impinging photons ω_2. The finite level widths lead to the fact that instead of the δ-function corresponding to the law of energy conservation in the process amplitude shows a Lorentzian profile of the finite width, and the resonance denominators $(E_{l'} - \omega_{2,1} + E_n)$ and $(E_l - \omega_{1,2} + E_n)$ acquire the imaginary additions (here E_l and $E_{l'}$ are the energies of finite particles, and E_n is the energy of a virtual particle, see [5]).

If the resonance denominator in the cross section, e.g.,

$$(E_l - \omega_{1,2} + E_n),$$

goes to zero (in this case the total "two-photon law of energy conservation" can be violated), the interaction at the first vertex can be considered as a birth of a real pair, and in the second as an excitation of one of the born particles. By analogy with the evolution of an atomic system, it can be shown that in the absence of interference, the resonance part of K can be represented as

$$k^{(2)}_{res} = \frac{\gamma_1 |H_{1/0}H_{2/1}|^2 N(\omega_2)dq}{[(E_l + E_{l'} - \omega_1 - \omega_2)^2 + \gamma_l^2/4]\cdot[(E_l + E_n - \omega_1)^2 + (\gamma_n + \Gamma_n)^2/4]} \qquad (4)$$

where γ_l and γ_n are radiative widths of levels E_l and $\overline{E_n}$ of the final and intermediate states, Γ_n is the spread of the intermediate level due to the process of the one-photon pair generation, $H_{1/0}$ is the matrix element of the process of one-phonon generation of a pair and $H_{2/1}$ is the matrix element of the absorption of photon ω_2 (see, e.g., [7]). The nonresonance part of $K^{(2)}$ corresponds to pulses q for which the second denominator in (4) is nonzero. For some K_2 there exist such q that both denominators in (4) go to zero simultaneously. In this case one can consider the resonance in the cross section of 2-photon generation. In [5] the authors considered only the case where none of the photons in one-photon process could form a pair. In the nonresonance case the results stated in [5] are also applicable, since the integration over the resonance region yields

$$k^{(2)}_{\text{res}} = \int \frac{\gamma_l \, |H_{1/0}|^2 \, |H_{2/1}|^2 N(\omega_2) dg}{(\gamma_n + \Gamma_n) \, [(E_{l'} - E_n - \omega_2)^2 + \gamma_l^2/4] \cdot |qE_n + PE_l|} = \frac{\Gamma_n}{\gamma_n + \Gamma_n} \frac{|H_{1/2}|^2}{|qE_n + PE_l|} \qquad (5)$$

and with due regard for the determination of the collisional width

$$\Gamma_n = \int \frac{\gamma_l \, |H^2/1|^2 \, N(\omega_2) \, d\, q}{[(E_{l'} - E_n - \omega_2)^2 + \gamma_l^2/4]} \qquad (6)$$

and over the region q corresponding to the complete law of energy conservation yields a two-photon cross section calculated in [5].

It can readily be seen that the sum of (5) and the first terms from (3) yield the one-photon generation coefficient and, therefore, the one- and two-photon processes in the nonresonance case are additive.

In resonance, if $\gamma_l \sim \gamma_n$, only numerical study of the dependence of the total absorption coefficient on the physical conditions of the reaction is possible. If a photon ω_1 generates a pair in such n and l states that $\gamma_n \ll \gamma_l$, then the region of the minimum of the first cofactor in (4) includes much smaller region of the second cofactor and a qualitative analytical study of this expression is possible. Using the sharp Lorentz profile of the second denominator, it is possible to substitute it by the δ-function, and integral (4) at the interval $\omega' - \omega \geq \gamma$ can also be represented in form (5). The rejected interval $\omega' - \omega \leq \gamma_n$ gives an addition due to its own two-photon process proportional to the concentration of photons ω_2. Thus, in case of low $N(\omega_2)$ consentrations, in resonance it is also possible to represent the absorption coefficient ω_1 photons as a sum of one- and two-phonon generation.

The specific behavior of the absorption coefficient in a strong magnetic field was studied numerically. The coefficient (3) can be conveniently represented in the form

$$k(\omega_1) = k^{1\Phi} \, (1 + N_2(\omega_2) / N_k(\omega_1, \omega_2)) \qquad (8)$$

The calculation of quantity $N_* = k^{1\Phi} / (\delta^{2\Phi}(1 - \cos(\vec{k}_1, \vec{k}_2)))$N = K /(s (1-cos(K ,K))) of dimensionality cm^{-3} allows the determination of such photon concentration $N_2(\omega_2)$, at which the contribution of two-photon process into the total coefficient are noticeable.

3. The ratio of the contributions of one- and two-photon processes

Figures 1 and 2 show critical concentrations N_* (sign , along the Y-axis decimal logarithm is plotted) for the case of impinging photons normal to the magnetic field B = 1 and 0.3, respectively. One of the photons has a fixed energy of 715 keV and cannot generate a pair in one-photon process. Along the X-axis the energy of other photons in mc^2 is plotted. The same figures also show the curves (sign)obtained for two-photon birth by the classical formula (B = 0). For any given concentration N_2, Figure 1 allows one to determine the prevalent process - one-photon ($N_* > N_2$) or nonresonance ($N_* < N_2$). From Fig.1 it is also seen that with a decrease of the field strength the concentration that makes the two-photon generation the main process also decreases. Our consideration is limited to N_* values not exceeding the concentration N_0 above which one has to take into consideration also induced processes.

Figure 3 (field B = 1) corresponds to the photon propagation across the field: ω_1 (of the fixed energy 1.28 MeV) and ω_2 photons (their energy in mc^2 units is plotted on the X-axis). In this case, the latter photons cannot form a pair at any energy. The values calculated with the aid of the formula for two-phonon birth at B = 0 (sign) have a threshold at ω_2 = 200 keV higher than for B = 1. The cyclotron resonanses in Fig.3 are the regions where N_* values approach, the concentration N_0 at which the absorptional and radiational widths become equal. For energies within several widths around the resonance, the concentration values where the two-photon birth exceeds one-phonon generation correspond to the induced mode and are not considered here.

□

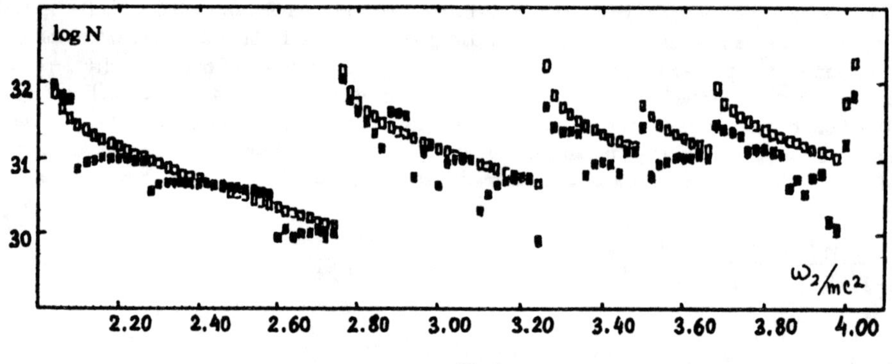

Fig.1.

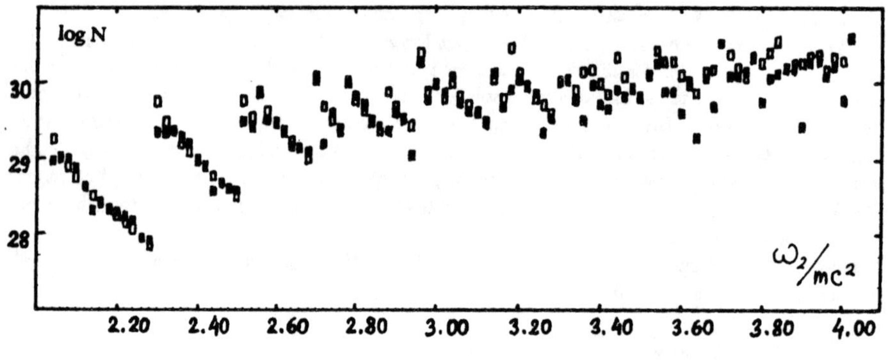

Fig.2.

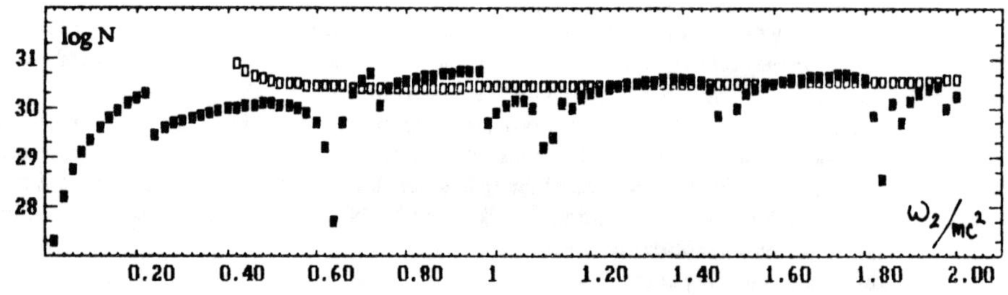

Fig.3.

References
1. AIP Conf. Proceed.141, Stanford, California, 1984.
2. N.P. Klepikov, ZhETF, 1954, 26, 1, 19.
3. J.K. Daugherty and A.K. Harding, Positron-electron Pairs in Astrophysics, M.L. Burns, A.K. Harding and R.Ramaty eds., AIP{ Conf. Proc., No.101, 1983, p.163.
4. A.I. Akhiezer and V.B. Berestetskii. (Kvantovaya elektrodinamika (Quantum electrodynamics) Moscow: Nauka, 1981.
5. A.A. Kozlenkov and I.G. Mitrofanov, ZhETF, 1986, 91, 6, 1978.
6. H. Herold, H. Ruder, and G. Wunner, Astron. Astrophys., 1982, 115, 90.
7. I.G. Mitrofanov and A.S. Pozanenko, ZhETF, 1987, 93, 6, 1951.
8. V. Geitler, Kvantovaya teoriya izlucheniya (Quantum theory of radiation) Moscow: IL, 1956 (in Russian).

Physics of Neutron Stars
Edited by A. M. Kaminker

ISBN 1-56072-164-2

ON SYNCHROTRON GAMMA RADIATION OF NEUTRON STARS

I. G. Mitrofanov and A. S. Pozanenko
Institute of Space Research of the USSR Academy of Sciences, 117810 Moscow

Electron transitions from excited states in a strong magnetic field $B \sim B_c$*/ have been considered. Transition times, mean square recoil pulses, and radiation spectra have been calculated numerically.

1. Introduction

At present it it is accepted that the synchrotron mechanism is one of the most important processes of gamma-ray generation by neutron stars (see, e.g., [1]). For a superstrong magnetic field the process should be considered in a relativistic quantum mechanical limit. Below some features of the synchrotron radiation for such a case are considered, and the laws that should be taken into account in the construction of physical models of sources are established.

2. Spectrum of Synchrotron Radiation in Decay of Excited States and Transitions to the Ground State

An excited state of an electron ζ_n is characterized by the principal quantum number n and the projections of the spin s and pulse p onto the direction of the magnetic field $\vec{B}$. Let $R(\zeta_n \to \zeta_m)$ be the probability (per unit time) of spontaneous transitions from the state ζ_n into the state ζ_m,

$$R(\zeta_n) = \sum_{m<n} R(\zeta_n \to \zeta_m) \qquad (1)$$

is the total velocity of spontaneous departure from the excited state. The probability of the implementation of a certain quantum transition in the decay of the state ζ_n is

$$W_{nm} = R(\zeta_n \to \zeta_m)/R(S_n) \qquad (2)$$

The quantities $R(\zeta_n \to \zeta_m)$ and $R(\zeta_n)$ were studied numerically for the case where the longitudinal component of the pulse in the initial state is zero, p = 0; obviously that the probability (2) is independent of this pulse.

One should distinguish between the spectra of photons emitted during the departure from the given level (decay spectrum) and the spectra of photons emitted in the transition into the ground state through an intermediate state (the transition spectrum). If the number of particles in the excited states $N(\zeta_n)$ in a radiating ensemble is constant,then the decay of this state makes the following contribution into the spectral density of the flux:

$$\frac{dN}{d\omega dt} = F_n(\omega) = R(\zeta_n) \sum_{m<n} \int d\Theta\delta(\omega - \omega_{nm}(\Theta))\, P(\Theta) \qquad (3)$$

where $P(\Theta)$ is the angular radiation distribution. In order to calculate the spectrum of the stationary synchrotron radiation (SSR) one has to integrate (3) over the whole ensemble.

The complete spectrum of photons emitted in the transition into the ground state for time interval from t = 0 to t = ∞ can be conveniently presented as a sum of the transition spectra for separate channels, if a channel is considered as a certain sequence of quantum transitions from the initial to the ground state [2]. For $n \gg 1$, the total number of channels is rather large (3^{n-1}) and the summation over channels cannot be performed.

But it should be taken into account that the probability M(K) of the implementation of channel K can be represented as a product of the probabilities of individual transitions in this channel:

---

*/ Below the following units are used: $E \to E/(m_e c^2)$, $B \to B/B_c$, where $B_c = m^2_e c^3/(e\hbar) = 4.4 \cdot 10$ G.

$$M(k) = \prod W_{ni} \qquad (\forall \zeta_i \in K) \qquad (4)$$

In addition, it is also possible to use the approximation of a partial allowance for a recoil pulse (PAR) when a recoil pulse is taken into account in photon emission but the accumulation of the longitudinal pulses in the intermediate states is neglected (see below). In this case the expression for the transition spectrum can be written in the recurrence form

$$\frac{dN}{d\omega} = G_n(\omega) = \sum_{m<n} \left(\int d\Theta\delta(\omega - \omega_{nm}(\Theta))P(\Theta) + G_m(\omega)\right). \qquad (5)$$

If the condition for the applicability of this approximation is not fulfilled (see [2]), then expression (5) cannot be used to calculate the spectra of synchrotron radiation, and one should invoke the Monte-Carlo technique.
But the spectra of the transition to the ground state can also be calculated in an arbitrary case, if to consider the spectra not with respect to energies but with respect to the emitted harmonics $G_n(\nu)$, where $\nu = n - m$. The expression for the spectra with respect to the number of harmonics is obtained from (5) by the substitution of the δ-function by the Kronecker function $\delta_{n-m,\nu}$. The study of spectra with respect to harmonics allows one to establish the characteristic features of the transition spectra from level n to level $m = 0$ and their difference from the decay spectra.
The spectra of the initial state decay and the transition to the ground level are shown in Fig.1. These spectra demonstrate the characteristics studied in a quasiclassical case and with allowance for quantum corrections [3,4]. In a constnat field, an increase in the electron energy makes the spectra harder due to a larger relative contributions of higher harmonics. The spectra are terminated at harmonics that correspond to the energy of the initial state. The spectra of transitions (and decay) from the states with the spin $s = +1$ are softer than for $s = -1$ because in the former case the direct transitions to the ground state are forbidden.
The decay spectrum of the state is always more hard than that of transition to $m = 0$. This difference becomes obvious from the comparison of (3) and (5): in the transition spectrum the contribution from the low harmonics is more pronounced due to the radiation of intermediate states. For the constant energy of the initial electron state (curves C and D in Fig.1a and curves A and B in Fig.1b) an increase of the field intensity reduces the difference between the transition and decay spectra because of an increase of the relative contributions of the transitions with spin reorientation. A reduction of this difference with an increase of the electron energy is associated with higher probability of the implementation of the transition channels with emission of higher harmonics.

3. Lifetimes of Levels and Transition to the Ground State The decay of the state is characterized by the lifetime τ_1 (the quantity reciprocal to the natural level width, i.e., the total probability of all the spontaneous transitions (1)). The dependence of τ_1 on the level energy for different values of the magnetic field is illustrated in Fig.2. On the other hand, the rate of the particle transition from an excited to the ground state is characterized by the transition time τ_2 which is determined from the solution of a system of kinetic equations:

$$\dot{b}_{\zeta_m} = -R(\zeta_m)\, b_{\zeta_m} + \sum_{m<i<n} R(\zeta_i \to \zeta_m) b_{\zeta_i} m = 1, n, \qquad (6)$$

where $b_{\zeta_m}(t)$ is the probability of the state $\zeta_m z$.
System (6) has a simple solution only in the case $nB \ll 1$: $\tau_2/\tau_1 = n \cdot \ln(n)$. With an increase of the magnetic field intensity, the above ratio decreases. The exact solution of (6) for τ_2 under the PAR conditions is also shown in Fig.2.
The comparison of the characteristic time τ_0 of the exciting of levels $e^{\pm}$ in a magnetic field with the times τ_1 and τ_2 provides the establishment of the conditions for synchrophotons generation: at $\tau_0 \ll \tau_1$ the SSR is generated, if $\tau_0 \gg \tau_2$ the synchrotron cooling radiation (SCR) is generaged; in an intermediate case the calculation of the spectra requires the solution of system (6).
In the numerical calculations of the decay spectra by the Monte-Carlo method [5] the contribution of each transition into the complete decay spectrum related to the lifetime of the level from which the transition occurred. If the lifetimes of all the levels were equal, the spectra obtained in [5] would have had the same form as the spectra calculated by (5). But it is well knwon that this time depends on the number of levels, especially in the vicinity of the ground state (Fig.2). Therefore spectra [5] must considerably differ from those calculated by formulae (5): in the latter spectra the number of photons for low harmonics is much larger.
In order to estimate the mean rate of the photon generation in the transition to the ground state the spectrum (5) should be divided into τ_2. Since the lifetimes of all the levels (with $m \ll n$) are much less than $\tau_2(n)$ (Fig.2), the generation rate obtained in [5], is overestimated in comparison with $G_n(\omega) / \tau_2(n)$. It should be noted that the difference in the above indicated estimates increases for lower field intensities.

4. Applicability of the Approcimation of Partial Allowance for Recoil Pulse

The possibility to conduct numerical calculations of the main characteristics of the quantum synchrotron radiation by (5,6) is associated with the validity of the approximation of the partial allowance for the recoil pulse (PAR). Otherwise the Monte-Carlo method should be used. Below we shall consider the the applicability range of the PAR approximation.

If in the initial excited state ζ_n an electron was at rest ($p = 0$), then upon the transition into the ground state it acquires a pulse $q \neq 0$ because the recoil effects. In a magnetic field $B \ll 1$ this pulse is small, but, in the quantum region $B \sim 1$ it drastically increases. At an intermediate level, $m < n$, the recoil effect is characterized by the mean square recoil pulse $< q^2_m(n) >$:

$$< q^2_m(n) > = \sum_{\{K\}} < q^2 >_K \qquad (7)$$

the contributions $< q^2 >_K$ to which come from all the transition channels. Along the sequence of the intermediate states, the pulse increases in the downward direction, being naximal in the ground state. It can be shown that (7) admits also the recurrence notation similar to (5). Thus calculated mean square recoil pulse is shown in Fig.3 as lines of constant level. From Fig.3 it follows that the PAR approximation terns out to be quite applicable to the calculations of the transitions from sufficiently high quantum levels in intense magnetic fields.

If $nB \ll 1$ and $< q^2_{m=0}(n) > \sim nB^2$, an increase in the field and an initial level n decelerates $< q^2_{m=0}(n) >$. The main cause of a slow increase of $< q^2(n) >$ with an increase in n is the narrowing of the directional diagram with respect to the directions across the field in the emission of high harmonics.

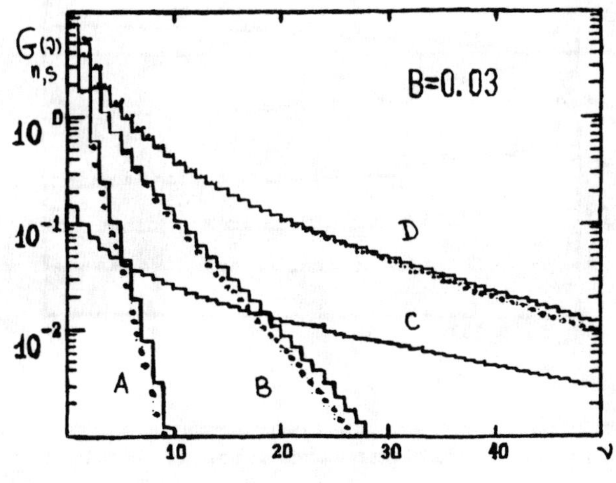

Fig.1.(a) Transition and decay spectra $G_{n,s}(\nu)$ by harmonic numbers: (A) m = 15, (B) n = 50, (D) n = 150 are the transition spectra; C n = 150 is the decay spectra.
Solid line s = -1, dash line s = + 1 (the decay spectrum is shown only for s = -1).

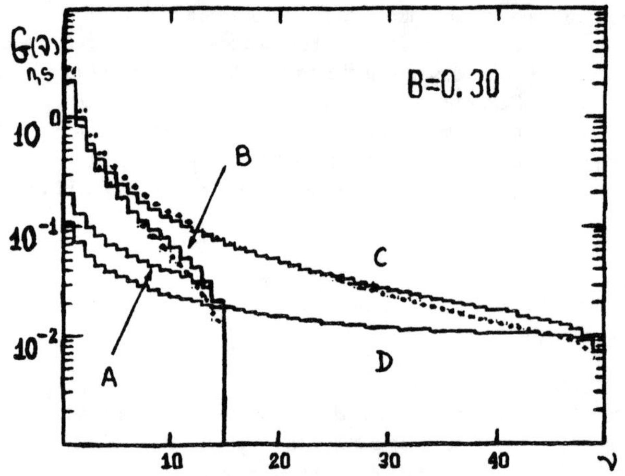

Fig.1

(b) The same as in (a). (A) n = 15, (C) n = 50 are the transition spectra, (B) n = 15 and (D) n = 50 are decay spectra. Solid line s = -1, dash line s = +1. (the decay spectra are shown only for s = -1).

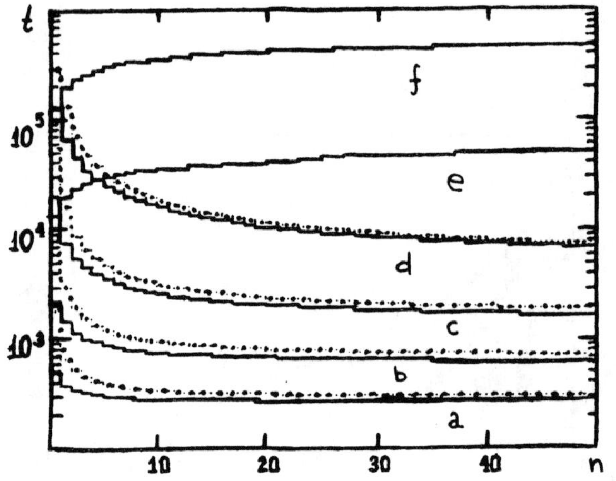

Fig.2. The lifetime of the initial state ζ_n (a, b, c, d) and the time of the transition into the ground state (e, f) in $h/(mc^2)$ units . Solid line is s + -1, dash line s = +1. The time of the transition is shown for s = m-1; B = 1 (a), 0.3 (b), 0.1 (c,e), and 0.03 (d,f).

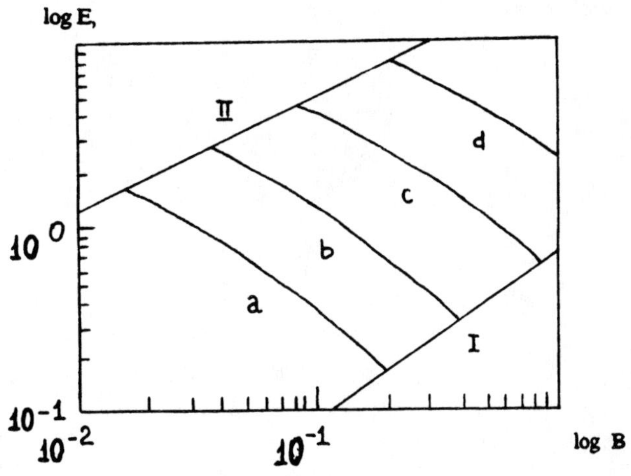

Fig.3. Constant-level lines of the mean square recoil pulse in the transition from the state ζ_n with energy $E = (1 + 2nB)^{1/2} - 1$ into an unexited state (averaging over the spins in the initial state was made). $<q^2> = 0.01$ (a), 0.03 (b), 0.1 (c), and 0.3 (d), I is the boundary of the first excited level, and II corresponds to n = 200.

References

1. I.G. Mitrofanov, Fizika neitronnykh zvezd. Pul'sary i barstery (Physicsof neutron stars. Pulsars and barsters) Leningrad: FTI, 1988, p.70 (in Russian).
2. I.G. Mitrofanov and A.S. Pozanenko, ZhETF, 1987, 93, 6, 1951.
3. A.A. Sokolov and I.M. Ternov. Relyativistskii elektron (Relativistic electron), Moscow: Nauka, 1983.
4. G.G. Pavlov, S.V. Golenetskii, Astrophys.Sp.Sci., 1986, 128, 341.
5. A. Harding and R. Preece, Astrophys.J., 1987, 939.

Physics of Neutron Stars
Edited by A. M. Kaminker

ISBN 1-56072-164-2

TWO-PHOTON ANNIHILATION RADIATION OF ELECTRON-POSITRON PLASMA IN MAGNETIC FIELDS OF NEUTRON STARS

A. D. Kaminker, P. G. Mamradze** and G. G. Pavlov**

* A. F. Ioffe Physical-technical Institute of the USSR Academy of Sciences, 194021 Leningrad

** Abastumani Astrophysical Observatory of the Academy of Sciences of the Georgian SSR, 380060 Tbilisi

The influence of longitudinal motion of particles of the electron-positron plasma in a strong magnetic field on the characteristics of two-photon annihilation radiation has been considered for Maxwellian and thermal e^- and e^+ distributions. The one-dimensional character of this motion results in the dependence of the annihilation radiation spectra on the direction of γ-quanta emission, an increase of the total annihilation rate, and a strong (about tens of percent) linear polirazation of the radiation. The $e^- e^+$ annihilation over a power spectra may be responsible for hard (of the order of several MeV) tails of the gamma-burst spectra.

Among the mechanisms responsible for the radiation of cosmic gamma-bursts of interest is the process of two-photon e^- and e^+ annihilation, which may take place close to the surface of neutron starts (NSs) with strong magnetic fields, $B \sim 10^{11}$ - 10^{13} G. The emission spectral features in the range $\sim$ 300-500 keV observed for $\sim$ 20% of γ-bursts [1] are naturally interpreted as annihilation lines shifted towards the red range due to the gravitational field of a NS. In [2] the annihilation radiation was invoked for the interpretation of the so-called hard component of the γ-burst radiation ($\geq$ 1 MeV) whose spectrum is of a power character (see also [3]). On the other hand, it is well known [4,5] that a strong magnetic field of a NS makes a substantial effect on the rate of the $e^- e^+$ annihilation and the properties of annihilation radiation. In particular, if $B \geq B_c = 4.41 \cdot 10^{13}$ G, a process of one-proton annihilation prevails in the $e^- e^+$ plasma studied in detail in [5-7]. The present study is devoted to the effect of the longitudinal motion of e^- and e^+ in a magnetic field $B \ll B_c$ on the characteristics of two-photon annihilation radiation.

If the rate of cyclotron cooling of the $e^- e^+$ plasma ($\propto B^2$) exceeds the rate of heating (i.e. of a Coulomb or a Compton type), which fills the excited Landau levels ($n_{\mp} = 1, 2, 3, ...$), then e^- and e^+ are located mainly on the ground Landau level, $n = 0$. In this case the particle motion is one-dimensional, and the spins of e^- and e^+ are oriented aniparallel and parallel to $\vec{B}$. respectively. Thus in the case of Coulomb heating of a moderately relativistic plasma such a process takes place at $B \geq 10^{10} (N/10^{24}\ cm^{-3})^{1/2}$ G. Under such conditions the plasma state is given by the one-dimensional distribution functions $\eta_-(p_-)$ and $\eta_+(p_+)$, where p_- and p_+ are the longitudinal pulses of e^- and e^+. In the limit of low longitudinal velocities of e^- and e^+, where $\eta_{\mp}(p_{\mp}) = \delta(p_{\mp})$, the two-photon annihilation was considered in more detail in [8], where, in particular, the linear polarization of the annihilation radiation was first analyzed. The effect of one-dimensional motion of the thermal $e^- e^+$-plasma on the intensity and polarization of the annihilation radiation was first considered in [9]. We shall continue the study of two-photon annihilation at the ground Landau level and consider two cases. In case (a), we assume a one-dimensional Maxwellian distribution of e^- and e^+ with (longitudinal) temperature $T = t \cdot mc^2$:

$$\eta_{\mp} = [2k_1(1/t)]^{-1} \exp(-\varepsilon_{\mp}/t) \qquad (1)$$

where $\varepsilon_{\mp} = (1 + p^2)^{1/2}$ is the energy of e^- and e^+, and K_1 is the Macdonald function. Hereinafter all the energies, pulses, and velocities are expressed in mc^2, mc, and c units, respectrively. We restrict ourselves to the condition $b \equiv B/B_c \ll \min(1;t)$. Note also that the condition $t \gg b$ does not contradict the assumption that $n_- = n_+ = 0$ if the transversal temperature $T_\perp = mc^2 t_\perp$ is much lower than the longitudinal temperature, so that the condition $b \gg t_\perp$ is fulfilled.

In case (b) we assume a power one-dimensional distribution of e^- and e^+:

$$\eta_{\mp} = \frac{\Gamma\left(\frac{k}{2}\right)}{\sqrt{\pi}\,\Gamma\left(\frac{k-1}{2}\right)}\,\varepsilon_{\mp}^{-k} \qquad (2)$$

consider relatively weak magnetic fields (see above), then for $d\sigma_\lambda/d\Omega d\omega$ the expression can be used obtained at B = 0 for e^- and e^+ (see, e.g., [10]) with oppositely oriented spins moving along the same line

$$\frac{d\sigma_\lambda}{d\Omega\,d\omega} = \left(\frac{d\sigma_\lambda}{d\Omega\,d\omega}\right)_0 \frac{d\Omega_0\,d\omega_0}{d\Omega\,d\omega} = \left(\frac{d\sigma_\lambda}{d\Omega\,d\omega}\right)_0 \gamma^{-1}(1-\beta c)^{-1}, \qquad (4)$$

$$\left(\frac{d\sigma_\lambda}{d\Omega\,d\omega}\right)_0 = \frac{\delta(\omega_0-\varepsilon_0)}{4\varepsilon_0\,|p_0|}\,\frac{\varepsilon_0^2+p_0^2[1+(-1)^\lambda s_0^2]^2}{(\varepsilon_0^2 s_0^2+c_0^2)^2} = \left(\frac{d\sigma_\lambda}{d\Omega}\right)\delta(\omega_0-\varepsilon_0), \qquad (5)$$

where the subscript o denotes the quantities in the system of center of masses (CMS), $\beta = (p_+ + p_-)/(\varepsilon_+ + \varepsilon_-)$ is the velocity of the CMS with respect to the laboratory reference system, $\gamma = (1-\beta^2)^{-1/2}$, $\omega = \omega_0/\gamma(1-\beta c)$, $c = \cos v$, v is the angle formed by $\vec{B}$ and $\vec{q}$, p_0 ($-p_0$) is the longitudinal pulse of e^-(e^+), $\varepsilon_0^2 = 1 + p_0^2$, $c_0 = \cos v_0$, $s_0 = \sin v_0$, the polarization index takes the values $\lambda = 2$ and $\lambda = 1$, when the vector of photon polarization either lies in the $\vec{B}\,\vec{q}$ plane or is normal to it, respectively.

Expressions (3)-(5) with due regard for (1) or (2) allow one to obtain the intensity of annihilation radiation I, the Stokes parameter Q, and the degree of linear polarization P:

$$I = A_1 + A_2, \qquad Q = A_2 - A_1 \qquad P = Q/I \qquad (6)$$

In this case the Stokes parameters U = V = 0, i.e., in particular, there is no circular polarization. One can also obtain an direction-integrated radiation spectrum $S(\omega)$, the total annihilation rate P, and the energy-integrated degree of radiation polarization $\wp$:

$$S(\omega) = \frac{1}{2}\int d\Omega\, I, \quad R = mc^2\int_0^\infty d\omega\, S \qquad (7)$$

(the factor 1/2 takes account of the equivalence of two photons)

$$\wp = \int_0^\infty d\omega\, Q \Big/ \int_0^\infty d\omega\, I \qquad (8)$$

(a) One-dimensional thermal distribution

The I, S, R, and $\wp$ values calculated by (3)-(5) with due regard for (1) are shown in Figs.1-4. Figure 1 demonstrates the dependence of the $I(\omega)$ spectra on temperature and angle v.

If $v = 0$, formulae (3)-(5) and (1) yield

$$I = I_0 \begin{cases} \dfrac{\omega^{3/2}}{\sqrt{\pi t}} \exp\left[-\dfrac{(\omega-1)^2}{\omega t}\right] & t \ll \min(1, \omega^{-1}) \\ \dfrac{\omega}{t^2}[\ln 4\omega t - 1 - c]\exp\left(-\dfrac{\omega^2+1}{\omega t}\right) & t \gg \max(1, \omega^{-1}) \end{cases} \qquad (9)$$

where C = 0.577 is the Euler constant. If $t \to 0$, the spectrum reduces to an infinitely narrow annihilation line $I \to I_0\,\delta(\omega - 1)$. With an increase in t this line broadens and shifts towards higher energies. For nonrelativistic temperatures $t \ll 1$, the broadening is caused mainly by the linear Doppler effect due to longitudinal motion of e^-e^+ pairs. The width of the (FWHM) line is determined by the formula $\Gamma = 2(\ln 2)^{1/2} t^{1/2}$, its maximum being shifted towards the blue range, $\omega = 1 + 0.75t$, because of the joint action of the quadratic Doppler effect and an increase in $\omega_0 = \varepsilon_0 = (1 + p_0^2)^{1/2}$ caused by a relative motion of e^- and e^+. At $t \geq 1$, the lines are transformed into an asymmetric continuum with $\Gamma \sim 2t$ and $\omega_m - 1 \sim t$. With an increase in v the annililation line becomes quite narrow. At $t \ll 1$, $|\cos v| \gg t^{1/2}$. t and we obtain

$$I = I_0 \frac{\omega}{\sqrt{\pi t}\,|\cos v|}\exp\left[-\frac{(\omega-1)^2}{t\cos^2 v}\right] \qquad (10)$$

i.e., the line is broadened due to the linear Doppler effect (the mean squared projection of the velocity onto the direction of the quantum emission $\sim t\cos^2 v$), and its width is equal to $\Gamma \approx 2(\ln 2)^{1/2} t^{1/2} |\cos v|$. Narrow peaks in Fig.1 indicate logarithmic divergence of the intensity I at $\omega \sin v = 1$, associated with singularity of the quantity $|v_- - v_-| \cdot (d\sigma_\lambda / d\Omega d\omega)$ due to one-dimensional character of the e^- and e^+ motion. The singularity also takes place for $p_- = p_+ = q_u = \gamma\beta = \omega \cos v = \cot v$; $\beta = \cos v, \gamma = \omega = (\sin v)^{-1}$; $p^0 = 0, \varepsilon_0 = 1, \cos v_0 = 0$.
If $v = \pi/2$ we have for the case $t \ll$? :

$$I = I_0 \begin{cases} \dfrac{\eta}{nt}\left(\ln\dfrac{t}{|1-\omega|} - c\right) & |\omega - 1| \ll 1 \\ \left[\dfrac{2\omega^3}{nt\,(1-\omega^2)}\right]^{1/2} \exp\left[-\dfrac{2(1-\omega)}{\omega t}\right] & 1-\omega \gg t \\ \left[\dfrac{2}{nt\,\omega^9\,(\omega^2-1)}\right]^{1/2} (3\omega^2-2)\exp\left[-\dfrac{2(\omega-1)}{\omega t}\right] & \omega - 1 \gg t \end{cases} \qquad (11)$$

$I(\omega)$ has a logarithmic singularity at $\omega = 1$. The line is much narrower than at $v = 0$, since at $v = \pi/2$ there is no longitudinal Doppler effect, and the line broadens due to the quadratic Doppler effect (the red wing) and an incrrease in $\omega_0 = \varepsilon_0 = (1 + p_0^2)^{1/2}$ because of the relative e^- and e^+ motion (the blue wing).
As is seen from Fig.1, the annihilation continuum at $t \geq 1$ "lowers down" with an increase in t for all the directions, $I \propto t^{-2} \ln t$.
The direction-integrated spectra $S(\omega)$ strongly differ from the spectra of an isotropic thermal plasma (Fig.2). For any t, $S(\omega)$ shows a sharp maximum at $\omega = 1$, whereas in an isotropic case an increse in t only displaces the maximum of $S(\omega)$ towards the blue range.
The total annihilation rate R decreases with an increse in t (Fig.3). For the R(t) dependence one can obtain the following approximate expression

$$R(t) \approx R_0 \left[K_0(t^{-1})\right]^2 \left[K_1(t^{-1})\right]^{-2} \xrightarrow[t \gg 1]{} R_0 \left(\ln 2t - c\right)^2 t^{-2} \qquad (12)$$

where $R_0 = 2\pi$ with $(e^2/mc^2)\, N_- N_+ = N_- N_+ (1.5 \cdot 10^{-14}\ cm^3/s)$, K_0 and K_1 are the Macdonald functions. From Fig.3 it is seen that for all t values, R considerably exceeds the total annihilation rate in an isotropic plasma:

$$R_{is}(t) = \frac{R_0}{2} K_0(t^{-1}) \left[K^2(t^{-1})\right]^{-1} \xrightarrow[t \gg 1]{} \frac{R_0}{4} \left(\ln 2t - c\right) t^{-2} \qquad (13)$$

Note also that if $t \to 0$, $R_{is} \to R_0/2$. This is associated with the exclusion of the annihilation of e_- and e^+ with parallel-oriented spins (see curve R_{par} in Fig.3) in the limit of low velocities of particles.
The energy-integrated degree of linear polarization (see Fig.4) at $t \ll 1$ is equal to $\wp \approx t \sin v$; it increases with t, at $t \approx 10$ $\wp$ attains the maximum, $\wp_{max}(v = \pi/2) \approx 0.28$, and then decreases at $t > 10$ for most directions $\sin v > t^{-1}$, $\wp \propto (\ln t)^{-1}$. The polarization vector is in the $\vec{B}\vec{q}$ plane ($\wp > 0$).
(b) One-dimensional power distribution
Figure 5 shows the $I(\omega)$ spectra of the annihilation radiation as a function of K of power distribution of particles (2) and angle v. With an increase in K the spectra change qualitatively similar to variation with a decrease in t in the thermal case (cf. Fig.1). Of interest is the case $v = 0$ to which there correspond the hardest spectra. Thus in the high-energy wing $\omega \gg 1$ we have

$$I(\omega) \propto \begin{cases} \omega^{-(k-1)} & \text{if } v = 0 \\ \omega^{-(2k+3)} & \text{if } v = 90^\circ \end{cases} \qquad (14)$$

As in case (a), I experiences logarithmic divergence at $\omega \sin v = 1$.
The direction-integrated spectra $S(\omega)$ at different k are shown in Fig.6. For the high-energy wing $\omega \gg 1$ we have

$$S(\omega) \approx \frac{4\sqrt{\pi}\,\Gamma^3(k/2)}{\Gamma\left(\dfrac{k+1}{2}\right)\Gamma^2\left(\dfrac{k-1}{2}\right)}\,\omega^{-(k+1)} \qquad (15)$$

Figure 7 shows the dependece of the total annihilation rate on K for different spin orientations - antiparallel spins (R), poarallel spins (R_{par}), and nonpolirized e^- and e^+ (R_{NP}) (cf.Fig.3).
The energy-integrated degree of linear polarization $p(v)$ (Fig.8) for the above considered values k = 2, 3, 4 in the wide angular range attains the value ~ 10 - 20% (cf.Fig.4).
The obtained results can be applicable, in particular, to the interpretation of the annihilation radiation from γ-burst sources. For assumed magnetic fields of these sources (B ~ 10^{12} G) the observed broadening of annihilation singularities is most probably caused by the longitudinal motion of e^- and e^+. From the present consideration it follows that the width of annihilation singularity is substantially dependent on the angle formed by the magnetic field and the line of sight. In other words, the conlusions about the parameters of the radiating plasma depend on the directional diagram of the emitted radiation. Thus, if the annihilation radiation is directed along $\vec{B}$, then the linewidths Γ ~ 100-200 keV obtained in [1] lead to unrealistically low plasma temperatures T ≤ 30 keV. On the other hand, the linewidths Γ ~ 350-500 keV obtained for another fitting model [2] lead to the values T ~ 100-200 keV that are in accord with the typical values of the fitting temperature for the soft continuum component (E ≤ 1 MeV). On the contrary, if radiation is emitted mainly normally to $\vec{B}$, then the agreement with the temperature of the soft continuum component is attained just for the linewidths ~ 100-200 keV.
Hard (seems, power with the exponent 2-3) tails of the γ-burst spectra observed on SMM [3] can hardly be consistent with the assumption on the thermal distribution of e^- and e^+. Since the power tails (see (b)) of the annihilation radiation arise in the power distributions of e^- and e^+, it is natural to draw a conclusion that such distributions are really attained in the emitting range (at least for the low-energy range of the spectrum of e^- and e^+). Thus, if the emitting regions close to a NS are rather extended and radiations emitted at different angles to the magnetic field reach the observer simultaneously (i.e., $S(\omega)$ is recorded), then the observed hard tails can be explained by the annihilation radiation of e^- and e^+ with the spectrum index k ~ 1-2 (see (15)). If the directional diagram is narrow (along the field), then the observed emission spectra correspond to softer e^- and e^+ spectra (K ~ 3-4), see (14)).
An important conclusion seems to be is that about a strong (tens of percent) linear polarization of the annihilation radiation in a magnetic field. Since polarization provides an important information, e.g., about the directional diagram and the geometry of the NS magnetic field it seems rather important to include polarization measurements into further observations of cosmic gamma-bursts.

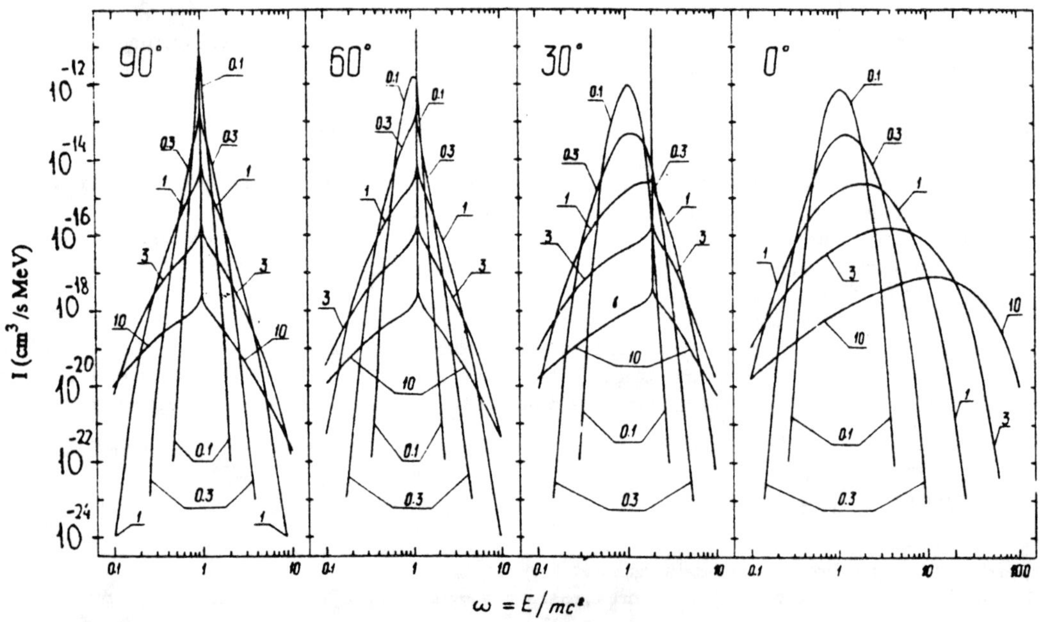

Fig.1. The spectra of intensity I (divided into $N^- N^+$) for different y and t (the numbers at the curves). For curves t = 0.1, 0.3, 1, 3, 10 the factors 10^{-2}, 10^{-1}, 10^{0}, 10^{1}, and 10^{2} were introduced.

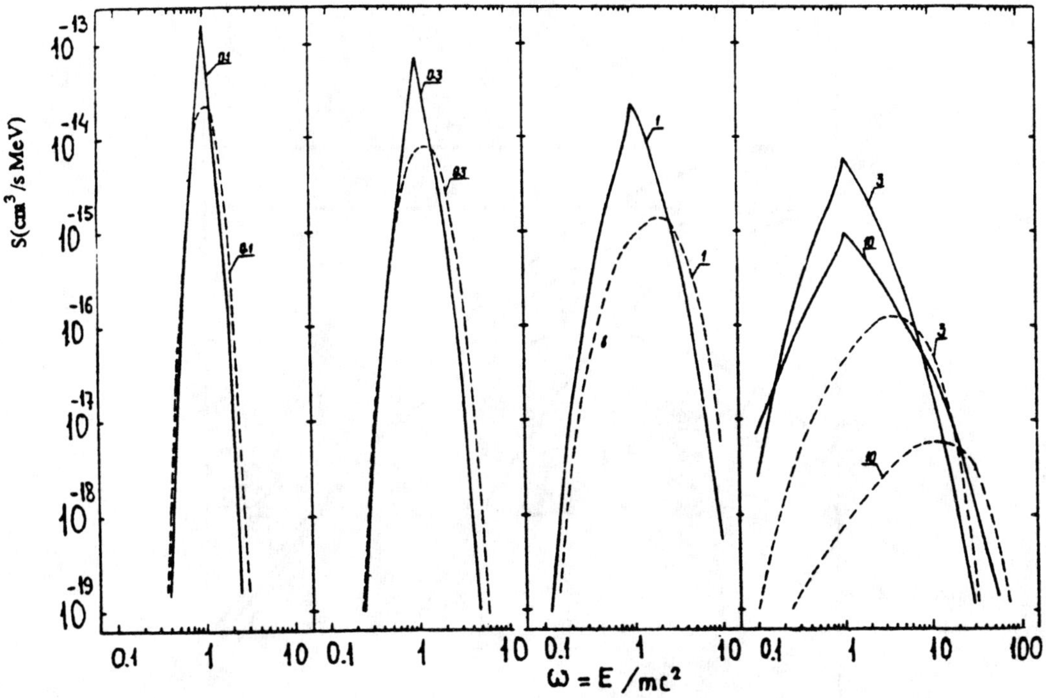

Fig.2. Spectra of direction-integrated intensity S (divided into $N^- N^+$) for sifferent t (numbers at the curves). The dash lines indicate the S_{is} spectra in the isotropic case (B = 0)

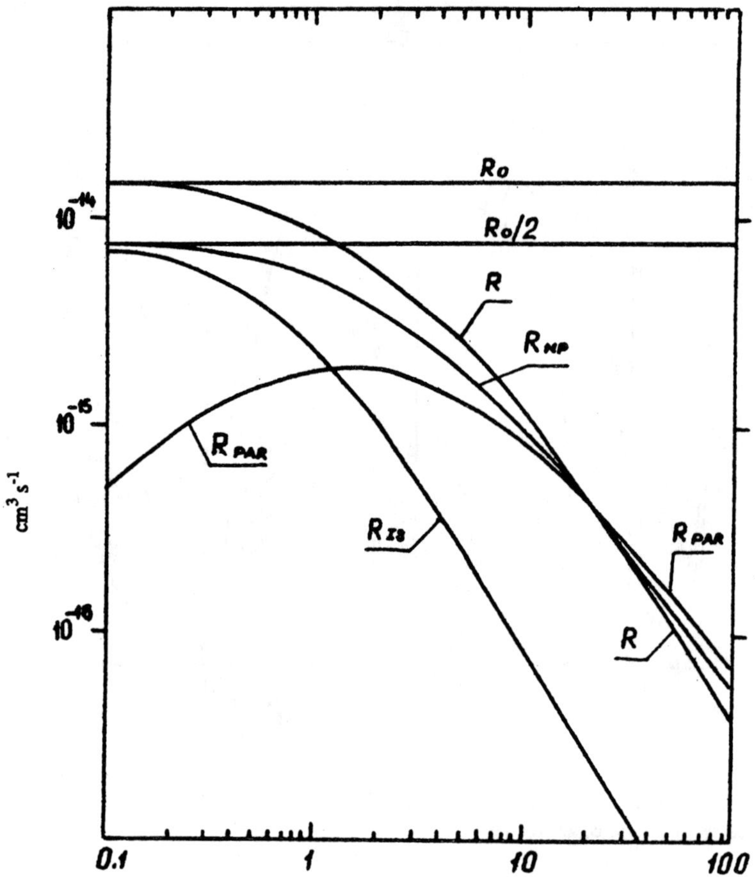

Fig.3. The total annihilation rate R (divided into $N^- N^+$) as a function of the parameter $t = T/mc^2$. The curve R_{IS} corresponds to the isotropic case (B = 0), R_{par} corresponds to parrallel spins in one-dimensional motion of e^- and e^+ , R correspond to unpolarized e^- and e^+ .

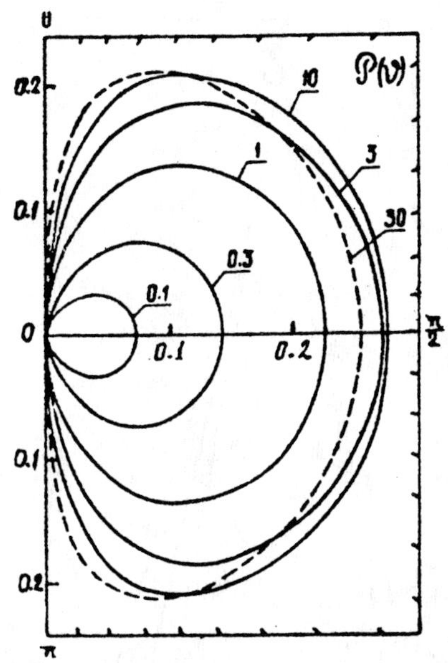

Fig.4. Angular dependence P(y) of the integral polarization for different t (numbers at the curves).

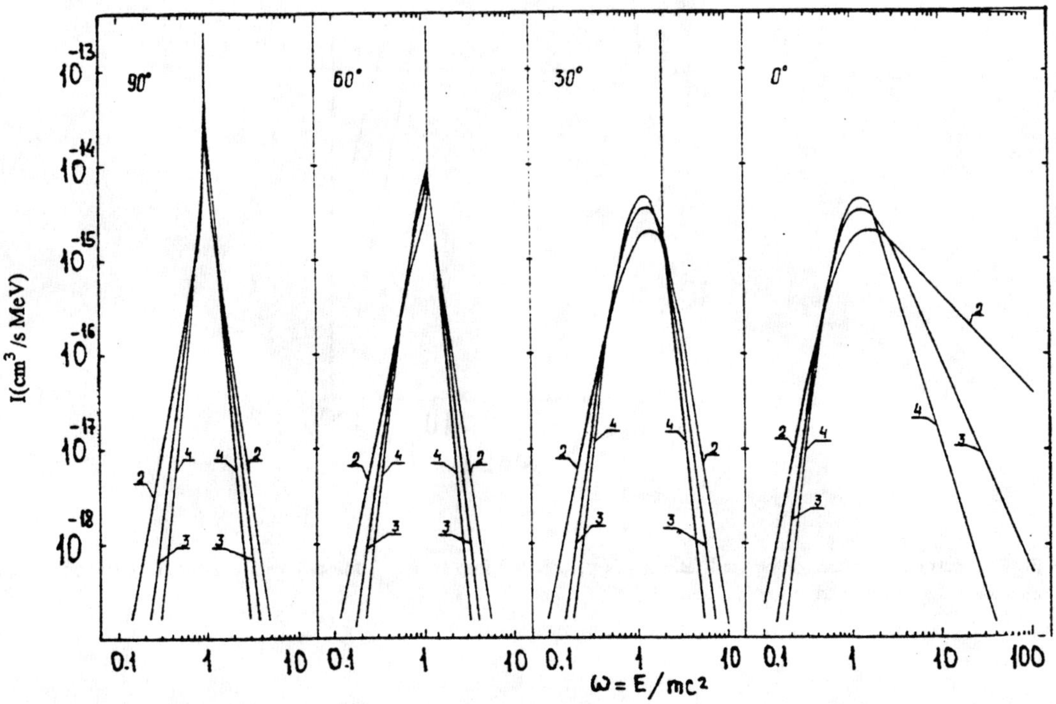

Fig.5. Intensity spectra I(w) (divided into $N^- N^+$) for various y and exponents k (numbers at the curves) of the power distribution of e^- and e^+ .

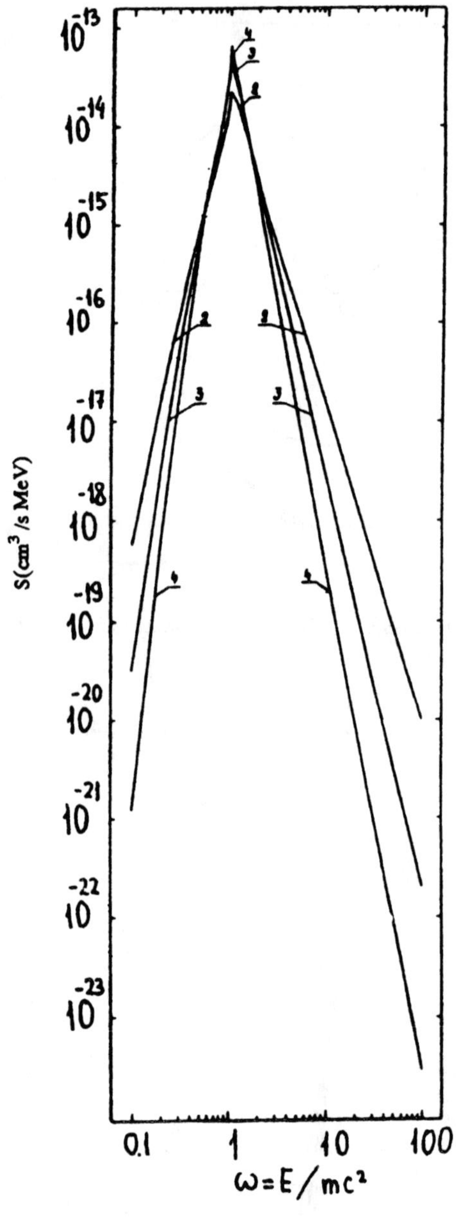

Fig.6. Spectra of direction-integrated intensity S(w) (divided into $N^{-}N^{+}$) for different exponents k (numbers at the curves) of the power distribution of e^{-} and e^{+} .

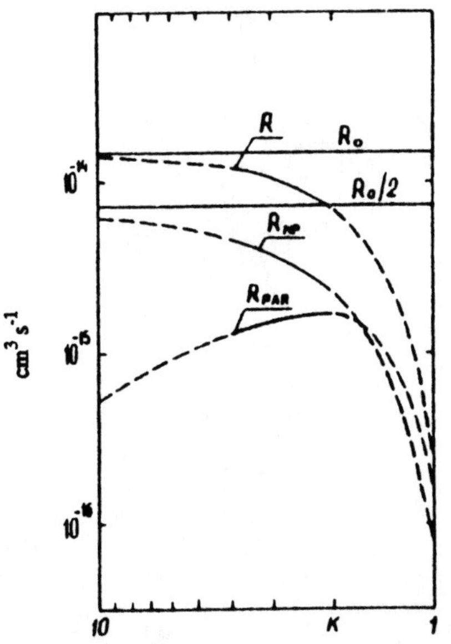

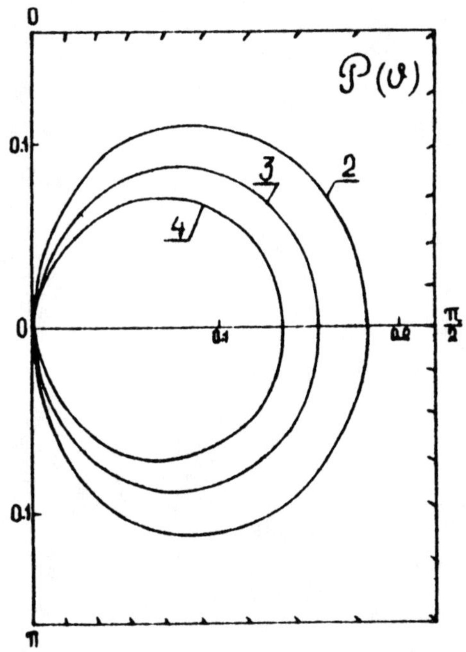

Fig.7. Total annihilation rate R (divided into $N^- N^+$) as a function of the exponent of the power distribution of e^- and e^+.

Fig.8. Angular dependence of the integral polarization P at different exponents k (the numbers at the curves) of the power distribution of e^- and e^+.

References

1. E.P. Mazets, S.V. Golenetskii, R.L. Aptekar, et al., Nature, 1981, 290, 378.
2. S.V. Golenetskii, E.P. Mazets, R.L. Aptekar, Yu.A. Guryan, and V.N. Ilynskii, Astron. Space. Sci., 1986, 124, 243.
3. P.L. Nolan, G.H. Share, S. Matz, et al. High-Energy Transients in Astrophysics, S.E. Woodsley, ed., AIP Conf. Proc. No. 115, 1984, 399.
4. G. Wunner, Phys. Rev. Lett., 1979, 42, 79.
5. J.K. Daugherty and R.W. Bussard, Astrophys. J., 1980, 238, 296.
6. A.K. Harding, Astrophys. J., 300,, 167.
7. G. Wunner, J. Paez, H. Herold, and H. Ruder, Astron. Astrophys., 1986, 170, 179.
8. A.D. Kaminker, G.G. Pavlov, and P.G. Mamradze, Space.Sci., 1987, 138, 1.
9. A.D. Kaminker, G.G. Pavlov, and P.G. Mamradze, Proc. of the Joint Varenna-Abastumani Intern. School & Workshop "Plasma Astrophysics", ESA, 1986, 393.
10. A.I. Akhiezer and V.V. Berestetskii, Kvantovaya elektrodinamika (Quantum electrodynamics), 1981, Moscow: Nauka (in Russian).

Physics of Neutron Stars
Edited by A. M. Kaminker

ISBN 1-56072-164-2

CYCLOTRON RADIATION OF PLASMA WITH ANISOTROPIC TEMPERATURE IN MAGNETIC FIELDS OF NEUTRON STARS

V. G. Beschastnov and G. G. Pavlov
A. F. Ioffe Physical-Technical Institute, 194021, Leningrad

It has been shown that for an anisotropic distribution function of plasma particles nonequidistance of Landau levels in a strong magnetic field may result in moticeable oscillations of the spectra of cyclotron radiation absorption and emission.

As is well knwon, the quantum relativistic effects strongly influence the plasma radiation from the surface layers of neutron stars possessing magnetic fields $B \sim (10^{-2}\text{ -}1)B_c$, where $B_c = m^2c^3/e\hbar = 4.41 \cdot 10^{13}$ G (the respective electron gyrofrequency is in to the X-ray range $\hbar\Omega_B = \hbar eB/(mc) \sim$ (S - 500) keV). On the other hand, radiating particles are not necessarily in a thermal equilibrium. Thus, in accretion on the magnetic poles of a neutron star [1] or in the generation of electron-positron pairs by high-energy photons in a superstrong magnetic field $\vec{B}$ [2] the conditions may arise such that the degrees of freedom of the particles transverse with respect to $\vec{B}$ (high Landau levels) are excited much more easier than longitudinal. The nonequidistance of the Landau levels for such anisotropy can result in noticeable oscillations of the spectra of cyclotron radiation and absorption.

Let us consider the formation of spectra by way of example of an anisotropic Maxwell distribution

$$f_n(p_z) = \frac{N\, th\,(b/2t_\perp)}{\pi b\sqrt{2\pi t_{||}}} \exp\left(-\frac{nb}{t_\perp} - \frac{P_z^2}{2t_{||}}\right) \qquad (1)$$

where n is the number of the Landau level, p_z is the longitudinal electron pulse in mc units, N is the particle concentration, $b = B/B_c = \hbar\Omega_B/(mc^2)$, $t_\perp = T_\perp/(mc^2)$ and $t_{||} = T_{||}/(mc^2)$ are the parameters that, being much smaller than unity, have the sense of dimensionless trnasverse and longitudinal plasma temperatures.

Figure 1 depicts the energy levels (in mc^2 units) of a relativistic electron in a quantizing magnetic field (Landau levels):

$$\varepsilon = \varepsilon_n(p_z) = (1 + 2nb + p_z{}^2)^{1/2} \qquad (2)$$

Interacting with the radiation, electrons pass from one levels to another Dimensionless frequencies of absorbed and emitted photons in such a process, $\omega = \hbar\Omega/(mc^2)$, are determineed by the laws of energy and longitudinal pulse conservation:

$$\omega = \varepsilon_{n+s}(p_z + q_z) - \varepsilon_n(p_z), \quad s = 1, 2,... \qquad (3)$$

for absorption and by

$$\omega = \varepsilon_n(p_z) - \varepsilon_{n-s}(p_z - q_z), \quad s=1,2,..., n \qquad (4)$$

for emission ($\vec{q}$ is the photon wavevector). For simplicity, consider the radiation along the magnetic field. The selection rules in this case allow the transitions between the neighboring levels (s = 1) only. In a sufficiently rarefied plasma, $q_z = \omega$, and, in virtue of (2)-(4), an electron that transits from the state (n, p_z) either absorbs or emits photons having the same frequencies:

$$\omega_n(p_z) = b/[\varepsilon_n(p_z) - p_z]. \qquad (5)$$

To the most filled state with $p_z = 0$ (transitions from such states are depicted by arrows) there correspond the following photon frequencies:

$$\omega_n = \omega_n (0) = b\,(1+2nb)^{-1/2} \qquad (6)$$

If only such states were filled ($t_{||} \to 0$), then the absorption and emission spectra would have been represented by sets of infinitely narrow lines with decreasing sequence of frequencies because of nonequidistance of the levels ($\omega_{n+1} < \omega_n$). For small but finite $t_{||}$ an electron on the nth level has longitudinal pulses $|p_z| \le \sqrt{2t_{||}}$. It can readily be seen that the quantity

$$\gamma_n = \frac{1}{2}\left[\omega_n(\sqrt{2t_{||}}) - \omega_n(-\sqrt{2t_{||}})\right] = b\sqrt{2t_{||}}\,(1+2nb)^{-1} \qquad (7)$$

has the sense of a characteristic linewidth. The transitions that are responsible for the spectral lines with allowance for the thermal scatter in p are depicted by hatched zones in Fig.1. It is seen that if a region of filled p is sufficiently narrow, such zones for several low Landau levels do not intersect each other, i.e., the respective spectral lines do not overlap. This may occur if the condition $\omega_n - \omega_{n+1} \ge \gamma_n$ is fulfilled. At $n = 0$ it yields $b \ge 2\sqrt{2t_{||}}\,(\sqrt{1+2t_{||}} + \sqrt{2t_{||}})$. On the other hand, the lines may have comparable intensities if the level occupancy only slightly depends on n, i.e., $b \le t_\perp$. Thus, individual lines are clearly seen if

$$t_\perp \ge b \ge 2\sqrt{2t_{||}}\,(\sqrt{1+2t_{||}} + \sqrt{2t_{||}}) \qquad (8)$$

These conditions are satisfied only in case of strong anisotropy, when $t_\perp \ge 2\sqrt{2t_{||}}\,(\sqrt{1+2t_{||}} + \sqrt{2t_{||}})$.
The quantum relativistic expression for the absorption coefficients μ and the emissive power j of a plasma with the anisotropic distribution function (1) has the form [3]:

$$\mu = \frac{4\pi^{3/2}\alpha\lambda_c^2 N}{\sqrt{2t_{||}}\,\omega}\, th\left(\frac{b}{2t_\perp}\right)\left\{\exp\left[-\left(\frac{\omega^2-\omega_0^2}{2\omega\gamma_0}\right)^2\right] + 2\sum_{n=1}^{\infty}\exp\left[-\left(\frac{\omega^2-\omega_n^2}{2\omega\gamma_n}\right)^2 - \frac{nb}{t_\perp}\right]\right\}$$

$$j = \frac{\alpha mc^2 N\omega^2}{2\pi^{3/2}\sqrt{2t_{||}}}\, th\left(\frac{b}{2t_\perp}\right)\sum_{n=1}^{\infty}(2n-1)\exp\left[-\left(\frac{\omega^2-\omega_n^2}{2\omega\gamma_n}\right)^2 - \frac{nb}{t_\perp}\right] \qquad (9)$$

where $\alpha = e^2/(\hbar c)$, $\lambda = \hbar/(mc)$. The spectral line profiles are described by the exponents in the sums over the Landau levels and are of the Doppler type close to the central frequencies ω_n. For $b \ll \min(\sqrt{t_{||}}, t_{||}, t_\perp)$ the summation over n can be substituted by integration, the lines are smoothened, and expression (9) is transformed into the classical expressions:

$$\mu_{cl} = \frac{\pi\Omega_p^2\Omega_B}{2c\,\Omega^2 t_\perp}\exp(y^2 - z^2)\,erfc\,(y),$$

$$j_{cl} = \frac{\Omega_p^2\Omega_B m\sqrt{2t_{||}}}{16\pi^2 c\,t_\perp}\exp(-z^2)\left[\frac{1}{\sqrt{\pi}} - y\exp(y^2)\,erfc\,(y)\right], \qquad (10)$$

where $\Omega_p = (4\pi Ne^2/m)^{1/2}$ is the plasma frequency,

$$z = \frac{\Omega^2 - \Omega_B^2}{2\Omega\,\Omega_B\sqrt{2t_{||}}}, \quad y = z + \frac{\Omega_B}{\Omega}\sqrt{\frac{t_{||}}{2t_\perp{}^2}}$$

$$erfc\,(y) = \frac{2}{\sqrt{\pi}}\int_y^{\infty}\exp(-u^2)\,du \qquad (11)$$

Figure 2 shows the spectra calculated by formulae (9)-(11). In the ultraquantum case (a), $b > t_\perp$. Higher spectral lines are strongly suppressed, therefore the absorption spectrum is determined mainly by the line due to the transition $0 \to 1$, and the emission spectrum by the line due to the transition $1 \to 0$. The second line is noticeably shifted relative the first one towards the red range becuase of a rather pronounced nonequidistance of the levels. The case (b) corresponds to the optimum conditions (8) when a large number of amplitude-comparable lines appear. In the case (a), $b = 2\sqrt{2t_{||}}$, so that the oscillations become very small and the spectra approach the classical ones. The classical profiles of the absorption coefficient and radiating power are strongly asymmetric: if $\omega > b$, they are broadened by the linear Doppler effect caused by the thermal motion of electrons along the magnetic field, and if $\omega < b$, the broadening is determined by the squared Doppler effect associated mainly with motion of electrons across the field.

We have considered the cyclotron absorption and the photon emission by the plasma electrons along the magnetic field. A similar consideration can be made also for the radiating positron transitions. The only difference consists in that the direction of the circular polarization of absorbed and emitted photons would be opposite to the direction of the circular polarization of radidation interacting with electrons (in accordance with the opposite directions of the Larmor precession for electrons and positrons).

In conclusion it should be said that the effect of quantum relativistic oscillations in cyclotron absorption and emission in superstrong magnetic fields of neutron stars can manifest itself for various anisotropic distribution functins of particles and for radiation propagation at arbitrary angles v to the magnetic field. Taking that $q_z = \omega \cos v$ and solving equations (3) and (4), it is easy to obtain that at

$v \neq 0$ oscillations in the absorption and emission spectra take place at different frequencies, ω_a and ω_e,

$$\omega_{a,e} = \left\{ \pm (\varepsilon - p_z \cos v) \pm \left[(\varepsilon - p_z \cos v)^2 \pm 2 s b \sin^2 v \right]^{1/2} \right\} / \sin^2 v \qquad (12)$$

At $\omega \approx \omega_e$, the conditions may arise such that the processes of forced emission would prevail over absorption processes, which would result in negative absorption - a maser increase in emission (at $v = 0$ this effect is impossible because $\omega_a = \omega_e$). Also, at $v \neq 0$, the spectra may have some additional features associated with the singular behavior of the electron-state density in a quantizing magnetic field (see [4] and the examples of the spectra of thermal plasma given there). This the investigation of a more general case, which we hope to carry out in future, is of obvious interest.

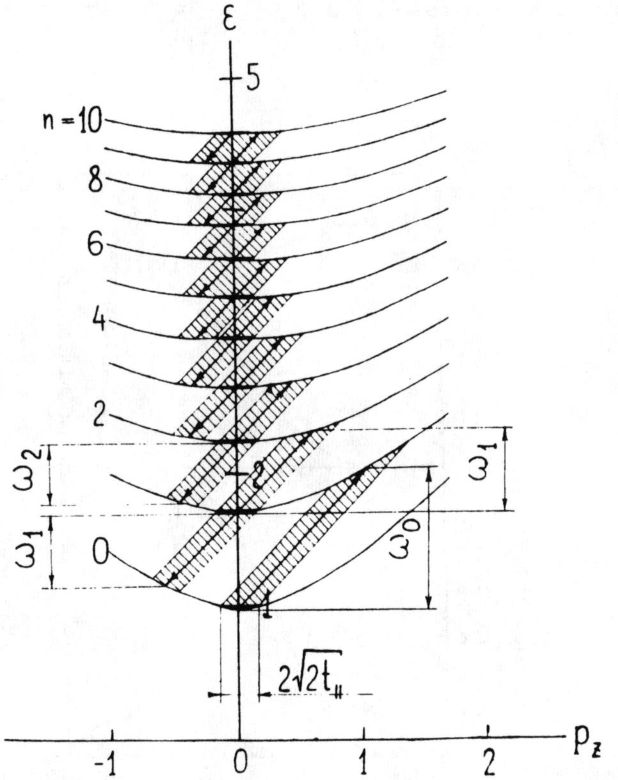

Fig.1. Transitions between the Landau levels associated with absorption and emission along the magnetic field for strongly anisotropic particle distribution when typical longitudinal pulses are small $\left(|p_z| \leq \sqrt{2t_{||}} \ll b \right)$ and many levels are filled $\left(t_{\perp} \gg b \right)$. The transitions occurring in nonintersecting dashed zones cause quantum oscillations of the absorption and emission spectra)

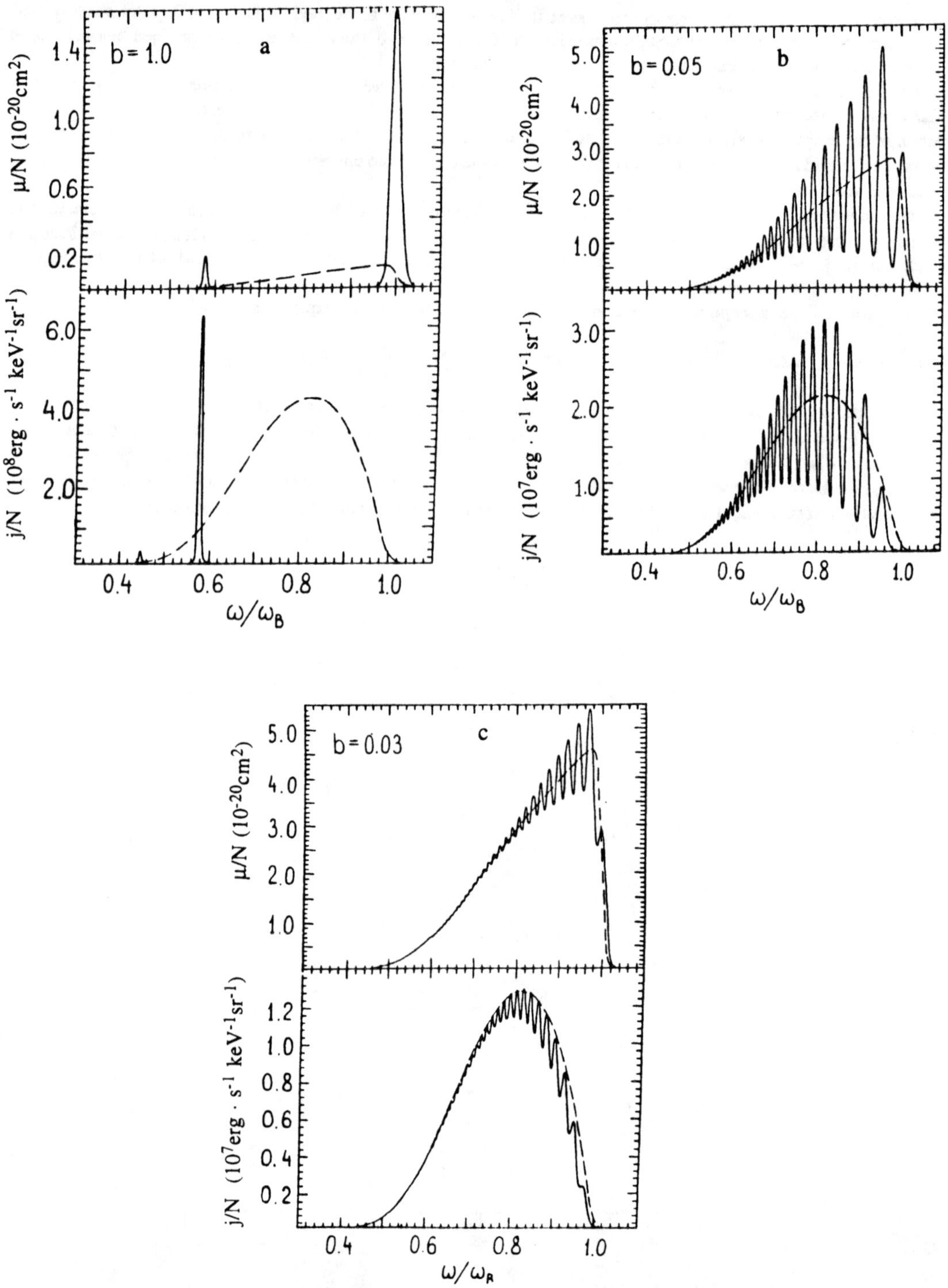

Fig.2. The spectra of cyclotron absorption and emission along the magnnetic field in a plasma with the temperature anisotropy ($t_\perp = 0.3$, $t_{||} = 10^{-4}$) for different values of b. With an increase in b, oscillations are smoothened and the spectra approach the classical ones (depicted by dashed lines).

References
1. Yu.N. Gnedin, G.G. Pavlov, K.M.V. Apparao, and S.M. Chitre, Astrophys. Space.Sci., 1981, 76, 105.
2. V.S. Beskin, Astrofizika, 1982, 18, 439.
3. V.G. Beschastnov and G.G. Pavlov, Preprint FTI 1196, Leningrad, 1988 44 pp. (in Russian).
4. G.G. Pavlov, in: Proc. of the Joint Varenna-Abastumani Intern. School & Workshop Plasma Astrophysics, Sukhumi, 19-28 May, 1986. Europ. Space Agency.

Physics of Neutron Stars
Edited by A. M. Kaminker

ISBN 1-56072-164-2

INTERACTION OF RADIATION AT CYCLOTRON HARMONICS IN ACCRETING PLASMA OF X-RAY PULSARS

A.A. Litvinchuk and V.V. Chugurin
Institute of Applied Physics of the USSR Academy of Sciences,
603600 Gorky

An equation for cyclotron radiation transfer with allowance of the wave mutual conversion at the first and the second gyrofrequency harmonics has been derived. Its solution was obtained in the limit of a strongly inhomogeneous magnetic field of the accretion plasma of X-ray pulsars. It has been shown that for X-ray luminosity lower than 10^{37} erg/s the cyclotron line in the emission spectrum of a hot spot disappears upon the passage of the second gyroresonance layer because of the radiation interaction at cyclotron harmonics.

1. The formation of the observed radiation of X-ray pulsars has been considered in numerous articles (see review [1]). The attention is focused on the radiation generation in the deceleration range of the accretion plasma (in a hot spot or an accretion column [2]). At the same time, scattering of the hot-spot radiation by an accreting plasma in an inhomogeneous magnetic field can also have a substantial effect on the formation of the observed radiation [3]: the optical thickness of gyroresonance layer in the cyclotron scattering at the first gyrofrequency harmonic, $\tau_1 \approx \pi^2 e^2 NL_B/(3mc\omega_B)$ ($L_B = |B^{-1}$ $dB/dz|^{-1}$) exceeds unity at the accreting plasma concentration $N \geq 10^{16}$ cm^{-3} (this corresponds to the X-ray pulsar luminosity $1_x \geq 10$ erg/s). The radiation scattering at the first gyrofrequency harnomic by a plasma in an inhomogeneous magnetic field was considered in [4,5]; in [6] it was shown that such a process may be responsible for a decrease of the modulation depth and an increase of the pulse width for lower freequencies and also for the appearance of interpulses in the radiation of X-ray pulsars at low frequencies. In these studies radiation transfer at higher harmonics was neglected. But, according to [7], the optical thickness, $\tau_s \sim (B/B_{cr})^{s-1}\,\tau_1$, of the gyroresonance layer for the radiation at higher hyrofrequency harmonics may also exceed unity (s is the harmonics number, $B_{cr} = 4.4 \cdot 10^{13}$ G). In the present study we considered the radiation at the second gyrofrequency harmonic in a collisionless plasma.

2. The transfer equation is obtained by a method conventional for the theory of star atmospheres without LTR. Let us fix the radiation intensity and determine from the balance equations the population of the Landau levels. The obtained data are then used for the determination of the transfer coefficients. Since under the conditions of X-ray pulsars the population of the Landau levels rapidly decreases with a growth in the level number [3], we restrict ourselves to the consideration of the first theree levels (Fig.1, the level numbers are indicated, thin arrows show spontaneous transition, double arrows show induced transitions, collisions are neglected). In a similar way [8], one can determine the level populations and then, using the method of the Einstein coefficients, can also calculate the radiative power and the absorption coefficient of the cyclotron radiation at the first and the second gyrofrequency harmonics. Substituting thus obtained values into the transfer equation, we obtain (in the Schwarzschiel-Schuster approximation) the following set of equations for the radiation with both polarizations

$$\frac{dI_1^+}{dy} = \tau_1 \frac{4A_2}{2A_1 + A_2}\left[\frac{1}{8}I_2^+ - \frac{4\pi^3 c^2}{h\,\omega_B^3}\left(I_1^+\right)^2\right], \quad \frac{dI_1^-}{dy} = -\tau_1 I_1^- \qquad (1)$$

$$\frac{dI_2^+}{dy} = \tau_2 \frac{16A_1}{2A_1 + A_2}\left[-\frac{1}{8}I_2^+ + \frac{4\pi^3 c^2}{h\,\omega_B^3}\left(I_1^+\right)^2\right], \quad \frac{dI_2^-}{dy} = -\tau_2 I_2^- \qquad (2)$$

In this set of equations

$$I_s^{\pm} = \frac{1}{2}\left(I_s^{(1)} \pm I_s^{(2)}\right), \quad I_s^{(1)} = \frac{1}{2\pi}\int_{\cos\alpha>0} I_\omega(z,p_{||},\alpha,s)\,d\Omega, \quad I_s^{(2)} = \int_{\cos\alpha<0} I_\omega(s)\,d\Omega \quad (3)$$

$$I_\omega(z,p_{||},\alpha,s) = \int I_\omega(z,\omega,\alpha)\,\delta\left\{\omega - s\omega_B\left(1 - p_{||}\cos\alpha/mc\right)^{-1}\right\}d\omega \quad (4)$$

is the intensity of the radiation at the sth harmonic of gyrofrequency which is in resonance with that of the particles with the given longitudinal (with respect to the magnetic field) pulse $p_{||}$, α is the angle formed by $\vec{B}$ and the wavevector of a photon. The transition to (4) allows one to take account of the Doppler effect in the cyclotron scattering [4,9], the recoil effect is neglected. The operator of the radiation transfer, $I_\omega(z, p_{||}, \alpha, s)$ consists of two terms and in the case where the magnetic field is normal to the plasma layer boundary has the form

$$\cos\alpha\frac{d}{dz}I_\omega(z,p_{||},\alpha,s) - \frac{mc}{\omega_B}\frac{d\omega_B}{dz}\frac{d}{dp_{||}}I_\omega(z,p_{||},\alpha,s) \quad (5)$$

where the first term describes the three-dimensional transfer and the second one the frequency of a photon leaving the line [5]. Under the conditions of an accreting plasma over a hot spot of X-ray pulsars, the second term prevails, i.e., it is the case of strongly inhomogeneous magnetic field [5,6] analogous to the Sobolev approximation of the theory of moving star shells; therefore the space derivative in (1) and (2) is rejected.Further on, we have in (1) and (2)

$$y = (2\pi m æ T_{||})^{-1/2}\int_{-\infty}^{p_{||}} \exp\left(-p_{||}^2/2mæT_{||}\right)dp_{||} \quad (6)$$

and

$$\tau_1 = \frac{8}{3}\frac{\pi^2 e^2}{mc\,\omega_B}N L_B, \qquad \tau_2 = \frac{8}{5}\frac{\pi^2 e^2}{mc\,\omega_B}\frac{\hbar\omega_B}{mc^2} \quad (7)$$

denotes the optical thickness of the gyroresonance layer (with the characteristic size $\beta_{T||} L_B$ along the height [3]) on the first and the second gyrofrequency characteristics, respectively ($\beta_{T||} = (æT_{||}/mc^2)^{1/2}$, $æT_{||}$ is the longitudinal temperature). The probabilities of the $1 \to 0$ and $2 \to 0$ cyclotron transitions are denoted by A_1 and A_2, respectively (the probability of the $2 \to 1$ transition is twice higher):

$$A_1 = \frac{4}{3}\frac{e^2\omega_B^2}{mc^3} \qquad A_2 = \frac{32}{5}\frac{e^2\omega_B^2}{mc^3}\frac{\hbar\omega_B}{mc^2} \quad (8)$$

The right-hand parts of the first equations in (1) and (2) describe the wave interaction at the first and the second gyrofrequency harmonics. In this case an electron absorbs two photons having frequency ω_B ($0 \to 1$ and $1 \to 2$ transitions) and emits one photon having frequency $2\omega_B$; a photon is absorbed absorption at the second harmonic with the transition of an electron to the second excited Landau level with the subsequent cascade emission of two photons at the first gyrofrequency harmonic. In the equilibrium, $I_2^+ = (32\pi^3c^2/\hbar\omega_B^3)(I_1^+)^2$, which corresponds to the equality of the effective radiation temperatures T_b at the first and the second gyrofrequency harmonics (in the Wien limit $\hbar\omega_B \gg æT$, the efective radiation temperature is determined from the relationship $I_\omega = (\hbar\omega^3/4\pi^3c^2)\exp(-\hbar\omega/æT_b)$. We should like to note that systems (1) and (2) give the Manly-Row relationship $I_1^+ + 2I_2^+ = \text{const}$ that reflects the conservation of the total line energy in the coalescence and decay of quanta (cf.[10).

3. In the case where an incident radiation of intensity $I^{in}(\omega)$ falls onto the plasma from one side only, the boundary conditions have the form (cf.[5])

$$I_1^{(1)}(y=0) = I^{in}(\omega_B) \quad I_1^{(2)}(y=0) = 0$$
$$I_2^{(1)}(y=0) = I^{in}(2\omega_B) \quad I_2^{(2)}(y=0) = 0 \quad (9)$$

The solution of equations (1) and (2) with the boundary conditions (3) shows that if a layer is optically thick for scattering of the cyclotron radiation at the second gyrofrequency harmonic ($\tau_1 > \tau_2 > 1$), then the intensity of the scattered radiation is

$$I^{sc}(\omega_B) = I_1{}^{(1)}(y=1) \approx \frac{1}{2} I^{in}(\omega_B) + I^{in}(2\omega_B),$$
$$I^{sc}(2\omega_B) = (32\pi^3 c^2/\hbar\omega_B^3)\left(\frac{1}{2} I^{in}(\omega_B) + I^{in}(2\omega_B)\right)^2 \quad (10)$$

Under the conditions of X-ray pulsars the intensity of the incident radiation rapidly decrease with an increase in frequency: $I^{in}(\omega_B) \gg I^{in}(2\omega_B)$; therefore the intensity of the radiation that leaves the gyroresonance layer both at the first and second gyrofrequency harmonics is determined only by $I^{in}(\omega_B)$:

$$I^{sc}(\omega_B) \approx \frac{1}{2} I^{in}(\omega_B), \qquad I^{sc}(2\omega_B) \approx (8\pi^3 c^2/\hbar\omega_B^3)(I^{in}(\omega_B))^2 \quad (11)$$

4. Now let us assume that the radiation spectrum of a hot spot of a X-ray pulsar contains a cyclotron line at the frequency ω_{B0} (B_0 is the magnetic field at the star surface). Then, upon the scattering by the second gyroresonance layer the intensity of the transmitted radiation at the height over the star surface $Z \sim L_B = 3 \cdot 10^5$ cm (where $\omega_{B0} = 2\omega_B(z)$) and at frequencies close to ω_{B0} is determined by $I^{in}(\omega_{B0}/2)$, i.e., smoothly varies with the frequency and has no singularity. Thus, the singularity at the frequency ω_{B0} disappears because of the radiation interaction at the cyclotron harminocs Wupon at the scattering by the second gyroresonance layer, and therefore the observed spectrum of the X-ray pulsar has no cyclotron line. Such a situation is preserved if the optical density of the accreting plasma is low with respect to the Thomson scattering, i.e., at $L_x < 10^{36}$-10^{37} erg/s. For high luminocities, the Thomson scattering becomes essential (since the reaching plasma concentration is high) and the radiation from the hot spot passes through the side boundary of the accreting plasma not attaining the second gyroresonance layer. In the latter case, the effect of mutual wave conversion at the gyrofrequency harmonics produces no effect on the formation of the observed radiation, and the radiation spectrum of the X-ray pulsar with luminocity $L_x \geq 10^{37}$ erg/s may also have a cyclotron line. The latter conclusion agrees with the well known observations data (see review [2]) where the cyclotron line was recorded in the radiation spectra of two X-ray pulsars (Her X-1 and 4U 0115+63) with $L_x \sim 10^{37}$ erg/s.

References

1. P. Meszaros, Space. Sci. Rev., 1984, 38, No.2, 325.
2. P. Jose amd S. Rappaport, Ann. Rev. Astron. Astrophys., 1984, 22, 537.
3. V.V. Zheleznyakov, Itogi Nauki i Tekhniki. Astronomiya, (Advances in science and technology. Astronomy) 22, 1983. Moscow: VINITI, p.135 (in Russian).
4. A.A. Litvinchuk, Preprint IPF AN SSSR No.159, 1987, Gorky (in Russian).
5. A.A. Litvinchuk, Preprint IPF AN SSSR No.160, 1987, Gorky (in Russian).
6. V.V. Zheleznyakov and A.A. Litvinchuk, In: Plasma Astrophysics, 1986, ESA SP-251, p.375.
7. V.V. Zheleznyakov, Astrophys. Space Sci., 1982, 83, No.1, 81.
8. V.V. Zheleznyakov and A.A. Litvinchuk, Astrophys. Space Sci., 1984, 105, No.1, 73.
9. V.V. Zheleznyakov and A.A. Litvinchuk, Astron. Zh., 1987, 64, 2, 306.
10. S.A. Akhmanov and R.V. Khokhlov, Problemy nelineinoi optiki (Problems of nonlinear optics) 1964 Moscow: VINITI (in Russian).

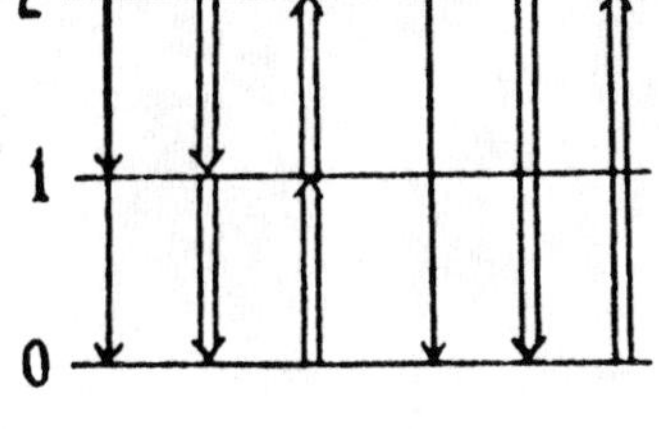

Fig.1.

Physics of Neutron Stars
Edited by A. M. Kaminker

ISBN 1-56072-164-2

PROPERTIES OF ATOMS AND MOLECULES IN MAGNETIC FIELDS OF NEUTRON STARS

V.K. Khersonskii
Leningrad Division of Special Astrophysical Observatory of the USSR Academy of Sciences, Pulkovo, 196140 Leningrad

The literature data on the properties of atoms and molecules in a strong magnetic fireld ($B \leq 10^{13}$ G) characteristic of a neutron star have been briefly reviewed. Spectroscopic characteristics, quantum transitions, ionization and dissociative equilibria, and some other related problems are being discussed.

1. Introduction

The existence of strong magnetic fields (of the order of 10^{11} -10^{13} G) around neutron stars confirmed by the observed cyclotron lines [1] has triggered numerous studies of the properties of atomic and molecular systems in such fields. Therefore the study of a system of charged particles in a magnetic field, which by itself is one of the most interesting problems of the quantum theory, has received a new impetus for its development in the context of astrophysical investigation of neutron stars.

In the magnetic fields under consideration electron shells of atoms and molecules turned out to be strongly deformed. The motion of an electron in the direction transverse to the field is controlled by the Lorentz force and is characterized by the so-called magnetic length $a_M = \sqrt{2\hbar c/e B}$ r . The dimensions of the atomic system along the magnetic field (and also in the absence of such a field) turn out to be of the order of the Bohr radius $a_0 = \hbar^2/m_e e^2$. A magnetic field for which both these dimensions turn out to be equal is called critical:

$$B_0 = 2e^3 m_e^2 c/\hbar^3 = 4.701 \cdot 10^9 G$$

A magnetic field is often characterized by its magnitude expressed in B_0 units:

$$\beta = B/B_0 = (a_0/a_m)^2$$

This quantity is called the degree of the deformation of an atomic system, since it characterizes the change in the system dimensions in the longitudinal and transverse directions with respect to the magnetic field.

If $B \sim 10^{11}$ -10^{13} G (β = 500-50000), the deformation of electron shells becomes so essential that the properties of the atomic systems change in a drastic way. Since at present it is impossible to create such strong magnetic fields under the laboratory conditions, the only way for studying the above properties is the construction of the theory and performance of the respective calculations. On the other hand, it is just such investigations that are necessary for the creation of the basic set of data on elementary processes in atoms and molecules in the magnetic fields under consideration. Such data sets can then be used for the development of spectroscopic methods for studying gases and plasma at neutron stars.

The aim of the present review is to summarize the known properties of atoms and molecules in magnetic fields $B = 10^{11}$ -10^{13} G; a brief consideration of the atomic and molecular spectra the quantum transitions there is also presented together with ionization and dissociation, some other related problems.

2. Atomic Spectroscopy

By present the most precise results on the structure of atomic spectra in 10^{11} -10^{13} G magnetic fields have been obtained for simplest hydrogen- and helium-like systems [2-7]. These results are obtained for a nonrelativistic interpretation of the problems, which is sufficiently accurate for magentic fields $B \ll 5 \cdot 10^{13}$G when $\hbar\omega_B \ll m_e c^2$ ($\omega_B = eB/m_e c$ is the cyclotron frequency). Numerically, the problem was studied in [8]. The consideration of the known data on atomic spectra is expedient to start with the those obtained for a hydrogen atom.

2.1. Energy spectrum of a hydrogen atom. In the nonrelativistic approximation the complete set of quantum numbers determining the state of a hydrogen atom includes the following quantities. The transverse motion of en electron is characterized by the number n of the Landau level and the projection m of an orbital moment. The longitudinal motion is characterized by the quantum number ν (the number of the wavefunction nodes in the direction of the magnetic field) that determines the parity of the state $\pi = (-1)^{\nu}$. The total energy is also characterized by the projection σ of an electron spin on the direction of the magnetic field.

To the ground state of a hydrogen atom there correspond the quantum numbers n = 0, m = 0, ν = 0, and σ = -1/2. The state with σ = 1/2 (or n = 1 or n = 1) lies higher by $\hbar\omega_B = \beta \cdot 54.4$ eV than the ground state. There exists a one-to-one correspondence between the quantum numbers $nm\nu$ and $n_p lm$ (n_p is the principal quantum number, l is the eigenvalue of the operator of the orbital moment) of the hydrogen atoms in the limit $\beta \to 0$ [9]. The energy spectrum consists of two groups of levels: (a) hydrogen-like levels with $\nu \geq 1$ (levels corresponding to the longitudinal motion form in the limit b $\to \infty$ a conventional discrete spectrum corresponding to the Coulomb potential) and (b) closely related levels with ν = 0. In the limit $\beta \to 0$, these levels correspond to the values m = -l, and with an increase in β their energy varies in a logarithmic way. The energy of levels with m > 0 drastically increases with an increase of the field (as has already been noted, by $m\hbar\omega_B$). The exact values of the energy levels for different fields are tabulated in [3], but the authors neglected the effects associated with the finiteness of the proton mass.

As is well known, in the presence of a magneteic field one cannot distinguish between the motion in the system of the center of masses from the motion of the whole system as a wholein the Hamiltonian of the systems of charged particles. Therefore if the system of particles moves in such a way that the transverse component of its velocity is nonzero, the exact calculation of the energy levels is rather a cumbersome problem. The indicated effects can be neglected if

$$K_{\perp}a_0 \ll 10\left(\frac{a_0}{a_m}\right)^2\frac{m_e}{m_p} = 10\,\beta\frac{m_e}{m_p}$$

where $\vec{K}_{\perp} = \vec{P}_{\perp}/\hbar$ is the projection of the generalized pulse P onto the plane normal to the magnetic field. If $K_{\perp}$ = 0, then the allowance for the finitness of the proton mass in the calculations of energy levels reduces to allowance for a correction dependent on the quantum number m [4]:

$$\Delta E = 4\beta E_H\frac{m_e}{m_p}|m| = 0.0296\,\beta\,|m| \quad eV$$

where $E_H = e^2/2a_0 = 13.6$ eV.

2.2.Lamb shift. The energy shift of the same order of magnitude arises due to an anomalous magnetic field of an electron. The resulting shift of the states corresponding to the excited Landau levels n ≥ 1 is ~ $(\alpha/2\pi)\,\hbar\,\omega_B \approx \beta\cdot 63$ eV.

2.3.Bound-bound transitions in a hydrogen atom. The detailed calculation of the oscillator forces $f_{\tau\tau'}$ for such type of transitions (τ is a set of quantum numbers, whicn determine the system status) were performed and tabulated in [10] including the data on the transitions between the lower levels at values of β ranging between $10^{-4} \leq \beta \leq 10^{3}$. The change in the character of the given dependences $f_{\tau\tau'}(\beta)$ from the region where the structure of an atom in determined by the Coulomb interactions between an electron and the nucleus ($\beta \approx 0$) to the region where it is determined by the magnetic field occurs at $\beta \sim 0.1$. Some transitions supressed at $\beta \to 0$ turned out to be allowed at $\beta \approx 1000$ and vice versa.

2.4. Hydrogen ionization and recombination. First, this problem was considered in [11]. Then the problem was studied in [12-14] whose authors calculated the photoeffect cross-section as a function of the energy of incident photons for different incidence angles and polarizations of these photons. It is also noted that on the whole the cross-section increases with an increase of the field. The cross-section has characteristic peaks corresponding to the excitation of states with n = 1 and σ = -1/2, and also the states with n = 0 and σ = 1/2. If $\vec{e}_k \perp \vec{B}$ (photons are polarized normally to the magnetic field), the ionization cross-section is equal to zero; if the photon energy is lower than the excitation threshold for the first Landau level, then n = 1, σ = -1/2, since in this case the transitions from the ground Landau state give no contribution into the process cross section.

In [14] the cross-section and the recombination rate were obtained (with the use of the principle of detailed equilibrium). It was shown, in particular, that for low electron energies $E^{'}$ $\omega_{rec} \propto 1/\sqrt{E}$ the ratio has the value $E^{'}/E_H \approx 100$ (for $\beta \approx$ 1000) and then drops down to zero with a decrease in $E^{'}$. The recombination rate drastically increases if the electron energy is such that recombination can proceed through excited Landau levels, since in this case recombination occurs with the participation of such fast processes as the transitions between the Landau levels.

2.5.Positronium is one of the most interesting hydrogen-like systems. The detailed calculations of quantum transitions and annihilation performed in [5]. The authors also determined the level lifetimes with respect to the spontaneous dipole transitions and 2g- annihilation. One-quantum annihilation at $\vec{K} = \vec{P}/\hbar = 0$ (see Sect.2.1) is strictly forbidden, which is the consequence of the conservation laws for energy, photon pulse, and the generalized pulse of the system, $\hbar\vec{K}$. The allowed transitions are those from the odd to the ground (tightly bound) states of opposite parity. The transitions from even states can occur only to positron-like (odd) states. The probabilities of such transitions are relatively low. Therefore the even states decompose directly along the 2γ-annihilation channel, whereas the decomposition of odd states occurs mainly via the dipole transition into a tightly bound state with the subsequent 2γ-annihilation.

2.6. Hydrogen-like ions. The level energies and oscillator forces for hydrogen-like ions with the nucleus charge Z can be readily determined by the scale recalculation taking into account the analogous quantities for a hydrogen atom (see, e.g., [6]):

$$E(Z,B) = Z^2 E(Z=1, B/Z^2)$$
$$f_{\pi\pi'}(Z,B) = f_{\pi\pi'}(Z=1, B/Z^2)$$

These relationships follow from the analysis of the Hamiltonian and are valid in the approximation of an infinitesimal proton mass. In order to take into account the finiteness of the proton mass, it is necessary to perform an additional scale recalculation (see, e.g., [6]):

$$E_m(m_e, m_p, B) = \lambda E_m(m_e, m_p \to \infty, B/\lambda^2) - \hbar m e B/m_p$$
$$f_{\pi\pi'}(m_e, m_p, B) = \lambda^{-1} f_{\pi\pi'}(m_e, m_p \to \infty, B/\lambda^2)$$
where $\lambda = m_p/(m_e + m_p)]l = m/(m + m)$.

Note also the scale charge transformation of the photoionization cross-section [14],

$$\sigma(Z, B, E, E' \vec{k}) = Z^{-1}\sigma(Z=1, B/Z^2, E/Z^2, E'/Z^2, \vec{k}/Z)$$

where E is the energy of the state from which photoionization occurs, E' is the total energy of an emitted electron, and $\vec{k}$ is the wavevector of an incident photon. In [14], this expression is used, in particular, for obtaining the photoeffect cross-section in a hydrogen-like iron ion, Fe^{25+}.

2.7. Helium and helium-like ions. By present, the most studied problem is that of the energies of the ground state for systems of the isoelectron series of helium [2, 15]. These calculations were performed mainly by the Hartree-Fock method and embrace all helium-like ions up to Fe^{24+}. The obtained results indicate that the energy of the ground state for strong fields can reach the values from several kiloelectronvolts up to several tens of kiloelectronvolts. Thus if $B = 10^{13}$ G, the energy of the ground state attains the value ~ 1.2 keV for He, ~ 6.2 keV for C^{4+}, ~ 19.3 keV for Si^{12+}, and 42.2 keV for Fe^{24+}. It is of interest that the energy of the ground state of H^- for such fields turns out to be also rather high, ~330 eV.

3. Molecular Spectroscopy

By present, the best studied states are the low electron states of a molecular H_2^+ ion, which in the case of a weak field correspond to the terms $1\sigma_g$, $1\pi_g$, $1\pi_u$, $2\pi\sigma_u$, $2d\sigma_g$ and $2p\pi_u$ [16-24]. These calculations were performed mainly for the fields $B < 10^9$-10^{10} G. The electron terms at $B > 10^{11}$ G were calculated only for three first indicated states [18-23], most of these results relating to the calculations of the potential curves determining the dependece of the energy of interaction between the H atom and H^+ ion on the internuclear distance R. The potential surface determining the dependence of the interaction energy not only on R but also on the angle formed by the direction of the magnetic field and internuclear axis Θ, was first studied in detail in [20]. The discussion of this problem was continued then in [18, 23]. The vibrational-rotational spectrum of H_2^+, the probabilities of rotational transitions, and some related problems were considered in [21, 22, 25-27].

3.1. Potential surface of H . The calculations performed in [18, 20, 23] revealed a series of important features of the potential surface H_2^+ in $1\sigma_g$ state in a strong magnetic field. First, the equilibrium internuclear distance at B ~ 10^{12}-10^{13} G substantially decreases down to a value of 0.17-0.08 Å. At the same time, the dissociation energy increases attaining ~ 200 eV at $B = 10^{13}$ G. Secondly, the potential well is formed only for the small range of angles in the vicinity of the value $\Theta \approx 0$ ($\Theta \leq 15 - 20°$). For large angles, the potential surface takes the values above the dissociation threshold.

3.2. Vibrational-rotational spectrum. The analysis of the the potential surface as a function of R and Θ leads to a conclusion [22] that two types of motion are possible in the above indicated potential well: (a) the precession of the molecular axis about the direction of the magnetic field, which is characterized by two quantum numbers - N (the number of a Landau level) and Λ (the projection of the angular momentim of the nuclear motion onto the direction of the magnetic field) and (b) the vibrations of nuclei relative the center of gravity characterized by the quantum number ν. The resulting spectrum is a set of vibrational bands each of which is determined by its own N and Λ values. Since the potential well is of a finite depth, the number of levels is also finite.

The analysis of the statistics of nuclear spins indicates that in a strong magnetic field three different H_2^+ modifications exist, each corresponding to one of three possible values of the total projection Σ of nuclear spins onto the direction of the magnetic field. The total energy of the vibrational-rotational level included an additive term proportional to Σ. Moreover, the spin modifications with $\Sigma = 0$ have rotational levels corresponding to any integer Λ, whereas if $|\Sigma| = 1$, Λ can take odd values alone.

3.3. Vibrational-rotational transitions. The wavelengths of the vibrational-rotational transitions are within a wide spectrum range (from optical to X-ray). To calculate the probability of these transitions one should know the dependence of the quadrupole moment of H_2^+ on R and Θ. Such dependence have been calculated in [21].

The probabilities of vibrational-rotational transitions in the field range$B = 10^{12}$ - 10^{13} G were determined in [26]. The authors have also studied the polarization characteristics and angular radiation distributions. Among most general features of these curves one should note that at any polarization of an emitted quantum the radiation distribution has a minimum along the magnetic field direction.

Electron transitions in molecules for the above considered magnetic fields have not been considered.

The problem of molecule photodissociation in a strong magnetic field is open. Partly, the problem was considered in [25] where it was shown that the magnetic field produces a stabilizing effect on photodissociation. But this conclusion is of a preliminary character, since it has been made on the basis of the calculations limited to the above considered case $\Theta = 0$.

3.4. Other molecular systems in strong magnetic fields almost have not been studied. Some potential curves for H_2 molecules for the case $\Theta = 0$ were calculated, e.g., in [28] where, in particular, it has been shown that the internuclear distances of a neutral hydrogen molecule in the above considered magnetic fields decrease even more, whereas the dissociation energies increse to a larger extent than in a molecular H_2^+ ion.

One should also mention an attempt of calculating the potential surface of an H ion [29] which showed an essential changes in the ion structure in fields $B > 10^9$ G.

3.5. Nuclear tunneling in molecules. As has already been noted, the internuclear distances in molecules in the direction of a strong applied field substantially decrease, which may results in nucleus tunneling from the molecular potential well into a nuclear one with the subsequent nuclear reaction. This catalytic effect was discussed in detail in [27] where, in particular, the probabilities of the reactions in hydrogen were calculated and it was established that with an increase of the field from $B = 10^{12}$ G up to $B = 10^{13}$ G the reaction probability increases by more than ten orders of magnitude.

4. Ionization and Dissociative Balance

The development of the methods for calculating the ionization equilibrium in a medium at a sufficiently high temperature and strong magnetic field was first discussed in [11]. The authors obtained an approximate expression for the constant of ionization equilibrium of hydrogen-like ions in the temperature range where the proton motion can be considered as quasiclassical. At lower temperatures, $10^5 \leq T \leq 10^6$ K this expression was refined in [30], where an additional allowance for the internal structure of a hydrogen-like ion was made. The problem of dissociative equilibrium was discussed first in [31] (the dissociative equilibrium of an H_2^+ ions was studied). The authors also performed an illustrative calculation of the ionization and dissociative balance of H atoms and H_2^+ ions in the temperature range 10^5 -10^6 K at the gas concentrations 10^{20} -10^{22} cm^{-3}.

4.1. The ionization equilibrium constant in the above considered magnetic fields strongly depends on the temperature. With the temperature rise from 10^5 up to 10^6 K this value varies by almost 75 orders of magnitude. A magnetic field influences the constant of ionization equilibrium in two ways. On the one hand, it increases the ionization energy, i.e. "strengthens" atoms and ions. On the other hand, it decreases the phase volume per quantum state, i.e., promotes ionization. The temperature at which both factors become equal is $T_c = 5 \cdot 10^5$ K. At $T < T_c$ an increase in the magnetic field shifts the equilibrium towards the formation of neutral ions, and at $T \geq T_c$ the same increase in the field shifts the equilibrium in the reverse direction. As has already been noted, the molecular and atomic spectra are influenced by the motion of a system as a whole (thermal motion) across the magnetic field direction. But if the requirements to the accuracy of the determination of the ionization equilibrium constant are not very severe, this perturbation in the spectra can be neglected up to the temperatures $3 \cdot 10^6$K [11,30].

4.2. The constant of dissociative equilibrium of H_2^+ is qualitatively similar to the above considered constant of ionization equilibrium (temperature dependence, the effect of a magnetic field) but quantitatively differs from it since it additionally depends on the total projection Σ of nuclear spins onto the direction of the magnetic field. In other words, each spin modification of H_2^+ has its own constant of dissociative equilibrium. The temperature T_c (see Sect.4.2) is different for different Σ. For $|S| = 0$, $T_c \sim 1 \cdot 10^5$ K. For $\Sigma = 0$, $T_c \sim 10^6$ K.

4.3. Ionization and dissociative balances. The illustrative calculations performed in [31] show a strong dependence of the content of H atoms and molecular H_2^+ ions on the magnitude of the magnetic field, the field influence showing itself in full accordance with the effects discussed earlier in Sects.4.1 and 4.2. If the total gas concentration is lower or equal to 10^{22}

cm^{-3} , the maximum value of the relative H_2^+ content is 10^{-4} - 10^{-7} (depending on the total gas concentration and the magnetic field value) in the temperature range T ~ (2-3) · 10^5 K.

5. Conclusion

The investigations in physics of atoms and molecules in a strong magnetic field are carried out on an ever increasing scale. This is associated with the development of new concepts on the properties of strongly magnetized matter and with the applications of the results to asptophysics and solid state physics. As to the neutron star radiation, the combination of the data on the properties of atoms and molecules in magnetic fields of the above considered magnitudes (including spectroscopy and elementary processees) with the models of real gas distribution, radiation, and magnetic fields in the vicinity of an object allows the development of new spectroscopic methods, which, as a rule, turn out to be more informative than those based on the radiation observations in continuous spectra.

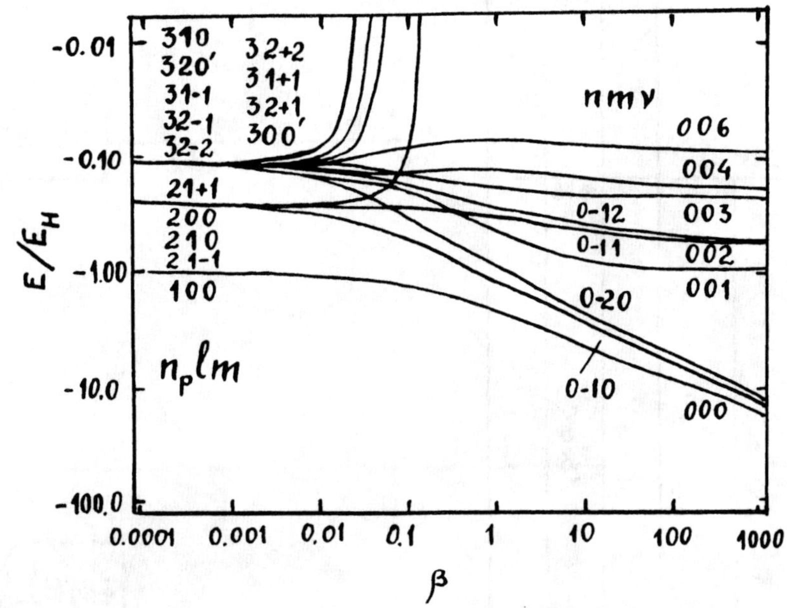

Fig.1. Energies of lower states of a hydrogen atom ($n_p \leq 3$) as functions of the magnetic field [3].

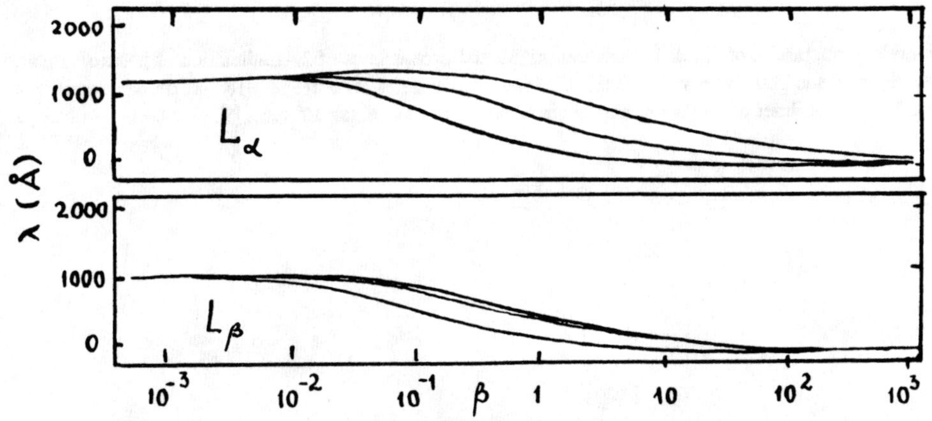

Fig.2. Wavelengths in L_α and L_β triplets of a hydrogen atom as a function of the magnetic field [3].

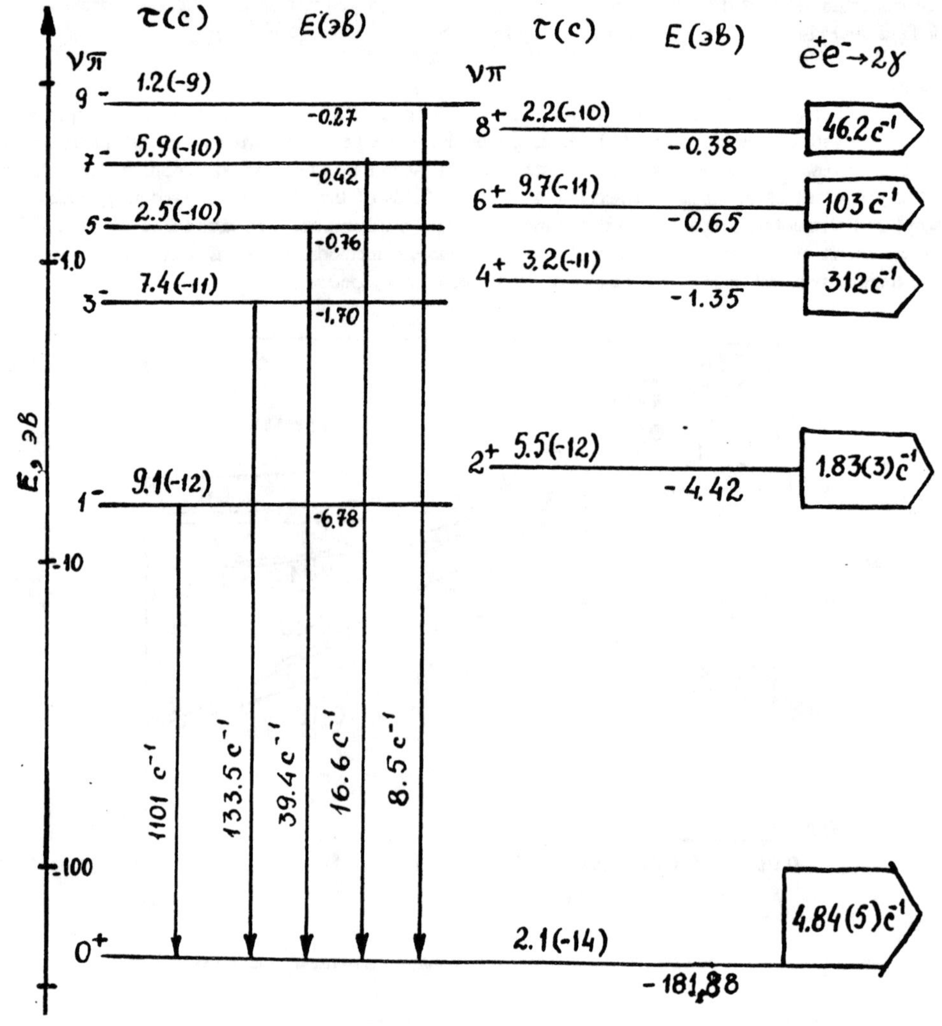

Fig.3. Energy levels, probabilities of dipole electric transitions, and probabilities of 2γ-annihilation of positronium at ρ = 1000 and K = 0, π is the state parity, τ are the lifetimes of the states with respect to the dipole transitions and 2γ-annihilation. The probabilities of transitions and 2γ-annihilation are given in the 10^8 units [5].

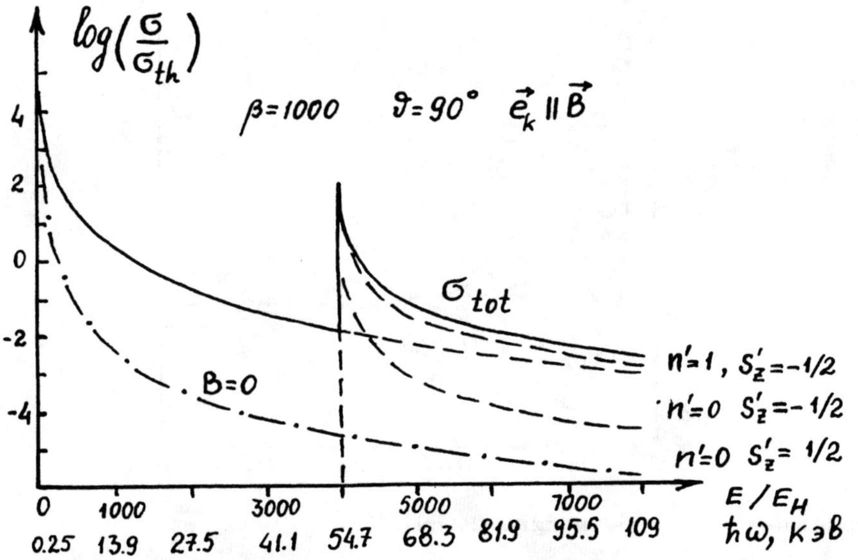

Fig.4. Photoionization cross-section of a hydrogen atom in the case where an incident (in the direction normal to the field) photon is polarized parallel with the magnetic field. The cross section is given in the Thomson cross-section units. The contributions of different Landau levels are shown. For comparison the values of log (σ/σ_{th}) at B = 0 are also given.

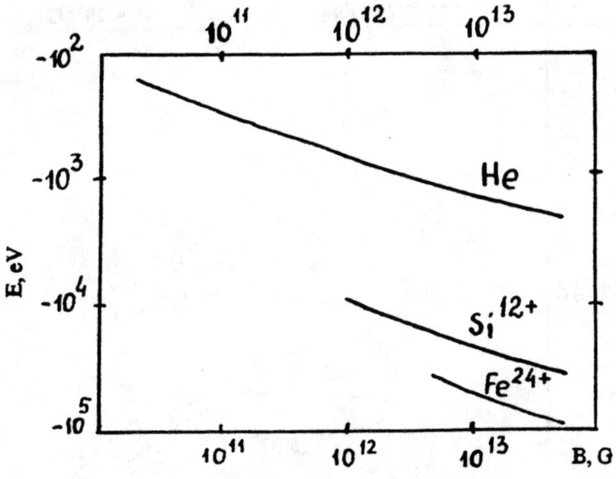

Fig.5. Energies of the ground states of a helium atom and helium-like ions as functions of the magnetic field [15].

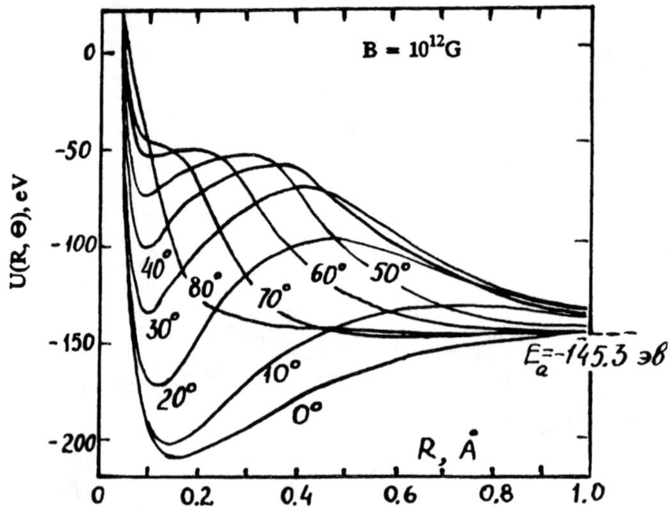

Fig.6. Potential surface of a H_2^+ ion in a strong magnetic field [20].

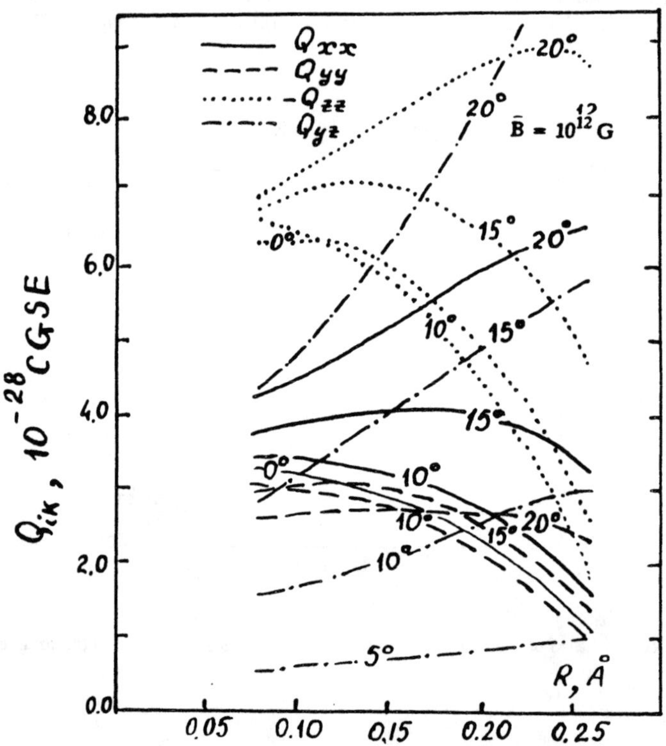

Fig.7. Components of the quandrupole-moment tensor of H_2^+ ions in a strong magnetic field [21].

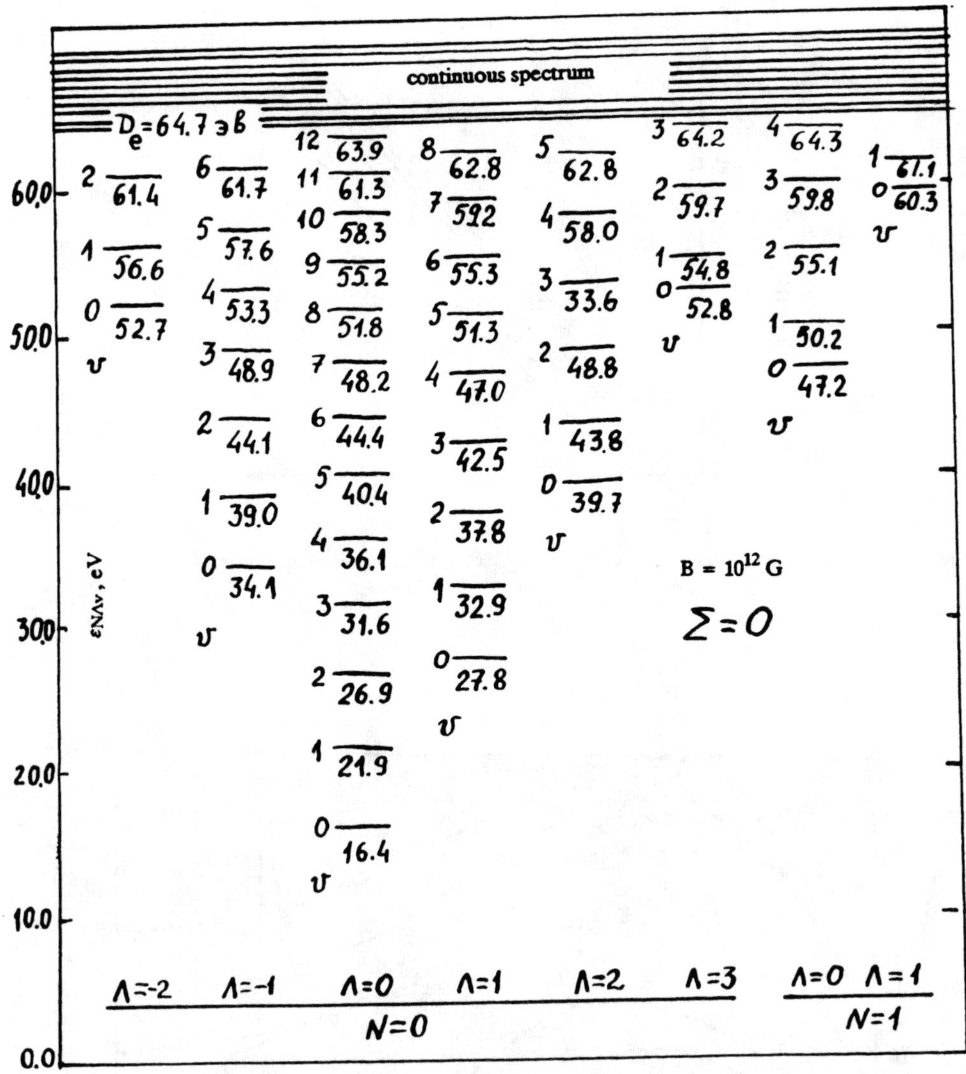

Fig.8. Vibrational-rotational energy spectrum of H_2^+ ions in a strong magnetic field. The numbers under the bars symbolizing levels indicate the level energy in eV.

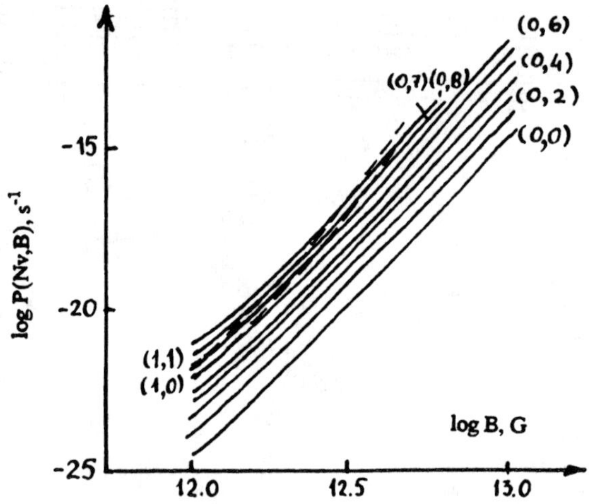

Fig.9. Probability of the nuclear reaction p + d → HD+ → He3 + γ in a strong magnetic field.

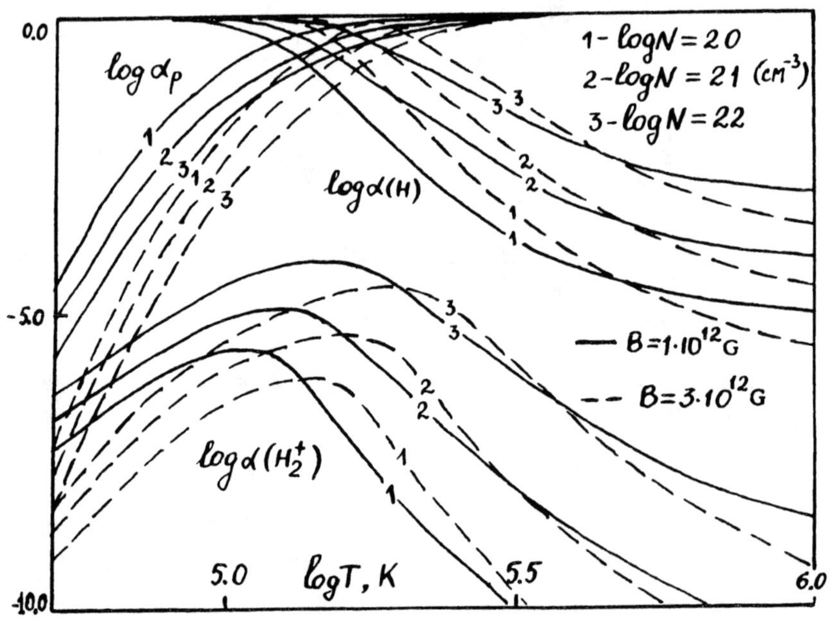

Fig.10. Relative content of protons (α_p), hydrogen atoms (α(H)), and H_2^+ ($\alpha(H_2^+)$) ions at different values of the total gas concentrations in different magnetic fields.

References

1. J. Trtmper. W. Pietsch, C. Reppin, W. Voges, R. Staubert, and E. Kendziorra, Astrophys. J., 1978, 219, L105.
2. J.E. Skjervol abd H.E. ostgaard, Phys.Scripta, 1984, 29, 448.
3. W. Rosner, G. Wunner, H. Herold, and H. Ruder, J. Phys. B: Atom. Molec. Phys., 1984, 17, 29.
4. H. Herold, H. Ruder, and G. Wunner, J. Phys. B: Atom. Molec. Phys., 1981, 14, 751.
5. G. Wunner, R. Ruder, and H. Herold, J. Phys.B: Atom. Molec. Phys.
1981, 14, 765.
6. G. Wunner, H. Ruder, and H. Herold, Astrophys. J., 1981, 247, 374.
7. P. Proschel, W. Rosner, G. Wunner, H. Ruder, and H. Herold, J. Phys. B: Atom. Molec. Phys., 1982, 15, 1959.
8. B.G.S. Doman, J. Phys.B: Atom. Molec. Phys., 1981, 14, L405.
9. J. Simola amd J. Virtano, J. Phys.B: Atom. Molec. Phys., 1978, 11, 3309.
10. H. Forster, W. Strupat, and W. Rosner, J. Phys. B: Atom. Molec. Phys., 1984, 17, 1301.
11. Yu.N. Gnedin, G.G. Pavlov, and A.N. Tsygan, ZhETF, 1984, 66, 421.
12. S.M. Kara and M.R.C. McDowell, J. Phys. B: Atom. Molec. Phys.,
1981, 14, 1719.
13. W. Schmitt, H. Herold, H. Ruder, and G. Wunner, Astron. Astrophys., 1984, 94, 194.
14. G. Wunner, H. Ruder, H. Herold, and W. Schmitt, Astron. Astrophys., 1983, 117, 156.
15. P. Proschell, W. Rosner, G. Wunner, H. Ruder, and H. Herold,
J. Phys. B: Atom. Molec. Phys., 1982, 15, 1959.
16. J.M. Peek and J. Katriel, Phys. Rev., 1980, A21, 413.
17. J. Ozaki and Y. Tomishima, Phys. Lett., 1981, 82A, 449.
18. D.R. Bridgman and J.M. Wadehra, Astrophys. J., 1987, 317, 865.
19. V.K. Khersonskii, Optika i Spektroskopiya, 1983, 55, 825.
20. V.K. Khersonskii, Astrophys., Space. Sci., 1984, 98, 255.
21. V.K. Khersonskii, Astrophys., Space. Sci., 1984, 103, 357.
22. V.K. Khersonskii, Astrophys., Space. Sci., 1985, 117, 47.
23. U. Wille, J. Phys. B: Atom. Molec. Phys., 1987, 20, L417.
24. D.M. Larsen, Phys. Rev., 1982, A25, 1295.
25. V.K. Khersonskii, Izvestiya SAO, 1985, 19, 54.
26. V.K. Khersonskii, Izvestiya SAO, 1987, 24, 68.
27. V.K. Khersonskii, ZhETF, 1986, 91, 1172.
28. A.V. Turbiner, Pis'ma v ZhETF, 1983, 38, 510.
29. C.S. Wark and A.K. Dutta, Phys. Rev., 1977, A16, 1747.
30. V.K. Khersonskii, Astron. Zh., 1987, 64, 433.
31. V.K. Khersonskii, Astron, Zh., 1987, 64, 1233.

Physics of Neutron Stars
Edited by A. M. Kaminker

ISBN 1-56072-164-2

PULSAR IN CRAB NEBULA AGAINST LIGHT DIBARIONS

M.I. Krivoruchenko
Instutite of Theoretical and Experimental Physics, 117259 Moscow

From the data on glitches of the pulsar in the Crab nebula the the following constraint on the masses of dibarion resonances have been obtained: m(6q) ≥ 2180 Mev.

At high nuclear-matter densities the generation of dibarions becomes advantageous [1-3].

The effect of dibarions on the properties of a nuclear matter can be explained with the use of the following considerations. Let us fill a box (Fig.1) with neutrons. In virtue of the Pauli principle, netrons successively fill ever higher levels. When the energy of a neutron on the external shell (the Fermi energy ε_F) becomes as high as the half of the dibarion mass, $\varepsilon_F = m(6q)/2$, dibarions start being generated. Dibarions are Bose-particles, therefore they will be accumulated in the ground state forming a Bose-condensate. Particles in the ground state have zero-pulse, therefore dibarions are at rest and "do not impact the wall" and, therefore, give no contribution into prressure. Upon the attainment by the Fermi energy of the value m(6q)/2, neutrons stop filling higher energy levels and, simultaneously, the pressure becones constant although the density of the nuclear matter continues increasing (Fig.2). The nuclear matter with dibarion impurity is indifferent to both compression and extension and looses its elasticity.

Now let us assume that a dibarion condensate has been formed inside a neutron star (Fig.3). The Euler equation has the form

$$\rho \frac{d\vec{v}}{dt} = -\vec{\nabla} P - \rho \vec{\nabla} \Phi$$

In the inner region where the pressure is levelled due to dibarions, $\vec{\nabla} p = 0$. The matter is distributed continuously, therefore the gradient of the gravitation potential $\vec{\nabla} \Phi \neq 0$. Thus, $d\vec{v}/dt \neq 0$, and no static solutions exist. Neither exist stable neutron stars with a dibarion Bose-condensate [3].

The data on the glitches of the pulsar from the Crab Nebula provide the determination of the ratio of moments of inertia for charged and neutral components, $I_c/I_n \approx 0.05$ [4]. The pulsar consists of 95% of neutral particles, i.e., neutrons, therefore it is neither a quark nor a strange star (quarks are charged particles) but a conventional neutron star. Another argument here is the rate of pulsar cooling which agrees quite well with the traditional calculations; if to assume the existence of a quark matter, the agreement is not so good [4].

Glitches of the pulsar in the Crab Nebula are repeated in $\Delta t \approx 5$ years. In the TV model, 5 years is necessary for the pulsar mass to attain the value $\dot{M} = 1.3\ M_\odot$ [4]. This value is close to the Chandrasekhar limit 1.4 $M_\odot$ and agrees with a small value of I_c/I_n. Knowing the mass, we can also find the chemical potential at the neutron star center: $\mu = 1088$ MeV. The birth of dibarions is energetically disadvantageous if $2\mu < m(6q)$, whence the restriction on the dibarion mass m(6q) > 2176 MeV. Here the interactions between dibarions and bitween dibarions and neutrons were neglected.

The obtained restriction relates to the dibarion states of the quark nature (e.g., spherical 6q-MIT-sacks). The bound states of nuclons of the deutron-type are, of course, unessential, since nuclon pulses at the Fermi surface strongly exceed the characteristic pulses in deutrons. Note here that the dimensions of 6q-sacks do not exceed those of the Δ-isobar (see, e.g., [5]); therefore the assumption on the point-like dibarions at the nuclear densities would be valid as for conventional barions n, p, Σ^-, Δ^-, etc.

The most reliably established lightest dibarion resonanse LN(2130) [6] has the mass m(6q) = 2130 MeV. If to take into account the roughness of the estimates (the model of an ideal dibarion gas), then it can be taken that there is no serious contradictions with the above obtained restriction. The condition m(6q) > 2176 MeV seems to make the existence of the NN(1960) resonanse dubious [6]. A dibarion with such a mass should be condensed at the densities two times lower than that existing in the conventional nuclei.

Rather a light dibarion resonance may destabilize neutron stars before they attain the OV-limit. This phenomenon is illustrated by Fig.4. We consider the TV model, taking into account the possibility of the Bose-condensation of the LN(2130) resonance. The upper mass limit of neutron stars in this case is determined not by the "relativization" of the nuclear matter, but by the mass of the dibarion resonance LN(2130).

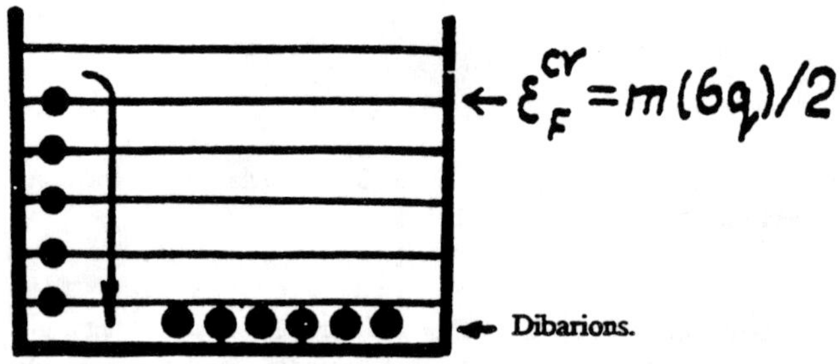

Fig.1.

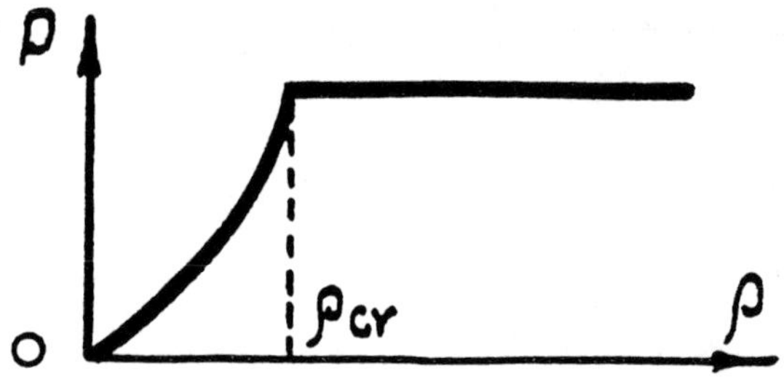

Fig.2.

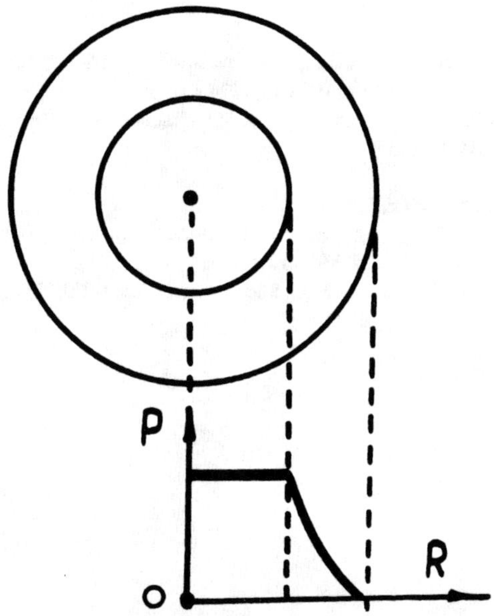

Fig.3.

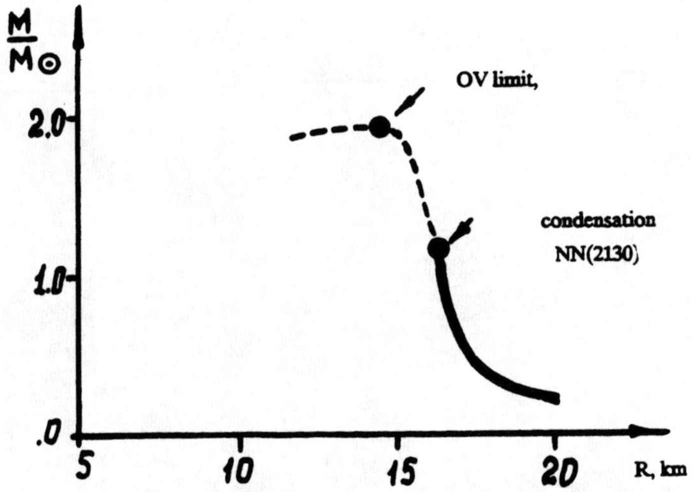

Fig.4.

References

1. A.M. Baldin, R.G. Nazmitdinov, A.V. Chizhov, A.S. Shumovskii, and V.I. Yukalov, Dokl. AN SSSR, 1984, 279, 602;
A.S. Shumovskii and V.I. Yukalov, EChAYa, 1985, 16, 1274.
i A.V. Chizhov, R.R. Nazmitdinov, A.S. Shumovsky, and V.I. Yukalov, Nucl. Phys., 1986, A449, 660.
2. St. Mrowczynski, Preprint E4 -85-855, Dubna;
Phys. Lett., 1985, B152, 299.
3. M.I. Krivoruchenko, Pis'ma v ZhETF, 1987, 46, 5.
4. S. Shapiro and S. T'yukolski, Chernye dyry, belye karliki i neitronnye zvesdy (Black holes, white dwarf stars and neutron stars) 1985 (Moscow: Mir) pp. 318, 320, 326, 354 (in Russian).
5. L.A. Kondratyuk, M.I. Krivoruchenko, and M.G. Shchepkin, Pis'ma v ZhETF, 1986, 43, 10.
6. Particle Data Group, Phys. Lett., 1986, B170.

Physics of Neutron Stars
Edited by A. M. Kaminker

ISBN 1-56072-164-2

ELECTRODYNAMICS OF NEUTRINO IN DISPERSIVE MEDIA

V.N. Oraevskii, V.B. Semikoz, and A.Ya. Smorodinskii
Institute of Earth Magnetism, Ionosphere, and Radiowave Propagation of the USSR of the Academy of Sciences,
142092 Troitsk, the Moscow District

It has been shown that the electromagnetic properties of a neutrino in a dispersive medium substantially differ from those in vacuum. Small medium-induced electric charge and normal magnetic moment of a neutrino (proportional to the product of the weak-interaction constant and the charge density) provide the consideration of a neutrino motion as a notion of an electron under the same conditions.

The application of the minimum standard model of electrically weak interactions ($SU_L(2) \otimes U(1)$) to dispersive media in the low energy range ($q^2 \ll M_w^2$, M_w is the mass of a W-boson*) leads to electrodynamics of continuous media with participation of neutrinos. In the above indicated energy range one can neglect the contribution of the self-action of the Young-Mills field into the medium polarization [1] which, in this case, is completely described by an Abelian U(1) part of the electrically weak model, i.e., quantum electrodynamics (QED) in the medium.

A neutrino, polarizing the medium by weak forces, acquires relatively small induced electric charge [2] and normal (not changing helicity) magnetic moment [3] and, therefore, in this sense, behaves in a medium like an electron.

A neutrino moving in a medium emits electromagnetic waves (polarization radiation in isotropic [4] and Vavilov-Cherenkov radiation in an anisotropic medium [5]), is scattered by electrons and ions of the medium by a Coulomb way (via the longitudinal plasmon [6]), and experiences the action of the external electromagnetic fields [7], etc.

Basic Polarization Effect

1.1. Polarization Origin of Neutrino Electric Charge in Dispersive Medium

Let us remind that the electromagnetic characteristics of neutrino in vacuum - mean-square electromagnetic radius $<r_\nu^2>^{1/2}$ and anomalous magnetic moment $\Delta\mu^{vac}$ - are associated with the interaction of the neutrino with the vacuum of vector W,Z bosons and leptons.

Having such characteristics, a neutrino, similar to a neutron, interacts with photons, electrons (electromagnetic νe-scattering), etc. In other words, the allowance for radiative corrections (RC) to a weak interaction calculated in one-loop approximation, allows us to calculate the electromagnetic structure of a neutrino in vacuum.

An electromagnetic structure of a neutrino in a medium is calculated in the same single-loop approximation [6]. In this case the self-action of the W, Z-fields does not make any contributions into electromagnetic form-factors of a neutrino, in distinction from the RC case in vacuum, where the self-action of vector bosons is always taken into account.* The main part is played by the polarization effects completely described in the low-energy range under consideration: (a) conventional polarization contribution of quantum electrodymanics (QED) in the medium, i.e., lepton loop, and (b) the simplest Lamb diagram with the W-boson exchange (see [6] and Fig.1).

Let us obtain an electromagnetic neutrino apex in a medium, not invoking the method of exact diagrams.

With this aim we use an analogy with the vacuum electromagnetic structure of an electroneutral particle. As is well known, in an electromagnetic 4-current $e\bar{\nu}\Gamma_\mu\nu$, where ν is a bispinor, part of electromagnetic vertex Γ_μ determined by the electric charge distribution, $\rho(r,t)$** is proportional to dimensionless formfactor $F(q^2)$ in the Lorentz structure $\Gamma_\mu \approx F(q^2)\gamma_\mu$.

---

* The following system of units is used $\hbar = c = 1$, the Feynman metrics $q^2 = q_\mu q^\mu = \omega^2 - \vec{k}^2$, $\mu = 0, 1, 2, 3$; the standard representation of the Dirac matrices with $\gamma_5 = \gamma_5^+ = i\gamma_0\gamma_1\gamma_2\gamma_3$.

** In the medium, there exists a natural termination of the contributions from large pulses of virtual particles because of the existence of the distribution functions that rapidly (exponentially, in the equilibrium medium) decrease with an increase of the pulse. Therefore no ultraviolet discrepancy diagrams arise with the self-action of the W, Z-fields that took place in the used unitary calibration in the case of vacuum RC. As a result, these contributions in the medium turn out to be of the oredr of $q^2/M_w^2 \ll 1$ with respect to the contributions from the main diagrams shown in Fig.1.

Such a form-factor in the lower 4-fermion apptoximation is proportional to the dimensional Fermi constant $G_F \approx 10^{-5}/m_p^2$ where m_p is the proton mass. This is a consequence of the νe interaction between the medium and electrons in the Born approximation. The formfactor $F(q^2)$ should not include any other binding constants.
Indeed, the charge e at the point 4-fermion vertex uniting the neutrino current and an electron - hole loop (see Fig.1) should be excluded from the constant of weak interaction G_F. This can be made in the polarization operator Π by dividing it into e or by writing $\Pi/e \sim e(\varepsilon - 1)/\alpha$, where $(\varepsilon - 1)$ is the polarizability of the medium and vacuum, e is the factor induced into the electromagnetic current $e\bar{\nu}\Gamma_{\mu\nu}$ and therefore it should not eneter the unknown formfactor $F(q^2)$, and $\alpha \equiv e^2$.
Finally, the dimensionless electric formfactor of a neutral particle is proportional, like in vacuum, to the square of the only 4-vector with the mass dimensionality $(\sim q^2)$.
Thus the electric formfactor of a neutrino in the medium should have the form

$$F(q^2) \sim G_F q^2 (\varepsilon - 1)/\alpha \qquad (1)$$

In the particular case of vacuum (the particle density $n_0 = 0$, $|\varepsilon - 1| = |\delta\varepsilon_{vac}| \sim \frac{\alpha}{3\pi} \ln q^2/m_e^2$, $q^2 \gg 4m_e^2$) one can readily estimate the contribution of vacuum polarization (QED) into a mean-square neutrino radius by the formula $F(q^2) = q^2 <r_\nu^2>/6$.

First of all, we should like to note that in a medium ($n_0 \neq 0$) the polarization of vacuum can be neglected in comparison with the polarization of the medium. In a medium, the polarizability $(\varepsilon - 1)$ can exceed unity. In the simplest case, where the weak dissipation is neglected, we have

$$\varepsilon = 1 - \frac{\omega_p^2}{\omega^2 - k^2 <v^2>}; \quad \omega_p = [4\pi\alpha n_0/m_e]^{1/2}$$

where w is the Langmuir frequency. Thus, for longitudinal plasmons whose spectrum is determined by the dispersion equation $\varepsilon(\omega, k) = 0$, the quantity under consideration, $|\varepsilon - 1|$ is eqact unity. In case of QED in vacuum, in order to make $\delta\varepsilon_{vac} \sim \frac{\alpha}{3\pi} \ln q^2/m_e^2$ equal to unity one has to pass to higher orders of the perturbation theory, i.e. to asymptotic values of the transferred pulse $q^2 \sim m_e^2 \exp(3\pi/\alpha)$, where the "zero-charge" of QED arises and the low-energy approximation becomes invalid.

$$|q^2|/M_W^2 \ll 1$$

The second remark about the difference of the formactor (1) from its value in vacuum is related to the magnitude of transferred pulse. The emission of a real plasmon by a plasma and the exchange with a resonance ($\varepsilon = 0$) virtual plasmon correspond to the existence of a mass of the medium "photon" $q^2 \neq 0$, i.e., in this case the second-order calibration invariance is violated. Under the nonresonance conditions ($\varepsilon \neq 0$), e.g., in case of the interaction with a quasistatic electric field ($w \ll k <v>$, $<v>$ is the mean velocity of medium particles) or in the static case of a massive neutrino ($m_\nu \neq 0$) at rest ($\omega = 0$), formula (1) directly yields the formula for an "electric charge" of a neutrino in the medium

$$e_\nu^{ind} \sim e G_F/\alpha \sqrt{}_D^2 \sim e(\lambda\!\!\!^-_W/\sqrt{}_D)^2 \qquad (2)$$

where $\lambda\!\!\!^-_W$l is the Compton wavelength of a W-boson, r_D is the Debye radius of the charge screening entering the well known formula for quasistatic dielectric constant: $\varepsilon(0, k) - 1 = (k\sqrt{}_D)^{-2}$.
Note also that a global calibration invariance (of the first kind) remains intact: at the boundary of a system (on the whole, electrically neutral) a compensating charge of the opposite sign is formed.

1.2. On Magnitude of "Electric Charge" of a Neutrino in a Medium
The term "electric charge" should be treated with a certain precaution. First, in distinction from an electron charge, this characteristic has no universal character, which manifests itself in the dependence of charge (2) on varying medium density and temperature. Secondly, neutrinos may have different "electrical charges" at the same medium temperatures and densities depending on the properties of external magnetic fields or the modes of the natural electromagnetic radiation of a moving neutrino.
Thus the qualitatively obtained charge in (2) [2],

$$e_\nu = -\frac{|e|\, G_v}{4\pi\alpha \sqrt{2}\, \sqrt{}_D^2} \qquad (2')$$

should be considered as quasistatic. It is with such a charge that a neutrino incteracts with the respective electric field. Here $G_V = G_F(1 + 4\xi)$ is the vector weak binding constant invariant with respect to the substitution of a neutrino (ν) by an antineutrino ($\tilde{\nu}$); $\xi = \sin^2\Theta_w$ is the parameter of the standard model determined by the Weinberg angle Θ_w.

A high-frequency electromagnetic field ($\omega \gg k<v>$ interacts with the neutrino via the charge*** different from (2') [8]

$$\tilde{e}_\nu = -\frac{|e|\, G_V \omega_p^2}{4\pi\alpha\sqrt{2}} \qquad (3)$$

and depending on another medium characteristic - Langmuir frequency ($\omega_p \sim <v>/\surd_D$).

It can readily be seen that the magnitude of charge (2) turns out to be small, of the order of $\sim 10^{-16}$e in a metal or 10^{-8}e under the conditions of collapsar with a density of $\rho \sim 10^{12}$ g/cm^3. But even under the laboratory conditions such a small quantity results in a much larger cross-section of "Coulomb" νe-scattering than the well known Born cross section calculated for the same standard model and averaged over the same state of electrons of the medium [6].

In order to obtain the energy range for neutrinos in which the late statement is valid, let us make the following estimate. The cross-section of "Coulomb" νe neutrino scattering can be obtained from the geometric one $\sigma_{coul} \cong \pi\surd^2_{min}$ L. Here L $= \ln(\surd_D/\surd_{min})$ is the Coulomb logarithm (L $\sim$ 10); $\surd_{min}$ is the distance of the maximum approach that can be estimated from the virial theorem:

$$e_\nu e / \surd_{min} \sim E_\nu$$

where E_ν is the neutino energy. Now let us comparé the obtained cross-section

$$\sigma_{coul} \sim \pi \left(\frac{e_\nu}{e}\right)^2 \alpha^2 L / E_\nu^2$$

with the standard cross-section of eleastic νe scattering in the ? system of an electron, $\sigma_{Born} = \sigma_0 E_\nu^{\ 2} / (m_e^2 + 2m_e E_\nu)$ where $\sigma_0 = 4G_F^{\ 2} m_e^2 / \pi \sim 1.7\cdot 10^{-44}$cm^2. It is easy to see that the earlier neglected electromagnetic reaction channel now becomes more essential than the standard Born contribution ($\sigma_{coul} \gg \sigma_{Born}$) for energies $E_\nu \ll E_0$, where

$$E_0 = \left[10^5 \pi\alpha L e_\nu / 2e\right]^{1/2} m_p$$

The constraint $E_\nu \ll E_0 \sim$ keV corresponds to the case of soft neutrinos ($E_\nu \ll m_e$) in the plasma of metals. For the conditions implemented in a collapsar ($E_\nu \gg m_e$) one can obtain another estimate of E_0 : $E_0 \sim$ 10 MeV.

In both cases the neutrino energy turns out to be lower than that of the Fermi pulse p_F ($p_F \sim$ keV in metals, $p_F \geq$ 10 MeV in a collapsar for a density $\rho \geq 10^{10}$ g/cm^3).

This signifies that the above discussed electromagnetic properties of a neutrino in a medium manifest themselves under the condition that the De Broglie wavelength $\lambda = p_\nu^{-1}$ strongly exceeds the mean interparticle distance $n_0^{-1/3} \sim p_F^{-1}$ (see Sect.2.1 below).

In the review, the interaction with unpolarized media of electron neutrino and antineutrino is being discussed. Neutrino of other aroma, whose microscopic interaction with electrons (and nuclei) is associated with neutral currents have the same electromagnetic characteristics in a medium as the electron neutrino with the only difference - the "large" value $1 + 4\xi \approx 2$ is substituted by a small one $4\xi - 1 \cong -0.08$, where the interaction parameter of electrically weak model $\xi = \sin^2\Theta w$ is approximately equal to $\xi \sim 0.23$ (from the experiments on accelerators*****). The collective energy losses of ν_μ and ν_τ neutrino in the substance drastically decrease by a factor of $[1 + 4\xi(1 - 4\xi)^{-1}]^2 \approx 625$. This may cause some differences in the transfer of different kinds of neutrinos in the formation of the spectra of neutrino radiation, etc.

*** The charge distribution density $e\rho(\vec{r})$ is determined in terms of the static form-factor $\rho(\vec{r}) = \int d^3k \exp(i\vec{k}\vec{r})\, F(o, k)\, (2\pi)^{-3}$. It can readily be seen that in vacuum there is no total charge of a neutral particle: $\int \rho(\vec{r})\, d^3r = 0$. Covariantly, this property is also reflected in the normalization of the vacuum formfactor $F^{vac}(0) = 0$ where $w = k = 0$; e is the electron charge ($e^2 = 1/137$).

**** Our attention to this form of the writing of high-frequency charge of a neutrino was drawn by V.N. Ursov.

***** The collective interaction of ν_μ and ν_τ neutrinos with the matter was first considered in {1, 9, 10]. The authors limited themselves to the contributions from neutral currents. Note also that the polarization losses of neutrinos in plasma were considered in the classical way using the Upov-Pointing flux. But in case of a small neutrino charge, the Bethe approximation is valid ($e_\nu e / \hbar c \ll 1$, cf. with Bethe formula $Z e^2 / \hbar v \ll 1$); the necessary quantum calculation was performed in [4].

For the parameter value $\xi = 0.25$, the interaction of $\nu_{\mu,\tau}$-neutrino with the matter does not change in comparison with the well known Born contributions in the standard model since $4\xi - 1 \cong 0$. In this case the differences in the interaction of $\nu_{\mu,\tau}$ and electon neutrino (ν_e) with the medium are especially well pronounced. Note here that in a homogeneous medium without space-time dispersion the same difference in the weak interaction constants for neutrino of different aroma leads to such effects as neutrino oscillations or to different Born cross-sections of ν_e-scattering.

1.3.Mott Scattering of Neutrinos by a Coulomb Center in a Medium

Using electromagnetic vertex of a neutrino in a dispersive medium [3,6], one can calculate the cross-section of neutrino scattering by a screened Coulomb center [11]. In a completely ionized plasma the screening is caused not by the atomic electrons but by the finite radius of Coulomb forces (potential Ze exp $(-\surd / \surd_D) / \surd$.****** Neglecting the contribution from the neutral currents the scatteing cross section acquire the form

$$\frac{d\sigma}{d\Omega} = \frac{Z^2\alpha^2(e_\nu/e)^2}{4v^2p^2}\,\frac{[1 - v^2\sin^2(\Theta/2) - m_\nu^2/2E^2]}{[\sin^2(\Theta/2) + (2p\surd_D)^{-2}]^2} \qquad (4)$$

where the neutrino charge e_ν is equal to (2'); m_ν is the meutrino rest mass that can be equal to zero (v = 0).
Results (4) differ from the case of electron scattering by the same screened potential by an inessential term $m_\nu^2/2E^2$ in the numerator of (4) which arise due to the presence of γ_5 matrix at the electromagnetic neutrino vertex, and, by a significant factor $(e_\nu/e)^2$ determined by formula (2').
If in the case of electron scattering by a Coulomb center in vacuum ($\surd_D \to \infty$) we obtain from (4) the standard Mott-Rutherford cross-section (neglecting the $(e_\nu/e)^2$ factor), then in the case of a neutrino the charge e_ν itself disappears in the limit $\surd_D \to \infty$ (see (2')). Therefore the "Coulomb" scattering can be compared only with the same electron scattering in the same medium.
The total cross-section $\sigma(E)$ in a medium obey the dependence $\propto E^{-2}$ neither for an electron nor for a neutrino due to the Debye screening of the potential which removes the Coulomb divergence at $\Theta \to 0$. But the estimates made in Sect.1.2 remain to be valid: in the energy range $E \ll p_F$ the cross-section calculated by (4) greatly exceeds the Born cross-section in the standard model determined by neutral currents alone.
Note also that everywhere in our review the medium is assumed to be infinite, isotropic (except of specially stipulated cases), and equilibrium. We consider mainly the neutrino (neutrino fluxes) passing through the electon plasma: degenerate electron plasma of metals, degenerate ultrarelativistic electron gas of collapsars, etc.

2. Single-particle Equation of Neutrino Motion in a Dispersive Medium

2.1. Wavefunction of Neutrino in a Medium

The most pictorial equation for the description of the neutrino interaction with the matter is a single-particle equation of neutrino motion in a medium (the Dirac equation). In order to derive this equation from the microscopic vacuum equation for the standard model used here it is important to explain how the statistical averaging of such an equation is understood.
It turns out that two conditions should be satisfied simultaneously. On the one hand, the statistical averaging (over the ensembles of equivalent systems) of the microscopic Maxwell equations in electrodynamics of continuous media is equivalent to averaging over the "physically infinitesimal volume" with such characteristic dimension l that $l \gg n_0^{-1/3}$, where n_0 is the mean density of the medium; at the same time, $l \ll \lambda$, where λ is the perturbation wavelength in the medium. The latter requirement means that the medium is a continuous continuum for a propagating wave. Thus, the size of "physically infinitesimal" volume is determined by the inequalities $n_0^{-1/3} \ll l \ll \lambda$.
On the other hand, in order to consider the neutrino coordinate "x" (in the wave function Ψ_ν (x,t)) as a "point" of a space filled with the medium with the given macroscopic characteristic, an additional condition should be fulfilled. The characteristic size l of the same "physically infinitesimal" volume should be much smaller than the experimental error of measuring the coordinate of a trial particle (neutrino) placed into the medium: $l \ll \Delta x$. Only in this case the medium can be considered as a continuous continuum for a moving neutrino. The uncertainty principle ($\Delta x \sim \hbar c/\varepsilon$) applied to a particle having no mass indicates that this error is of the order of a De Broglie wavelength $\hbar/p$ ($p = \varepsilon/c$).
Thus, an additional requirement

$$\hbar/p \ll l \qquad (5)$$

****** With due regard of neutral currents, there is a strong screening in the complete cross-section of neutrino scattering by nuclei in the plasma.

singles out the range of low neutrino energies, for which the description with the aid of the wave function Ψ_ν (x,t) is valid in a continuous homogeneous medium. The requirement (6) coincides with the criterion of the predominant contribution of the "Coulomb" cross-section in comparison with the Born cross-section (see Sect.1.2).
In the reverse limiting case of hard neutrino $p^{-1} \ll n_0^{-1/3} \ll 1$ (or a rarefied medium) the introduction of the above indicated wave function Ψ_ν (x,t) has no sense. The effects of collective interaction of neutrinos with the matter become small in comparison with the conventional two-particle interaction corresponding to the additive vacuum contribution. The latter is calculated with the aid of S-matrix elements of the standard vacuum electrically weak model (with the subsequent averaging over the particle distribution function in the medium) and turns out to be prevalent (in the $n_0 \to 0$ limit) in comparison with the contribution that comes from S-matrix elements of the collective processes with the participation of neutrinos.

2.2 Dirac Equation

Let us consider a single neutrino moving in an unbound weakly inhomogeneous medium. For simplicity, let us choose as a medium a binary completely ionized plasma consisting of one kind of ions ((Z, A) nuclei) and electrons with the equilibrium density $n_0^e = Zn_0^i$ where n_0^i is the ion density.
The microscopic (vacuum) Lagrange equation for neutrinos in the standard model of electrically weak interactions has the form:

$$\left[i\gamma_\mu \partial^\mu - \frac{e}{\sin(2\Theta_w)} \hat{Z}(\vec{x},t) \frac{(1-\gamma_5)}{2} - m_D^\nu \right] \Psi^\nu(\vec{x},t) + = \frac{e}{\sqrt{2}\sin\Theta_w} \hat{W}^+(\vec{x},t) \frac{(1-\gamma_5)}{2} \Psi^e(\vec{x},t) = 0 \tag{6}$$

Here the Dirac a neutrino mass m_D^ν (the term $m_D^\nu \overline{\Psi}^{(\nu)} \Psi^{(\nu)} = m_D^\nu (\overline{\Psi}_L \Psi_R + ?)$ was taken into account in the Lagrangian although all the results and conclusions of the present Section are applicable also to the mass-free case $m_D^\nu = 0$. The quantity e is the electron charge, $e^2 = 4\pi\alpha$, $\Psi_R = (1+\gamma_5)\Psi^{(\nu)}/2$, $\Psi_L = (1-\gamma_5)\Psi^{(\nu)}/2$.
The vector fields described by the Young-Mills equations - a neutral field $Z_\mu(\vec{x},t)$ and charged fields $W_\mu^\mp(\vec{x},t)$ - are determined by neutral currents of electrons and nuclons (of the medium) and the charged current $\bar{e}\nu$ of the interaction of a single neutrino with an electron of the medium.
The multiparticle character of the description is attained by secondary quantization of the electron and nuclon fields; in statistical averaging the background of low-mobile ions is assumed to be homogeneous and immobile simultaneously.
As an initial microscopic equation one can consider the Schwinger equation where the second term in (5) is substituted by the integral $\int \hat{M}(x,y) \Psi^\nu(y) d^4y$ with an operator $\hat{M}(x,y)$ determined (with the substraction of the rest mass contribution $m_D^{(\nu)} \delta^{(4)}(x-y)$) by the Green function of W-bosons[******]. One more variant of the initial microscopic equation adequate to the below used 4-fermion approximation is the Lagrange equation obtained with the aid of the approximate interaction Lagrangian (instead of the exact Weinberg-Salam Lagrangian used in the derivation of (5)):

$$L_{int} = \frac{G_F}{\sqrt{2}} \sum_j [\overline{\Psi}_j \gamma_\mu (c_L^{(j)}(1-\gamma_5) + c_R^{(j)}(1+\gamma_5) \Psi_j] \left(\overline{\Psi}_\nu \gamma^\mu \frac{(1-\gamma_5)}{2} \Psi_\nu \right)$$

where $c_{L,R}^j$ are the well known constant for fermions of the medium including quarks that constitute nuclons.
In the approximation of the point 4-fermion interaction, the statistical averaging (5) leads to the Dirac equation for a medium with weakly inhomogeneous electron distribution ($f_{rr'}^e(\vec{p},\vec{x},t) = (\delta_{rr'}/2) f_0^e(\vec{p}) + \delta f_{rr'}^e(\vec{p},\vec{x},t)$; $\delta f^e \ll f_0^e$; $\surd, r' = \pm 1$, the particle helicity):

$$\left[i\gamma_\mu \partial^\mu - V(\vec{x},t) \frac{(1-\gamma_5)}{2} - m_D^\nu \right] \Psi^\nu(\vec{x},t) = 0 \tag{7}$$

In a medium moving with the 4 -velocity $\Omega\mu$ ($\Omega^2 = 1$), the 4 potential $V_\mu(\vec{x},t)$ in the Dirac equation (7) consists of three terms ($V_\mu = \sum_{j=1} V_\mu^{(j)}$).

[******] The vacuum Schwinger equation for the Dirac and Majoran neutrino in external magnetic fields was studied at the Moscow University (see, e.g., [12]).

The equilibrium part $V_\mu^{(1)} = V_0 \Omega_\mu$ is determined by the homogeneous background of electrons and ions********

$$V_0 = \frac{(3Z - A)}{Z\sqrt{2}} G_F n_0^{(e)} \qquad (8)$$

The nonequilibrium parts $V_\mu^{(2)}$ and $V_\mu^{(3)}$ due to the averaged vector ($<\Psi_e(x)\gamma_\mu \Psi_e(x)>$) and pseudovector ($<\Psi_e(x)\gamma_\mu\gamma_5\Psi_e(x)>$) electron (and positron) currents have the form

$$V_\mu^{(2)}(\vec{x},t) = \frac{G_V}{\sqrt{2}e}\int \frac{d^4q}{(2\pi)^4} e^{-iqx} \prod_{\mu\nu}(\omega,\vec{k}) A^\nu(\omega,\vec{k}) \qquad (9)$$

$$V_\mu^{(3)}(\vec{x},t) = \frac{\varepsilon_{\mu\nu\beta\sigma} G_F}{2\sqrt{2}e}\int \frac{d^4q}{(2\pi)^4} e^{-iqx} A(\omega,\vec{k})[\mathcal{F}^{\rho\nu}\Omega^\sigma - \mathcal{F}^{\sigma\nu}\Omega^\rho] \qquad (9')$$

$\varepsilon_{\mu\nu\rho\sigma}$ is completely antisymmetric Levi-Civita tensor; $\Pi_{\mu\nu}(\omega,\vec{k})$ is the symmetric polarization tensor of statistical QED; $A(\omega,\vec{k})$ is the coefficient of the antisymmetric part of the polarization tensor, determining the induced magnetic moment of a neutrino [3]; $\mathcal{F}^{\rho\nu}(\omega,\vec{k}) = -i(q^\rho A^\nu - q^\nu A^\rho)$ is the electromagnetic field tensor in the medium. The contribution (9') is the spin ?, a correction to the main term (9) caused by the interaction (current ⊗ current) 4 -current of neutrino with the induced vector 4 -current of electrons.
To simplify the statement we consider the classical part of the main term in (9), which can be rewritten in the form of the 4 -current

$$\delta V_\mu^{(2)}(\vec{x},t) = \frac{G_V}{\sqrt{2}}\int \frac{d^3p}{\varepsilon_p} p_\mu \delta f^e(\vec{p},\vec{x},t) \qquad (9'')$$

with the Lorentz-invariant diatribution function $\delta f^e = \mathrm{Sp}\,\delta\hat{f}(\vec{p},\vec{x},t)$.
In order to exclude the distribution function $\delta f^e(\vec{p},\vec{x},t)$ in the 4 -potential (9") one can use the kinetic equation for electrons in the electrically weak model [13] and restrict oneself to the standard Vlasov approximation:

$$\frac{\partial\delta f^e(\vec{p},\vec{x},t)}{\partial t} + \vec{v}\frac{\partial\delta f^e(\vec{p},\vec{x},t)}{\partial\vec{x}} + e\left(\delta\vec{E}(\vec{x},t) + [\vec{v}\delta\vec{B}(\vec{x},t)]\right)\frac{\partial f_0^e}{\partial\vec{p}} = 0$$

As a result, it is possible to express the preserving ($\partial\delta V_\mu^{(2)} / \partial x_\mu = 0$) 4 -current in terms of the medium polarization with weak νe-interaction, determined by a conventional polarization vector $4\pi\delta P_i(\vec{x},t) = \delta D_i(\vec{x},t) - \delta E_i(\vec{x},t)$, $i = 1,2,3$.
From (9) we obtain [7]:

$$\delta V_0(\vec{x},t) = \frac{4\pi G_V}{\sqrt{2}e}\left(-\frac{\partial\delta P_k(\vec{x},t)}{\partial x_k}\right)$$
$$\delta V_i(\vec{x},t) = \frac{4\pi G_V}{\sqrt{2}e}\frac{\partial\delta P_i(\vec{x},t)}{\partial t} \qquad (10)$$

Whence it is seen that a neutrino in the medium moves in the field of neutrino-induced electric change field whose density, in accordance with electrodynamics of continuous media, is equal to $\rho_{ind}(\vec{x},t) = -\mathrm{div}\,\delta\vec{P}(\vec{x},t)$ (see the first of relationships (10)). In the theory of a linear response, the self-consistent fluctuation electric field $\delta A_\mu(\vec{x},t)$, the related strengths $\delta\vec{E}$ and $\delta\vec{B}$, and potential (10) were assumed to be weak. But a forced polarization of the medium under the action of strong external electromagnetic fields is also possible. A neutrino should interact with such induced currents, as a particle possessing, generally speaking, other electromagnetic characteristics than those for an isotropic medium.

2.3. Some Applications to Physics of Neutron Stars

From formulae (2') and (3) it is seen that in a dense medium the electric charge of a neutrino greatly increases. Therefore in the processes of supernova collapse and cooling of a hot neutron star the probabilities of elementary processes with the neutrino participation should be recalculated with due regard for the collective excitation modes of the medium, which, no doubt, would affect the collapse dynamics and the final characteristics of a neutron star.

******** Since the equilibrium plasma is electically neutral ($n_0^{(e)} = Zn_0^{(i)}$), then the potential (8) is exactly proportional to the amplitude of neutrino scattering by a neutral atoms [14] (with the compensation of the contributions of neutral vector currents of electrons and protons). If there is no contribution of the neutral neutron current (N = A - Z = 0), we obtain from (8) the potential $V_0 = \sqrt{2}\,G_F n_0^{(e)}$ that was used by Bethe for modifying of the mass matrix in the problem on neutrino oscillations [15]. The allowance for inhomogeneity of $n_0^{(e)}(x)$ made by Mikheev and Smirnov [16] corresponds to the introduction of slow (hydrodynamical) variables in distinction from the used here kintetic description with fast variables $\vec{x}$ and t in the additional contribution (9).

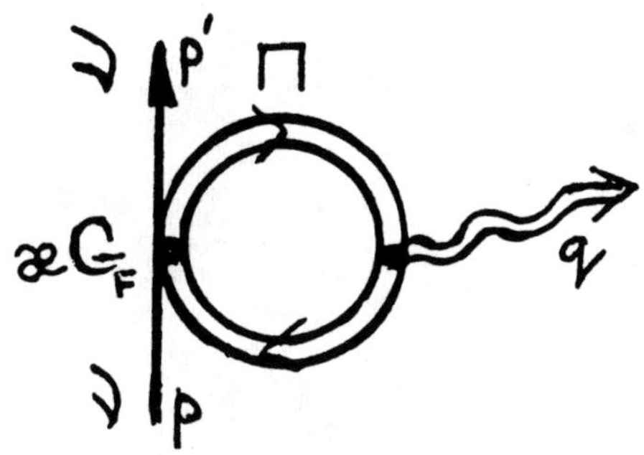

Fig.1. Electromagnetic apex of a neutrino in the point 4-fermion approximation; æ is the dimensionless coefficient depending on weak charges of polarizable particles of the medium (see below), q = p - p' is the 4-pulse of the irradiated photon in the medium (of a plasmon).

An increase of energy losses of a soft electron neutrino due to Cherenkov radiation in a magnetoactive plasma [5] can result in the redistribution of energy spectra of neutrinos of different aroma. Let us remind here that for $\nu_{\mu\tau}$ neutrino such losses are weakened by a factor of $[1 + 4\xi(4\xi - 1)^{-1}]^2 \approx 625$.

The probability of irradiation of $\nu\bar{\nu}$ pairs is modified in the decay of a transverse plasmon due to allowance for the induced magnetic moment of a neutrino [18].

The process of neutrino transfer in a completely ionized electon-nuclear plasma of a collapsar should be modified in the channel of elastic reactions of νA-scattering resulting in neutrino diffusion. The latter is essential where the free path with respect to the coherent scattering at nuclei l_{cr} is much less than the dimensions of a neutinosphere R ($l_{cr} \ll R$). But up to now this process has been calculated with the aid of the Born amplitudes in the standard electically weak model. The calculations [11] show that such amplitudes should be summed up with the "Coulomb" amplitudes of neutrino scattering by nuclei in the plasma, which leads (the neutral currents are neglected) to the Mott cross-section (4). With due regard for the neutral currents the screening in the plasma becomes essential and different for neutrino of different aroma. This results (for constant coherent-scattering energy) to different pulse duration of the neutrino radiation in the process of neutrino diffusion.

References

1. V.N.Oraevskii and V.B. Semikoz, ZeTF, 1984, 86, No.3, 796,
2. V.N.Oraevskii and V.B. Semikoz, Physica, A, 1987, 142, No.1, 135.
3. V.B. Semikoz, YaF. Pis'ma v redaktsiyu, 1987, 46, No.5, 1592.
4. V.N.Oraevskii, V.B. Semikoz, and Ya.A. Smorodinskii, Pis'ma v ZeTF, 1986, 43, No.12, 549.
5. V.N.Oraevskii and V.B. Semikoz, Proceedings of Joint Varenna-Abastumani Intern. School and Workshop. Sukhumi, USSR, 18-20 May, 1986, p.321.
6. V.N.Oraevskii, A.Yu. Plakhov, and V.B. Semikoz, and A.Ya. Smolukhovskii, ZeTF, 1087, 93, 1557.
7. V.B. Semikoz, Trudy IV Vsesoyuznoi shkoly "Chastitsy i kosmologiya" (Proceeding of the 4th All-Union School "Particles and cosmology"), BNO IYaI AN SSSR, 1987 (in Russian)
8.V.N.Oraevskii and V.N. Ursov, Phys. Lett., 1988.
9.V.N.Oraevskii and V.B. Semikoz, YaF, 1985, 42, 702.
V.N.Oraevskii and V.B. Semikoz, Trudy IV Vsesoyuznoi shkoly "Chastitsy i kosmologiya" (Proceeding of the 4th All-Union School "Particles and cosmology"), BNO IYaI AN SSSR, 1985, 145 (in Russian)
10. V.N.Oraevskii and V.B. Semikoz, Phys. Lett., 1384, 1398, No.1, 90-94.
11. L.B. Leinson, V.N.Oraevskii, V.B. Semikoz, and V.N. Usov, Phys. Lett.B, 1988; YaF, Pis'ma v redaktsiyu, 1988, 48.
12. A.V. Borisov, V.Vh. Zhukovskii, and A.I. Ternov, YaF, 1987, 46, 564.
13. V.B. Semikoz, Physica A, 1987, 142, No.1, 157.
14. B.K. Lewis, Phys. Rev., 1980, 21A, 663.
15. H.A. Bethe, PRL, 1986, 56, No.12, 1305.
16. S.P. Mikheev and A.Yu. Smirnov, YaF, 1985, 42, No.6, 1441.
17. E.S. Fradkin, Trudy FIAN, 1965, 29, 7.
18. L.B. Leinson, V.N.Oraevskii, V.B. Semikoz, PAZh, 1988, No.3.

Physics of Neutron Stars
Edited by A. M. Kaminker

ISBN 1-56072-164-2

ON THERMAL EFFECT OF ELECTRON CAPTURE REACTION IN DEGENERATE MATTER

M.A. Rudzskii and Z.F. Seidov
Scientific-Industrial Association of Space Research of the Academy of Sciences of the Azerbaidzhanian SSR, 370106 Baku

The changes in the entropy of a degenerate matter fully ionized caused by the reactions of electron capture by atomic nuclei have been studied.

1. Introduction

The reaction of electron capture (reverse beta-decay) in a stellar matter at not too high densities (lowerr than 10^{12} g/cm^3),

$$(A, Z) + e^- \rightarrow (A, Z-1) + \nu_e \quad (1)$$

are always nonequilibrium because a neutrino formed in this reaction can freely leave the star. In this case the kinetic (including thermal) energy of electrons decreases which, in turn, decreases the entropy (and temperature) of the matter. Along with such a nonequilibrium caused by the matter transparency for neutrino there takes place another nonequilibrium associated with the presence in the matter of electrons with energies greatly exceeding the threshold value for reaction (1). As has been shown in [1-3], in this case the reactions of electron capture at low temperatures lead (despite the "cooling" effect of the emitted neutrino) to an increase in the matter entropy and, at sufficiently high temperatures, to its decrease. In [4], the critical temperature above which an electron capture leads only to cooling for the chain of transformations $^{56}Fe \rightarrow {}^{56}Mn \rightarrow {}^{56}Cr$ was determined as $0.24\, m_ec^2$. The dependence of T_{cr} on the threshold energy δ and the chemical potential of an electron gas, μ, was briefly considered in [5]. In the present communication we shall consider in more detail the thermal effect of the electron capture reaction for one kind of nuclei and for the mixture of nuclides in the nuclear statistical equilibrium (NSE) at high temperatures. In addition to the transitions to the ground level of the final nucleus we shall also take into consideration the contribution from the Gamow-Teller resonance [6].

2. Reaction at One Nucleus

At low temperatures and high densities, when electrons are strongly degenerate whereas nuclei are nondegenerate, the change of the entropy in one reaction (1) is described by the expression [1-3, 7]

$$\bar{\epsilon} \equiv T\frac{dS}{dt} / W = \mu - \delta - \bar{\varepsilon}_\nu \qquad (2)$$

where $\bar{\varepsilon}_\nu$ is the average energy of emitted neutrino, $\bar{\varepsilon}_\nu = \varepsilon_\nu / W$. The total velocity of a decrease of the energy at the neutrino and the number of electron capture reactions at one nucleus per second are, respectively:

$$\varepsilon_\nu = \frac{\ln 2}{ft}\int_\delta^\infty \langle F\rangle\, (x^2-1)^{1/2} x\, (x-\delta)^3 \left[1 + \exp\left(\frac{x-\mu}{T}\right)\right]^{-1} dx \qquad (3)$$

$$W = \frac{\ln 2}{ft}\int_\delta^\infty \langle F\rangle\, (x^2-1)^{1/2} x\, (x-\delta)^2 \left[1 + \exp\left(\frac{x-\mu}{T}\right)\right]^{-1} dx \qquad (4)$$

The temperature T, chemical potential μ, reaction threshold δ, and ε_ν are expressed in the m_ec^2 units; m_e is the rest mass of an electron.
Obviously that $\overline{\in} > 0$ if the mean energy $\overline{E}_e$ of captured electrons is lower than the chemical potential of the electron gas and, vice versa, $\overline{\in} < 0$ if $\overline{E}_e > \mu$. The condition $\overline{\in} = 0$ determines T_{cr} for the given μ and δ. From (2-4) it is seen that if $\delta > 1$, the universal dependence of (T_{cr}/δ) on δ/μ takes place (Fig.1). The temperature is expressed in $T_9 = T/10^9$ K units. At high temperatures, $T_9 > 3$, there always exists a mixture of nuclei in the statistical equilibrium. Now we proceed to its consideration.

3. Mixture of Nuclei in Statistical Equilibrium

In the previous section we wrote the basic equations for the electron capture reaction in the case of the transition from the ground state of the initial nucleus to the ground state of the final nucleus (or to the first allowed state if the transition to the ground state is forbidden by the selection rules). At high T and large μ, when $E_e - \delta$ exceeds 2-3 MeV, the transitions to the states of the Gamow-Teller resonance are possible, which is taken into account by formulae [6]. Also, at $T_9 \geq 3$ one nucleus is replaced by a mixture of nuclei in the statistical equilibrium. In order to be able to take this fact into account we used the program from Ref. [8] that takes account of the concentrations of 29 nuclei, neutrons, and protons depending on T, m, and the mean $\overline{Z/A}$ value (where Z and A are the charge and the mass number of the nuclide). Figure 2 shows the temperature dependence of $\overline{\in}$ (the change in entropy) averaged over all the nuclei in the NSE for various values of μ (8, 10, 12, 14) and $\overline{Z/A} = 26/56$. The dash lines show such dependences for pure iron, ^{56}Fe; the solid lines show the same dependences for nuclides in the statistical equilibrium.
An interesting fact is the passage of $\overline{\in}$ through zero in the small temperature range at $\mu = 10$. Upon the attainment of NSE, the value of $\overline{\in}$ is always higher than its values in the reactions at one nucleus, and then, at sufficiently high temperatures, it decreases again and becomes negative. Note a strong dependence of $\overline{\in}$ on the parameter $\overline{Z/A}$. This is demonstrated by Fig.3 showing the temperature dependence of $\overline{\in}$ at μ, 8 and 10 for two values of the parameter $\overline{Z/A} = 0.46400$ (curves with subscript 1) and $\overline{Z/A} = 0.46429$ (curves with subscript 2).

4. Discussion

Upon the attainment of NSE the value of $\overline{\in}$ always increase with the temperature rise. As is seen from Fig.2 this effect is especially well pronounced for small μ. This is explained by the appearance in the NSE of nuclei with the electron capture threshold lower than that for the ^{56}Fe nucleus. An increase in $\overline{\in}$ provides an additional contribution into the effect of thermal spike (rapid growth of entropy upon the attainment of NSE) noted in the calculations of the initial stage of the collapse of a star with a small mass [9].
Another interesting field where a nonmonotonic temperature dependence of $\overline{\in}$ can play an important role is a slow star evolution due to cooling associated with the emission from the surface or, vice versa, heating due to the matter accretion.

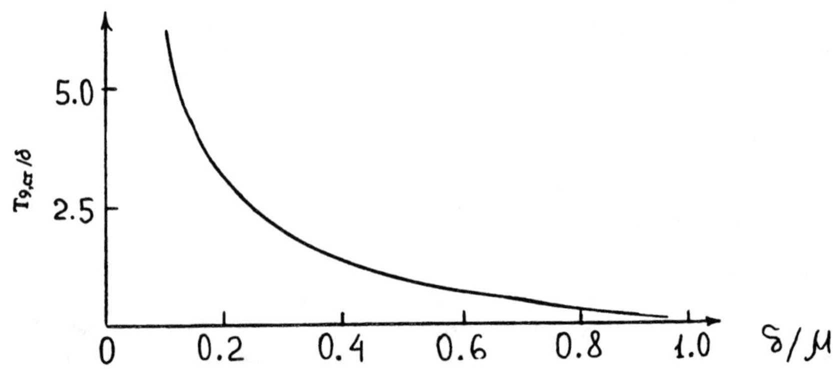

Figure 1

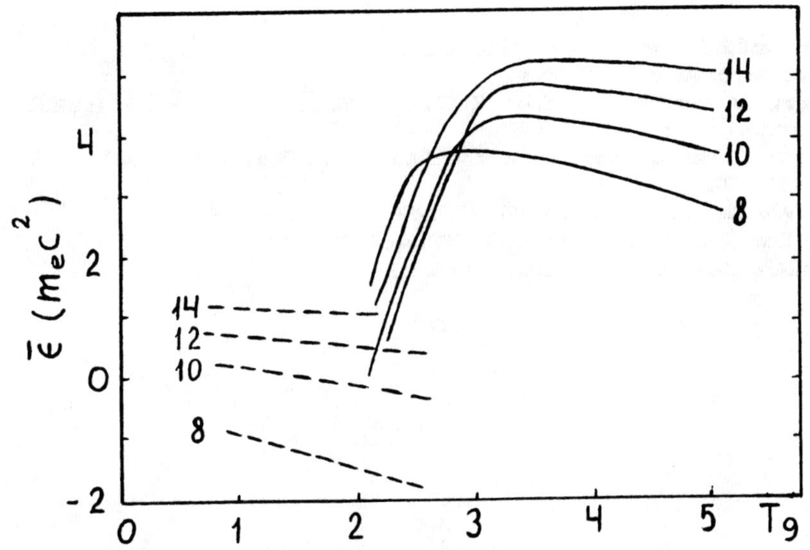

Figure 2

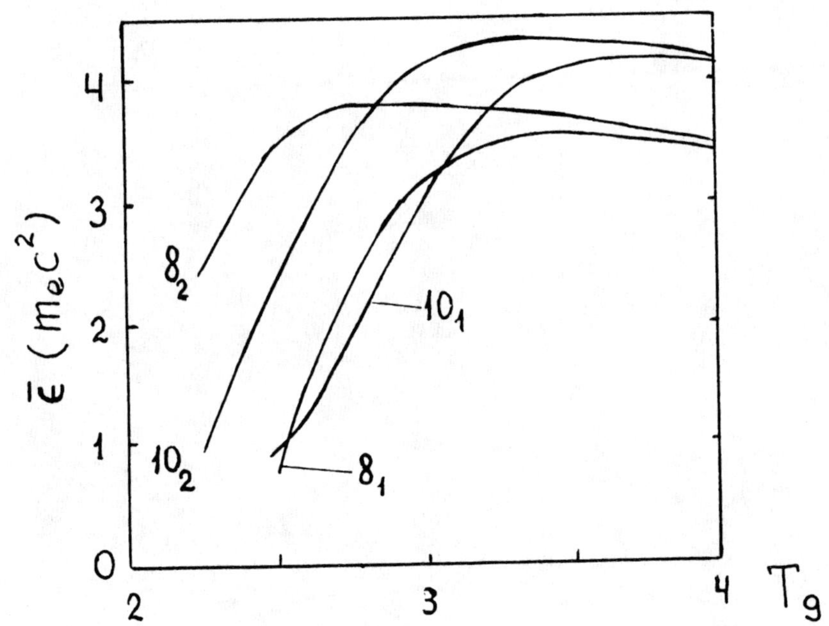

Figure 3

References

1. G.S. Bisnovatyi-Kogan and Z.F. Seidov, Aston. Zh., 1970, 47, 1, 139.
2. M.A. Rudzkii and Z.F. Seidov, Aston. Zh., 1974, 5, 936.
3. G.S. Bisnovatyi-Kogan, M.A. Rudzkii, and Z.F. Seidov, ZhETF, 1974, 67, 5, 1621. 4. M.A. Rudzkii and Z.F. Seidov, Izv. AN Azerb.SSR, Ser.fiz.-tekh. i matem. n., 1974, No.4, 89.
5. Z.F. Seidov, Summary of Doctoral Dissertation, Moscow, 1975. 6. G.H. Fuller, W.A. Fowler, and M.J. Newman, Astrophys.J., 1982, 252, No.3, 715.
7. V.S. Imshennik and V.M. Chechetkin, Astron. Zh., 1970, 47, 5, 929.
8. M.M. Baskom M.A. Rudzkii, and Z.F. Seidov, Astrofizika, 1980, 46, 4, 321.
9. S.I. Blinnikov and M.A. Rudzkii, Pis'ma v AZh., 1984, 10, No.5, 363.

Physics of Neutron Stars
Edited by A. M. Kaminker

ISBN 1-56072-164-2

ON POSSIBLE RESONANCE FREQUENCY (4.9 Hz) OF NEUTRON STARS

V.A. Kotov and V.P. Fomin
Crimean Astrrophysical Observatory of the USSR
Academy of Sciences, Nauchnyi, Crimea

The period distribution for 436 pulsars has been considered. It has been found that the frequencies of pulsar revolutionare statistically multiple to a certain characteristic frequency $\nu_0 = 4.93 + 0.04$ Hz. The reliability of the results reliability is $\approx$ 1-5 $\cdot 10^{-4}$. The frequency ν_0 is equal to a frequency of 4.94 Hz observed in a sequence of neutrino bursts from the Supernova 1987 A. It is assumed that ν_0 is a "resonance" frequency of natural oscillations of neutron stars.

The study of pulsar period distribution is important for understanding the process of pulsar formation, their evolution, and the effects associated with the gravitational radiation. Recently [1], it has been found that in such a distribution, a periodic modulation of the number of objects with a frequency of 10 Hz takes place which is a function of the period. Such an effect cannot be explained by observational selection, and up to now, no physical mechanism that can be responsible for such a periodicity has been suggested.

The method suggested in [1] is based on the analysis of the statistics of the ν_i/ν ratios, where ν_i is the frequency of a pulsar with the number i and ν_i is a trial frequency. The method may introduce a certain error within a factor of two in the calculated characteristic (of the highest multiplicity, commensurate)period. Therefore we have modified the method suggested in [1] - now the analysis is performed separately for "fast" ($\nu_i \geq \nu$) and "slow" ($\nu_i < \nu$) pulsars. This substantially increases the reliability of the obtained results. Note also that the set of pulsars under consideration (the number of objects is N = 436) exceeds by 32% that in [1] where the catalogue [2] for 330 pulsars have been analyzed.

The normilized function [3] has been calculated separately for "fast" and "slow" pulsars:

$$F_0(\nu) = (60 \cdot N)^{1/2} \left\{ 12^{-1/2} - \left[\frac{1}{N} \sum_{i=1}^{N} \Delta_i^2 \right]^{1/2} \right\} \qquad (3)$$

where $\Delta_i = x_i - \mathrm{INT}(x_i + 0.5)$, $x_i = \nu_i/\nu$ for $\nu_i \geq \nu$ and $x_i = \nu/\nu_i$ for $\nu_i < \nu$; i = 1, 2, ..., N. Upon averaging of two obtained $F_0(\nu)$ functions the following "resonance" spectrum was calculated

$$F(\nu) = F_0{}^3(\nu) \,/\, |\, F_0(\nu) \,|$$

The $F(\nu)$ spectrum for 436 pulsars is shown in Fig.1, where the only statistically valuable highest peak corresponds to the frequency $\nu_0 = 4.93 \pm 0.04$ Hz. Its formal significance is $\approx 4.4\ \sigma$; with due regard for the number of independent frequency ranges the probability of a random peak appearance is $\sim 5 \cdot 10^{-4}$.

We consider here two additional facts in favor of the existence of the found "specfic frequency" $\nu_0 \approx 5$ Hz.

1. Recently Hasinger [4] analyzed the distribution of fast quasiperiodic oscillations (QPO) of bright binary X-ray systems with small masses. For the so-called "normal branch" of the frequency source the QPO show a noticeable concentration in the vicinity of $\nu \approx$ 6 Hz. This brought Hasinger to a QPO mechanism based on a certain "fundamental" property of neutron stars and their environment.

Using the data [4], we constructed the distribution of nine QPO frequencies shown in Fig.2. It is seen that the prevalent frequency is closer rather to 6 than to 5 Hz.

2. The authors of Ref.[5] have analyzed 12 neutrino bursts from the Supernova 1987 Arecorded by a KAMIOKANDE detector. They have established (with the significance $\approx 3\sigma$) that the neutrino bursts followed with a frequency of 4.94 $\pm$ 0.04 Hz, which, within the error, coincides with the value of ν_0 obtained in our calculations.

We believe that the periodicity of $\approx$ 5 Hz revealed in all three cases may be due to the same mechanism caused by natural pulsations of neutron stars. By an unknown reason, the latter occur at a favorable "resonance" frequency ν_0.

The mechanism of excitation of a frequency of 5-6 Hz suggested by Hasinger [4] is based on an idea of a resonance sound wave generated in a disc accretion at the surface of a neutron star. The mechanism is rather simple and, probably, efficient in the case of QPOs of binary X-ray systems. The ν_0 -effect should be of a more general character, since, as follows from our analysis, the frequency ν_0 seems to be independent of radius R of the mass M and the period of a neutron star.

Acoustic oscillations of neutron stars and the so-called "shear waves" generated in the crust have periods $\leq$ 0.001 s [6]. Neither the hypothesis about g- and i-modes with periods $\geq$ 0.01-0.1 s can explain the ν_0 -effect: the spectrum of natural nonradial oscillations of a star is rather rich and strongly depends on M, R, and other parameters.

In this connection a new hypothesis may be put forward that the existence of ν_0 should reflect a certain property of the matter of neutron stars. It should manifests itself in a strong gravitational field, e.g., natural neutron oscillations.

In our opinion, the prevalence of frequency ν_0 in the pulsar distribution is of great interest for physics of neutron stars. In particular, the above indicated coincidence of ν_0 with the frequency of neutrino bursts may indicate that a pulsar that was formed at the moment of the explosion of the Supernova 1987 A oscillated and emitted a flow of neutrino with a frequency of "hypothetical" pulsations (neutron stars) equal to $\nu_0 \approx$ 4.9 Hz.

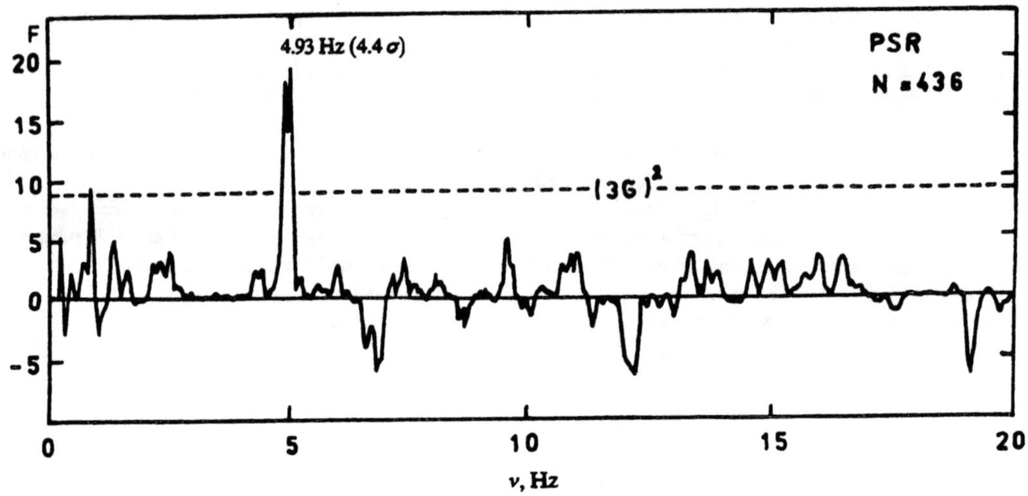

Fig.1. "Resonance" spectrum for 436 pulsars.

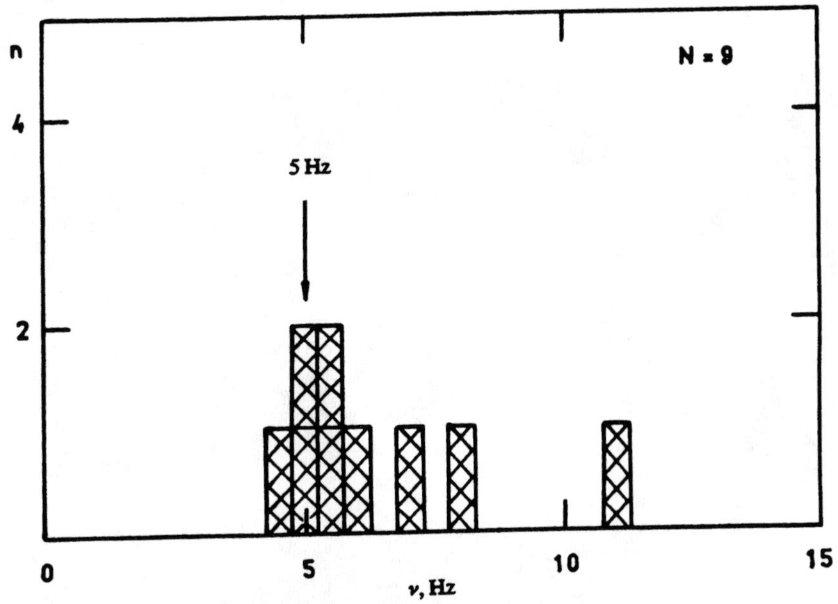

Fig.2. Distribution of nine QPO frequencies of binary X-ray sources (see [4]).

References

1. V.A. Kotov and B.M. Vladimirskii, Izv. Crym. Obs., 1987, 76, 93.
2. R.N. Manchester and J.H. Taylor, Astron. J., 1981, 86, 1953.
3. V.A. Kotov, Solar Physics, 1985, 100, 101.
4. G. Hasinger, Astron. Astrophys., 1987, 186, 153.
5. H. Ohelman and R. Buccheri, Astron. Astrophys., 1987, 180, L23.
6. P.N. McDermott, C.J. Hansen, H.M. Van Horn, and R. Buland, Astrophys.J., 1985, 297, L37.

Physics of Neutron Stars
Edited by A. M. Kaminker

ISBN 1-56072-164-2

NEUTRON STAR EVOLUTION IN MASSIVE BINARY SYSTEMS

V.M. Lipunov, K.A. Postnov, and M.E. Prokhorov
P.K. Shternberg State Astronomic Institute, 117234 Moscow

Theoretnical studies of evolution of normal and compact objects conducted in two last decades provided the restoration of the complete evolution scenario of a binary system with any initial parameters at any moment of time. The scenario is qualitatively complete, in other words, it explains whence and how all the known types of binaries, both often encountered and "ecsotic" ones, can appear.

Thus an inevitable question arises about the qualitative correspondence between the predictions of the evolution scenarioand real observations. Such an approach first suggested by Kornilov and Lipunov in 1983 emphasizes the scenario parameters which, leaving intact the main scheme of the evolution paths of binary stars, make a substantial effect on the occurrence frequency of various stages.

For the solution of such problems the distributions generated in a "standard" scenario for massive systems were simulated. The following schemario was considerred to be "standard":

the mass of initially more massive component of a binary system or a single star is distributed by the Solpeter law:

$$dN \propto M^{-2.35} dM, \quad 10\,M_\odot \le M \le 60\,M_\odot$$

The mass components in the system are related by a power ratio:

$$dN \propto q^2\,dq, \quad q = M_2/M_1$$

The orbit semiaxes are distributed uniformly with respect to the logarithm:

$$dN \propto d\,(lg A), \quad 10\,R_\odot \le A \le 10^7\,R_\odot$$

The stage with the shared shell starts at

$$q > Q_{ce} = 2.5$$

If the mass of a presupernova is above the critical $M_{cr} = 25\ M_\odot$, then the collapse results in a generation of a black hole with a mass approximately equal to the star mass, otherwise a neutron star is formed with the initial revolution period P_0 = 1 ms and a magnetic moment distributed by the law $dN \propto d(\log\mu)$, 10^{30} Gcm^3 $\le \mu \le 10^{32}$ Gcm^3 (accordingly, 10^{12} G $\le$ B $\le 10^{14}$ G). The Oppenheimer-Volkoff mass was taken to be equal to $M_{ov} = 2\ M_\odot$.

The star age was simulated in the range $0 \le T \le 30 \cdot 10^6$ years. It was taken that the number of stars in the Galaxy is equal to the number of stars in the binary systems. The general number of stars in the Galaxy is $N_G = 3 \cdot 10^{11}$ and its age is T = $15 \cdot 10^9$ years.

The "standard" scenario is well agree with a series of observational data, e.g.:

the burst frequency of supernovae is 1/55 year^{-1} ,

the origin frequency of neutron stars is 1/23 year^{-1} .

But there are also some substantial contradictions. Thus, e.g., the scenario "predicts" the existence of approximately 600 X-ray pulsars in the Galaxy (but there observed only 23) and 3% (~ 12) of potentially observable radiopulsars in pairs with nondegenerate stars (but at present such systems are not known).

In order to overcome the above indicated contradictions it was assumed that as a result of the asymmetry in the supernova burst the neutron star acquires a certain velocity ΔV in an arbitrary direction.

Figure 1 shows the dependence of a number of X-ray pulsars in the Galaxy (N_A) and the fraction of radiopulsars in pairs

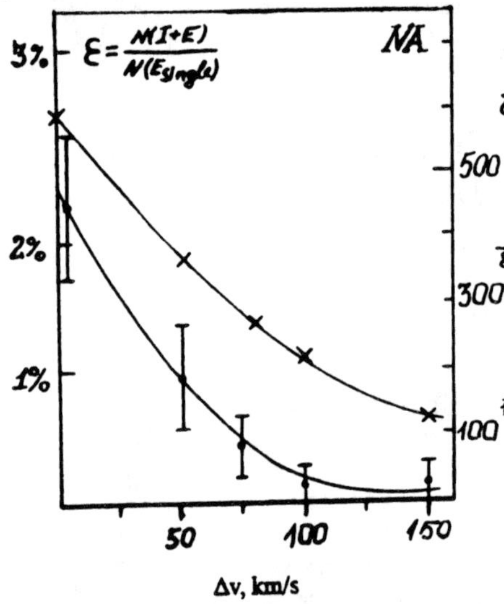

Figure 1

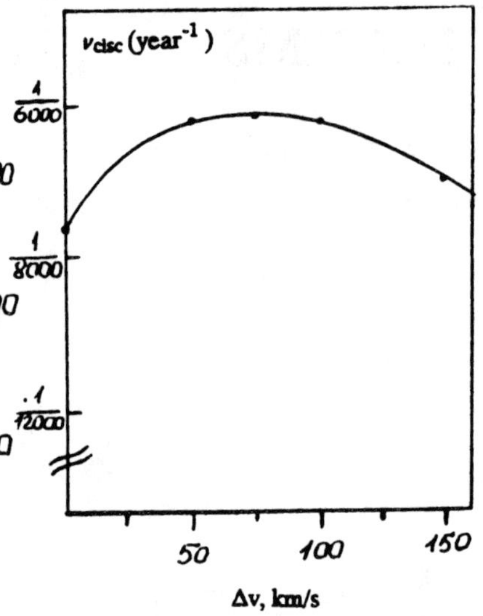

Figure 2

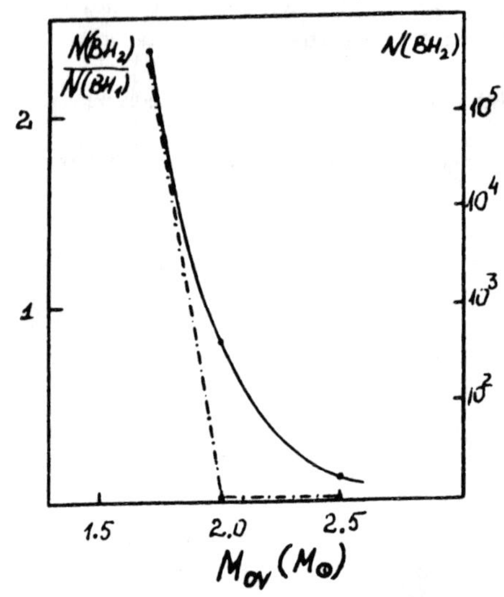

Figure 3

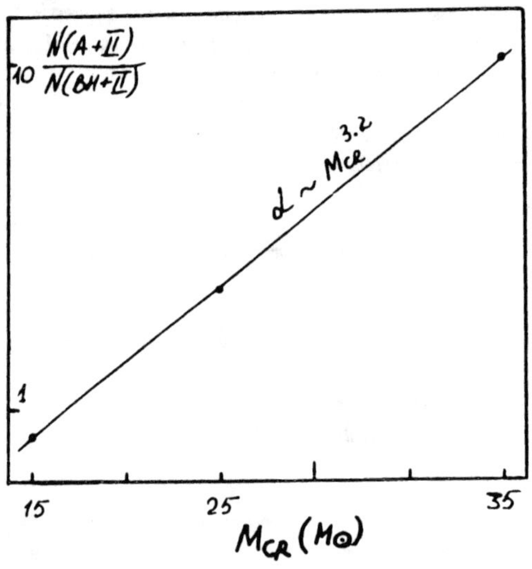

Figure 4

with nondegenerate stars (ε) as a function of ΔV. The contradiction disappears when $\Delta V \sim 100$ km/s. It should be noted that an increase in ΔV is accompanied by an increase in the frequency of neutron star merge (see Fig.2).

We also studied the dependence of the distribution on M_{cr} and M_{ov}. From Fig.3 it is seen that the number of binary X-ray sources with black holes decreases with an increase in M_{cr}. A result close to the observed ($N_{bh}/N_{ns} \propto 3/23$) yields $M_{cr} \sim 25 - 30\, M_\odot$. For smaller M_{cr} the number of such sources is too large.

The effect of the Oppenheimer-Volkoff mass is observed only for a number of "secondary" black holes formed in the course of the accretion upon the neutron star collapse (Fig.4). For $M_{ov} > 2M_\odot$ we have $N_{bh2}/N_{ns} < 1/10$, which does not contradict the known observations. But a rapid increase in N_{bh2} with a decrease of the mass M_{ov} does not allow one to "close" any neutron-star model prior the investigation of their initial mass spectra.

Physics of Neutron Stars
Edited by A. M. Kaminker

ISBN 1-56072-164-2

DISC ACCRETION OF A MAGNETIZED PLASMA ONTO A NEUTRON STAR WITH A STRONG MAGNETIC FIELD

G.D. Chagalishvili
Abastumani Astrophysical Observatory of the
Acedemy of Sciences of the Georgian SSR

Disc accretion onto a coaxial rotator is considered for the case where an accretion disc has rather a strong large-scale azimuthal magnetic field ($B\phi^2/8\pi \approx$ density of thermal energy of the medium), i.e., a field which, under certain conditions, is generated in the disc due to the action of a turbulent dynamo in the "$\alpha\omega$ approximation". It has been shown that the existense in the system of two magnetic fields of different origins results in an asymmetric accretion of the matter. Depending on the ratio of the radius of magnetic line closure to the Alfvwn radius, accretion usually proceeds selectively onto one of the magnetic poles. In such a case, physics of accretion substantially differs from physics of disc accretion for $B_\phi = 0$.

All the variety of the phenomena occurring in disc accretion onto neuton stars with magnetic fields have been studied since the early seventies. They are collected and classified in [1,2]. The completeness of these investigations is limited by the fact that only nonmagnetized discs were considered, in other words, only the cases where an accretion disc had no magnetic field that could influence the interaction processes between an accreting gas and the magnetic field of the neutron star. This relates, first and foremost, to the processes occurring on the Alfvwn surface. Indeed, it was shown [3,4] that under certain conditions a disc may possess a large-scale magnetic field whose generation is caused by the action of a turbulent dynamo. Thus generated field is, as a rule, of a toroidal configuration [4] and may attain an intensity at which the magnetic pressure becomes comparable with the gas pressure in the disc [3], i.e., on the Alfvwn surface it is comparable with the magnetic field pressure of a neutron star. It can readily be understood that this should drastically change the method and the path of the accreting gas penetration into the magnetic pole of a neutron star. To be more concrete, a strong toroidal magnetic field of the disc prevents the formation of the permulational instability and excludes the scenario of the plasma penetration inside the magnetosphere developed in [5,6].

The specific features of accretion for a magnetized disc are well seen in the simplest case where the rotation and the magnetic axes of a neutron star coincide. Let these axes be directed along the Z-axis of the reference system. We shall take that the magnetized accretion disc is in the equatorial plane of the star. Moreover, we also assume that the equilibrium revolution of a neutron star is almost attained [2] so that the accreting plasma on the Alfvwn surface is almost "short circuiting" with the force lines of the neutron star matnetic field. In other words, when the Alfvwn (R_A) and "short circuiting" (R_C) radii are close to one another, we have $R_A \approx R_C$. As is known [2], such a relationship between the radii is the most widespread. Moreover, if it is valid, then it is easy to describe the process of reclosing (recircuiting) of the force lines of the magneticc fields of the neutron star and the accretion disc.

Note also that accretion physics for a coaxial rotator in the disc accretion of the magnetized plasma is insensitive to the change of the magnetic dipole direction to the opposite one. In our case the existence in the system of two magnetic fields of different origins - a toroidal field of the accretion disc ($\vec{B}_\varphi$) and the field of a neutron star ($\vec{B}_N$) - results in the asymmetry with respect to the z axis determined by the mutual orientation of $\vec{B}_\varphi$ and $\vec{B}_N$. Accretion of the plasma mainly onto one magnetic field of the neutron star indicated in the abstract is just the manifestation of such an asymmetry. In accordance with the present study, the scenario of the phenomena occurring on the Alfvwn surface, is as follows. The magnetic fields of the neutron star and the accretion disc are crossed. This fact provides the reclosure of their force lines on the Alfvwn surface and, thus, the formation of some "paths" (Fig.3) along which the plasma "slips" mainly to one of the two magnetic poles of the neutron star.

Let us consider the process of reclosure of the force lines in more detail. In such situations the reclosure of magnetic force lines occurs due to the development of the tiring-mode instability [7-9].

In our case the dissipative effects are provided by the disc turbulence [3] which results in the magnetic Reynolds numbers $R_e \approx 10^2 - 10^3$. The mathematical methods of studying the tiring-mode instability are rather nontrivial. But the information necessary for the description of the qualitative behavior of force lines in their reclosure on the Alfvwn surface can be obtained from Refs.[7-9]. In accordance with these studies, the dynamics of the force lines can be schematically represented as shown in Figs.1, 2, 3.

Figure 1 shows separately the force lines of the magnetic fields of a neutron star ($\vec{B}_N$) and an accretion disc ($\vec{B}_\varphi$) on the Alfvwn surface. The heavy portions of the lines in Fig.1 are enlarged in Fig.2 to describe the physics of line "reclosure". The reclosure of the crossed force lines occurs at nodal points (Fig.2a) for a characteristic time $\tau \approx R_e^{1/2}/\Omega$ [7]. The further dynamics of the force forming the "paths", is illustrated by Fig.2a,b. Figure 3 illustrated the formation of one of the force tubes as a result of the above described processes. The configuration of the given force tube in the equator region is such that the matter in this tube acquires a pulse in the z-direction and 'slips" towards the upper magnetic pole, if the Kepler rotational velocity of the matter on the Alfvwn radius in the disc exceeds ΩR_A (where Ω is the angular velocity of the neutron star revolution). In other words, this occurs where $R_C/R_A > 1$. Naturally, if the condition $R_C/R_A < 1$ is fulfilled, the matter "slips" to the lower magnetic pole (Fig.3). We should like to emphasize here once more that we consider the case where R_A and R_C differ only slightly because it is only in this case that the Kepler rotation of the matter makes no effect on the physics of the instability development in the tiring-mode described in [7-9].

In the present study we have performed the qualitative analysis of the accretion onto a coaxial rotator, but it may readily be seen that the main characteristic of the accretion - its asymmetry - would be preserved only for accretion of a nonmegnetized plasma ($B_\phi = 0$) onto an inclined rotator.

Let us list the main observational manifestations resulting from the above described accretion asymmetry:

- if such a system can generate jets, they would be observed only in one direction;
- even a weak variation in the accretion rate may result in the change of the sign of the ratio $R_C/R_A - 1$. This results in the change of the magnetic pole for accretion which, in the final analysis, would produce a great effect on the radiation from the given source in the direction towards the observer.

The author expresses his gratitude to L.M. Zelenyi and A.L. Taktakishvili for fruitful discussions.

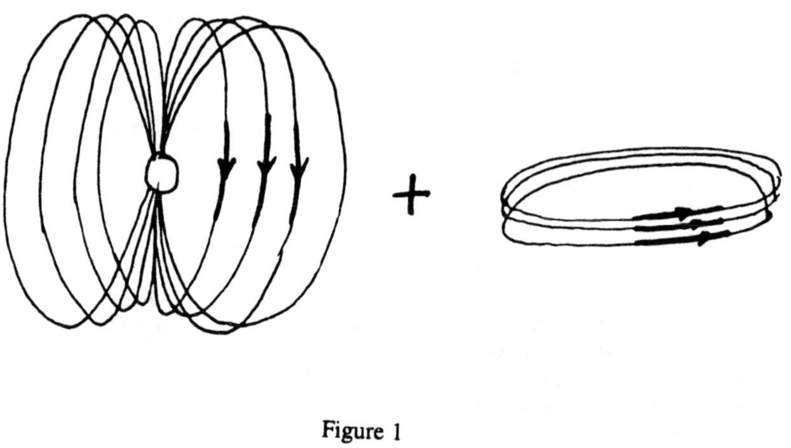

Figure 1

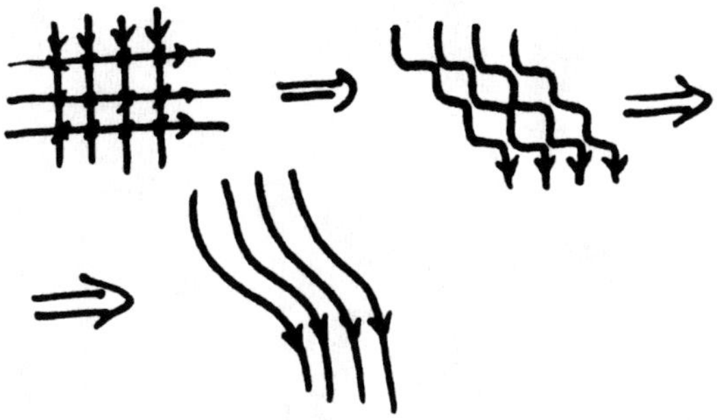

Figure 2

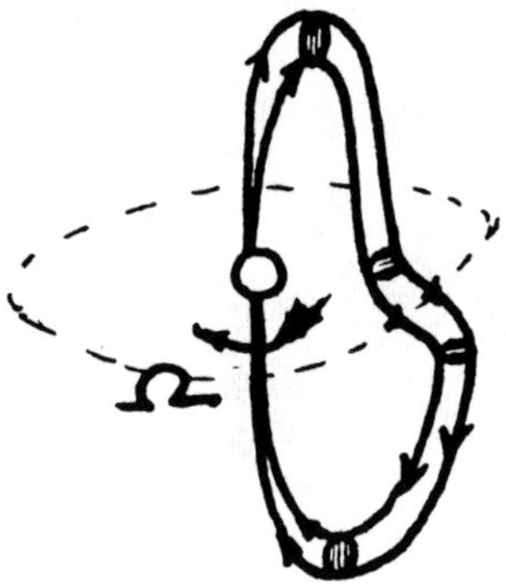

Figure 3

References

1. S. Shapiro and S. T'yukolski, Chernye dyry, belye karliki i neitronnye zvezdy (Black holes, white dwarf stars, and neutron stars) 1985 (Moscow: Mir) (in Russian).
2. V.M. Lipunov, Astrofizika neitronnykh zvezd (Astrophysics of neutron stars) 1987 (Moscow: Nauka) (in Russian).
3. G.D. Chagelishvili, R.G. Chanishvili, and J.G. Lominadze, Proc. Joint Varenna-Abastumani Internat. School and Workshop, Sukhumi, USSR 1986 (ESA sp-251, August 1986), 563.
4. G.D. Chagelishvili, R.G. Chanishvili, and J.G. Lominadze, Proc. Joint Varenna-Abastumani Internat. School and Workshop, Sukhumi, USSR 1986 (ESA sp-251, August 1986), 523.
5. P.Ghosh and F.K. Lamb, Ap. J., 1979, 232, 256.
6. P.Ghosh and F.K. Lamb, Ap.J., 1979, 234, 296.
7. H.P. Furth, J. Killen, and M.N. Rosenbluth, Phys. Fl., 1963, 6, No.4, 459.
8. L.C. Lee and Z.F. Fu, Geophysical Research Lett, 1985, 12, No.2, 105.
9. D.J. Southwood, M.A. Sounders, M.W. Dunlop, W.A.C. Mier-Jedrzejowicz, and R.P. Rijnbeek, Planet. Space Sci., 1986, 34, No.12, 1349.

Physics of Neutron Stars
Edited by A. M. Kaminker

ISBN 1-56072-164-2

ON POSSIBLE NATURE OF QUASIPERIODIC OSCILLATIONS OF LUMINOSITY IN X-RAY SOURCES

A.G. Muslimov
A.I. Ioffe Physical-technical Institute of the
USSR Academy of Sciences, 194021 Leningrad

A hypothesis is put forward that the oscillations of luminosity in X-ray sources may be caused by global Alfvwn oscillations of the plasma.

Recently quasiperiodic oscillations (QPO) of luminosity with the characteristic amplitude of 5-10% of the total flux and the frequency range 5-200 Hz [1] have been revealed for a series of Galactic X-ray sources. The theoretical models of this pnehomenon suggested by now are encounter serious difficulties [2]; thus it is impossible to explain fast variations in ν_{QPO} in the absence of any noticeable variations in luminosity (Sco X-1); in some X-ray sources (e.g., GX 17+2) at a certain phase of their activity the ν_{QPO} frequencies do not correlate with the total luminosity L_x [3].
Let us consider the disc accretion onto a neutron star (NS) in a low-mass binary X-ray system. We assume that the period of revolution of a NS is ≥ 10 ms, and the intensity of the magnetic field is ~ 10^9 -10^{10} G.
Due to instability (e.g. of Rayleigh-Taylor and Kelvin-Helmholtz type), the plasma of the internal part of the accretion disc is "frozen" into the magnetic field of the NS. Under certain conditions; then in this part of the accretion disc global magnetic hydrodynamic oscillations can be generated, in particular, nonlinear global Alfvwn (NGA) modes. These modes were first observed in the studies of an ideal magnetic hydrodynamic (MHD) spectrum of a toroidal plasma [4]. The detailed numerical investigation of NGA modes was performed in [5], later on, their existence was confirmed experimentally [6,7]. The NGA modes are natural modes lying under the lower boundary of the Alfvwn continuum, i.e., for these modes $w^2 < (w_A^2)$ where $w_A^2 = k_{||}^2 v_A^2$, $\vec{k}_{||}$ is the wave vector parallel to the equilibrium magnetic field, and v_A is the local Alfvwn speed. NGA modes have a substantial amplitude, $\delta B/B \approx 10\%$, their spectrum is of a discrete nature, the frequency of the fundamental mode is minimal, and that of a highly excited mode has the value exponentially close to (ω_A) .[8].
The basic statement consists in that the rate of the plasma accretion onto the NS surface can be modulated just with the frequency of the NGA model. For the lower boundary frequency of the Alfvwn continuum under the the above conditions we may write:

$$(\nu_A)_{min} \approx 340\, B_9^{-6/7} \dot{M}_{17}^{3/7} (M_\odot / M_{NS})^{5/7} R_6^{-18/7}$$

where $B_9 = B/10^9$ G, B is the intensity of the field at the NS pole, $\dot{M}_{17} = \dot{M}/10^{17}$ g $\cdot$s^{-1} , M is the accretion rate, $R_6 = R/10^6$ cm, and R is the NS radius.
Within the framework of the suggested model the above indicated difficulties in the interpretation of the QPOs can be overcome, in particular, it becomes possible to understand probable ν_{QPO} autocorrelation and total X-ray luminosity of the source.
I should like to express my gratitude to A.I. Tsygan for useful discussions.

References

1. M. van der Klis and P. Jansen, In: Proc. NATO-ATW Evolution of Galactic X-ray Binaries (Reidel NATO ASI Series C167), 1985, 129.
2. M. van der Klis et al., Astrophys. J., 1987, 316, 411.
3. L. Stella, A.N. Parmar, and N.E. White, Astrophys. J., 1987, 321, 418.
4. J.P. Goedbloed, Phys. Fluids, 1975, 18, 1258.
5. K. Appert, R. Grútber et al., Plasma Phys., 1982, 24, 1147.
6. R. Behn et al., Plasma Phys. Controlled Fusion, 1984, 26, 173.
7. T.E. Evans et al., In: Proc. of the Fourth Joint Varenna-Grenoble Internat. Symp. on Heating in Toroidal Plasma, Rome (Internat. School of Physics, Varenna), 1984, 1, 121.
8. S.M. Mahajan, D.W. Ross, and G.-L. Chen, Phys. Fluids, 1983, 26, 2195.

SUBJECT INDEX